SCIENCE AND CREATION

SCIENCE
AND
CREATION

From eternal cycles to an oscillating universe

by

STANLEY L. JAKI

Science History Publications
New York

Published in the United States by
Science History Publications
a division of
Neale Watson Academic Publications, Inc.
156 Fifth Avenue, New York, NY 10010

Library of Congress Cataloging in Publication Data

Jaki, Stanley L
        Science and creation: from eternal cycles to an
oscillating universe

        1.   Science—History.   1.   Title.
Q125.J27   509   73–22259
ISBN 0–88202–010–2

Printed in Great Britain by
R. & R. Clark, Ltd., Edinburgh

Contents

Introduction

With his penchant for startling dicta, Whitehead once defined European philosophical tradition as a series of footnotes to Plato. Whether this was the safest generalization to make on the topic is another matter, but it cannot be denied that the major themes of philosophy are as old as philosophy itself. Much the same holds true of scientific speculation. The word "atomic", which our age uses as its hallmark, has an ancestry leading back to the times of Pericles. Then and there it was clearly perceived that matter had to be either discrete or continuous. Decision on this represented the touchstone of truth for Democritus as well as for Aristotle. In the latter's words the verification of a strictly smallest quantity could be of such portent as to shake the very foundations of philosophy.

Ancient Greek philosophers showed equally keen interest in questions having to do with the very large. There again, a fundamental pair of alternatives was formulated with all possible clarity: the world could only be finite or infinite in extent. The counterpart of this along the parameter of time also received a most explicit attention. The classical Greeks' firm advocacy of an eternal world became a distinctive feature of their world view and science. Their concept of the eternity of a finite world, repeating itself in every Great Year, also anticipated to a surprising degree the idea of an oscillating universe, the favourite choice of many cosmologists of our day.

For them, as for their classical forerunners, fundamental philosophical considerations are at play in the acceptance of the idea of a universe that goes forever through a supreme cycle. It does not require a crystal ball to see that many decisive choices hang in balance with the truth or untruth of the notion of an oscillating universe. Its truth certainly would be a most palpable seal on the belief in a self-contained, non-contingent universe. Whether astronomy shall soon provide reliable support for the hypothesis of an oscillating universe remains to be seen. At any rate, life in our solar system would be extinguished long before the actual expanding phase had gone through much of the tens of billions of years assigned to it by brave calculators.

No such uncertainty characterizes, however, the history of speculations and contentions about a cosmos subject to the cosmic cycle of universal births, deaths and rebirths. Tracing out the historical record has an interest of its own, but the topic in question offers far more than opportunity for satiating scholarly preoccupations. This added attraction concerns a problem which should be of paramount interest to our scientific age. In a scientific perspective full maturity has certainly come to mankind during the second third of the twentieth century. The theoretical tools of exact physical science and its superb instrumentation permit for the first time a methodical exploration and control of the material world. Materials-science is perhaps the most telling proof of this, as practically any type of material can now be made to order.

Reaching full maturity always prompts a look backward. Young people do not care about genealogies. Family records begin to be of consuming interest only to those who have already arrived. The interest is particularly pressing when the beginning of the journey amounts to a miracle. The miracle is the emergence of a self-sustained type of the scientific endeavour. In a world history that had witnessed at least half a dozen great cultures, science had as many stillbirths. Only once, in the period of 1250–1650, did man's scientific quest muster enough zest to grow into an enterprise with built-in vitality.

A full account of all factors, the presence or the absence of which produced this startling singularity in history, will probably never be forthcoming. Portrayals of vast historical problems are never destined to be strictly complete. For all that, historians show unflagging interest in vast syntheses, fragile as they are bound to be. It should, therefore, seem strange that no major attempt has been made so far to come to grips with the problem of the stillbirths and birth of science.

This book is such an attempt. Its thesis can easily be summed up. Great cultures, where the scientific enterprise came to a standstill, invariably failed to formulate the notion of physical law, or the law of nature. Theirs was a theology with no belief in a personal, rational, absolutely transcendent Lawgiver, or Creator. Their cosmology reflected a pantheistic and animistic view of nature caught in the treadmill of perennial, inexorable returns. The scientific quest found fertile soil only when this faith in a personal, rational Creator had truly permeated a whole culture, beginning with the centuries of the High Middle Ages. It was that faith which provided, in sufficient measure, confidence in the rationality of the universe, trust in progress, and appreciation of the quantitative method, all indispensable ingredients of the scientific quest.

There is nothing original in this thesis, but it certainly needed to be told and documented in great detail. Not a few details of the documentation will probably strike as stark novelty a reader accustomed to the clichés of the historiography of science. He will be surprised at the unique features of the biblical world view, at the clash of early Christianity with paganism on the question of eternal returns, and at the theological reasons of the standstill of science among the Arabs. He will find novel the emphasis on the anti-scientific strain of Renaissance and Hegelianism (right and left), both saturated with the idea of eternal returns. Equally novel may appear to the reader the application of the theme to twentieth-century cosmology and science. He should, however, remind himself that the fundamental issue facing our scientific age is not about scientific know-how, but about the proper use of the scientific knowledge and tools. Constructive judgment on this point is simply impossible without defining individual and collective purpose. Yet, if man and mankind are only a chance ripple on dark, unfathomable cosmic waves, which swing cyclically back and forth, any reference to purpose will amount to mere equivocation and to a rank abuse of meaningful discourse. The future of man rests with that judgment which holds the universe to be the handiwork of a Creator and Lawgiver. To this belief, science owes its very birth and life. Its future and mankind's future rest with the same faith.

CHAPTER ONE

The Treadmill of Yugas

In the genesis of scientific ideas, as Duhem once noted, there is no well-defined starting point.[1] The concept of an oscillating universe forms no exception to this rule. It is the most recent type of cyclic cosmological models the origins of which go back to very early times, to the misty epochs of prehistory. There is a long way from prehistory to modern culture, and modernity is often the very antithesis to everything ancient let alone prehistoric. The educated man of the mid-twentieth century was, indeed, caught by surprise when scientific speculation presented him with the possibility of a universe subject to endless cycles of expansion and contraction. A conception of the cosmos along such cyclic lines stood in sharp contrast to the main features of scientific world pictures that had been in general acceptance since the rise of science in the seventeenth century. The Newtonian universe with its infinite space and infinite number of stars was certainly not conceived in a cyclic frame of reference and the same holds true of the Herschelian universe, in which the Milky Way stood for the universe at large. When the discovery of the expansion of the universe came in the late 1920's, the static aspect of former scientific world pictures had to yield. Still, for decades the expansion of the universe was mainly spoken of as being a linear, once-and-for-all process. In recent years, however, growing attention has been paid to the possibility that the present expansion of the universe was preceded by a contracting phase and shall be followed by it in an interminable sequence.

As can easily be seen, such a conception of the universe is a cyclic one, and its novelty for contemporary man, and for modern man in general, is undeniable. No less strikingly novel should appear the amount of time that may be as long as 85 billion years for a single expansion or contraction. The reasons for this surprise can easily be accounted for. During the last three hundred years modern mind has largely been shaped by the main trends of scientific thought, and until very recently none of those trends was congenial to the concept of a cyclic universe. In fact, scientific mentality coupled with other factors was particularly effective in discrediting speculations about perennially recurring cycles in the universe as a whole.

Ancient cultures were teeming with those speculations, which often contained quantitative estimates concerning the length of world cycles. Curiously enough, some of those estimates did not fall too short of the order of billions of years, and on occasion far surpassed it. A most telling illustration of this is found in the great Hindu classics of popular mythology and ethics, the Puranas, composed during the middle part of the first millenium. According to the Puranas, the physical universe entered, around the middle of the second millenium B.C., the last phase of a 4.3 billion-year-long cosmological process that will keep repeating itself. Undoubtedly, there is a marked uniqueness in the carefree delving into enormously large and precise figures that characterize the dicta about world cycles in the Puranas and

many other ancient Hindu literary classics. This uniqueness is a peculiar manifestation of a universal, primitive desire in man to escape the confines of temporal
existence.

Pre-modern cultures and societies were alike almost without exception in their
deep-seated aversion to time and history.[2] In building temples, or in staking out
sacred territories, primitive man longed to transfer himself to the mythical place
of timeless existence free of any change. The belief of primitive man in sacred
mountains and in sacred cities conveys the same motivation. Those mountains
and cities symbolized the mythical centre of the universe, or its absolute axis. They
represented the place free from the continuous changes taking place on the periphery, where man's actual existence unfolds as if he were in a state of exile. In
primitive man's belief it was at the central point that were located the tree of life
and the fountain of perennial youth, or immortality. Consequently, the mythical
centre was also the place where, in the beginning of time, the creation or the construction of the universe had occurred.

To secure a timeless character to his own constructions, primitive man, therefore,
saw to it that the foundation stone of a building erected by him be somehow
connected to the centre of the universe. Thus, in India it was the astrologer that
determined the exact spot where the cornerstone of a building was to be placed.
The mason in turn drove a little wooden peg into the ground in the belief that it
would hold the head of the snake motionless. The snake was the ancient Hindu
symbol of chaos and the mason's action repeated the feat of Indra who overcame
the serpent with his thunderbolt, securing thereby stability and timelessness to
what had been formed from the chaotic. This imitation by man of divine actions
seems to have been motivated by his belief that therein lay the only means to
overcome confinement to an existence strictly limited in time.

This preoccupation was echoed and honoured in all pre-modern cultures, and
nowhere asserted itself more forcefully than in the rituals of religious calendars.
Their usual cycle of one year symbolized the repeated immersion of man through
sacred rites into the presumed timelessness of the primordial cosmogonic act. In
view of the notorious volatility of man's attention, the repetition of symbolic
actions appeared from earliest days as a most natural means by which to achieve
some measure of steadiness through communing with the realm of the changeless.
The cycles of nature offered a most suitable framework, and were well exploited
by the resolve of primitive man to escape the irreversibility of time. The ultimate
meaning that underlies the various rituals performed at the onset of the new moon,
of spring, and of the new year almost invariably intimated some hostility toward
time and its once-and-for-all events. Through these rituals primitive man hoped to
devaluate time to the point of abolishing it on the subjective level. The periodic
recurrence of the phases of the moon and the four seasons was especially effective
in creating the belief that things, events, and human consciousness in particular
went by only to reappear later at regular intervals.

The principle of recurrence, when applied to individuals and society, had, of
course, to be based on cycles with periods much longer than 28 or 365 days.
Repetitions in the relative position of the planets promoted speculations for fixing
the length of cycles for the renewal of mankind and history as a whole. But the ties

between planetary cycles and cosmologico-historical ones remained tenuous at best. By and large, mythological beliefs, anthropomorphic imagination, and cabbalistic phantasies proved to be the decisive factors that determined ancient estimates of the cyclic ages of mankind and of the universe.

Such is certainly the case in the ancient Hindu speculations about the length of cosmological cycles. The following passage from the *Vishnu Purana* is representative of many similar utterances in ancient Hindu literature:

> Oh best of sages, fifteen twinklings of the eye make a Kashtha; thirty Kashthas, one Kala; and thirty Kalas, one Muhurtta. Thirty Muhurttas constitute a day and night of mortals: thirty such days make a month, divided into two half-months: six months form an Ayana (the period of the sun's progress north or south of the ecliptic): and two Ayanas compose a year. The southern Ayana is a night, and the northern a day, of the gods. Twelve thousand divine years, each composed of (three hundred and sixty) such [divine] days, constitute the period of the four Yugas, or ages. They are thus distributed: the Krita age has four thousand divine years; the Treta three thousand; the Dwapara two thousand; and the Kali age one thousand: so those acquainted with antiquity have declared. The period that precedes a Yuga is called a Sandhya, and it is of as many hundred years as there are thousands in the Yuga; and the period that follows a Yuga, termed the Sandhyansa, is of similar duration. The interval between the Sandhya and the Sandhyansa is the Yuga, denominated Krita, Treta, etc. The Krita, Treta, Dwapara, and Kali, constitute a great age, or aggregate of four ages: a thousand such aggregates are a day of Brahma ... At the end of this day a dissolution of the universe occurs, when all the three worlds, earth, and the regions of space, are consumed with fire ... When the three worlds are but one mighty ocean, Brahma, who is one with Narayana, satiate with the demolition of the universe, sleeps upon his serpent-bed ... for a night of equal duration with his day; at the close of which he creates anew. Of such days and nights is a year of Brahma composed; and a hundred such years constitute his whole life.[3]

The passage, although relatively free of morbid details, is clearly indicative of the pantheistic, animistic, and cabbalistic roots of the Hindu obsession with perennially recurring cosmic cycles which can be tabulated as follows:

1 divine day	1 year
1 divine year	360 years
12,000 divine years	
or a full cycle of four yugas	
or 1 mahayuga	4,320,000 years
1000 mahayugas	
or 1 kalpa	4.32×10^9 years
2 kalpas	
or 1 day of Brahma	8.64×10^9 years
1 year of Brahma	
or 360 days of Brahma	3.1104×10^{12} years
100 years of Brahma	3.1104×10^{14} years
or 1 life of Brahma	or 311 trillion years

Imagination naturally thrived on embellishing the foregoing scheme with sundry subdivisions. For the most part they were as arbitrary as the length of the "twinkling of the eye," of which, as indicated above, $15 \times 30 \times 30$, or 13,500, constituted a full day or a muhurtta. This meant that one "twinkling of the eye" was almost seven seconds long. But a muhurtta could also be only one-thirtieth of a full day, and a "twinkling of the eye" 360 times shorter than its foregoing measure. At the other end of the spectrum were such embellishments as the cosmic dawn (sandhya) and cosmic dusk (sandhyansa) which preceded and followed each yuga. Together with the yugas they constituted the following timetable:

	divine years	ordinary years
Kritayuga	4000	
dawn and dusk	800	
	4800	= 1,728,000
Tretayuga	3000	
dawn and dusk	600	
	3600	= 1,296,000
Dwaparayuga	2000	
dawn and dusk	400	
	2400 =	864,000
Kaliyuga	1000	
dawn and dusk	200	
	1200 =	432,000
	4,320,000	= 1 Mahayuga

The relative lengths of the four yugas correspond to the sequence of 4, 3, 2, 1, and these four numbers stand in close relation to the names of the four yugas.[4] As "four" is the symbol of perfection, so is Kritayuga the golden age and the longest of the four yugas. "Three" is already a regression, as a triad or a triangle is less than a square. Therefore, in the Tritayuga labour, suffering, and death appear. "Two" carries farther the descent on the ladder of perfection, and as a result the Dwaparayuga (or the age of "two") is an age of general decadence. "One" which characterizes the losing throw in a dice game becomes in turn the symbol of the culmination of evil, as "kali" also may have the connotations "black" and "time." In other words, the historical, actual time corresponds to a process which is inherently ruled by evil and irrationality.

That such a process should end in a global and cosmic disaster is only natural to expect. But the hostile attitude toward time imposes almost of necessity the endless repetition of a revengeful though disastrous victory over it in the form of ever recurring dissolutions of the cosmos. The collapse of all at the end of each Kaliyuga is the "mildest" of these pessimistic finales. More radical is the world conflagration at the end of each kalpa, and both are far surpassed, if this is physically possible at all, by the "annihilation" at the end of each life of the Brahma, who in turn will keep dying and rising for ever.

The same pessimism manifests itself in the notion of "divine day" and "divine year." They are constantly referred to in the Puranas which are unanimous in identifying the present age with the Kaliyuga, the worst of the four yugas. The Puranas, a faithful reflection of Hindu mentality in its full-blown form, are also in accord in assigning the onset of the Kaliyuga to the not-too-remote past, to the middle of the second millenium B.C.[5] For those who lived with such an outlook on history, both cosmic and human, the future must have appeared extremely bleak. After all, the woes and ills of their times signalled for them only the beginning of a moral and physical decay that was to increase steadily for another 430,000 years. This represented anything but an enticing prospect, and behind its acceptance there most likely lay some dispiriting experience.

It may, indeed, be argued with some plausibility that the time scale of cosmic and historic cycles was the outgrowth of a shattering disappointment with the turn of historic events. This also explains why in some older Hindu writings the four yugas have a total span of only 12,000 years, that is, the sum of 4800, 3600, 2400, and 1200 years.[6] Now, if the beginning of the last and worst of the four yugas, the Kaliyuga, had taken place somewhere around the middle of the second millenium B.C., it should have come to an end 1200 years later, or in times that constituted a truly turbulent phase in Hindu history. Yet, the golden age, or Kritayuga, failed to come even after the last Bactrian invaders had left India in 165 B.C. In addition, the next century saw Buddhism at its height and Brahmanism at its lowest. No wonder that ancient Hinduism sought some rationale in the face of this adverse turn of events. The explanation was sought in a myth that substituted a divine day for a human year with the result that the Kaliyuga was to last not 1200 years, but as many divine years, or 1200 × 360 years. The myth was rooted in the old Vedic belief which located the dwelling place of gods in the northernmost regions, where daylight and night last half a year respectively. Consequently, if one full day of the gods was equivalent to 360 successions of daylight and night as experienced close to the equatorial regions, a year of the gods corresponded to 360 ordinary years.[7]

Such may be the immediate reasons for the replacement of the short scale of cycles with a much larger one. These reasons are, however, of secondary importance when compared with that general outlook on world and existence which readily accommodated the patently wilful shift. The ultimate origins of that outlook are older than the Vedas, the oldest and most sacred writings of Hinduism. The phrasing of the Vedas took well over a thousand years, a period that ended not later than the seventh century B.C., with the composition of the Upanishads, the concluding parts of each Veda.

In the Hymns of the *Rig-Veda*, the most important of the Vedas, the "creation" of the world is described as a process subsequent to the rhythmic breathing of "that one Thing," the undifferentiated, eternal all.[8] It is the supreme, but also all-encompassing Being who undergoes a self-evolution which gives rise to all creatures including the gods:

> The Brahman was his mouth, of both his arms was the Rajanya made.
> His thighs became the Vaisya, from his feet the Sudra was produced.
> The Moon was gendered from his mind, and from his eye the Sun had birth;

Indra and Agni from his mouth were born, and Vayu from his breath.
Forth from his navel came mid-air; the sky was fashioned from his head;
Earth from his feet, and from his ear the regions. Thus they formed the worlds.[9]

The Hymns of the *Atharva-Veda*, composed somewhat later and in a more
popular vein, mention the "earth that was before this earth, which only the wisest
Sages know."[10] There one also catches a glimpse of the god Agni, "the all con-
sumer," who devours the earth and the heaven with his flame as he passes between
them.[11] Time, in turn, is personified as riding a chariot, the wheels of which are
made of "all the worlds of creatures."[12]

These distinct elements of a pantheistic, cyclic world conception find a thematic
expression in the Upanishads.[13] "Verily, this whole world is Brahma," states the
Chandogya Upanishad,[14] and the same is paraphrased in forceful terms in the
Mundaka Upanishad:

> Brahma, indeed, is this immortal, Brahma before,
> Brahma behind, to right and to left.
> Stretched forth below and above,
> Brahma, indeed, is this whole world, this widest extent.[15]

This universal being is absolutely eternal. Expressions like "in the beginning"
do not denote an absolute start for everything out of nothing. *The Chandogya
Upanishad* expressly faces the problem of coming into being out of nothing and the
pantheistic resolution of the question is unequivocal: "In the beginning, my dear,
this world was just Being (sat), one only, without a second. To be sure, some
people say: 'In the beginning this world was just Non-being (a-sat); one only
without a second; from that Non-being Being was produced.'—'But verily, my
dear, whence could this be?' said he. 'How from Non-being could Being be
produced? On the contrary, my dear, in the beginning this world was just Being,
one only, without a second.' "[16]

The immediate continuation of this passage is extremely instructive about a
fundamental tenet of the cosmology of Hinduism. The world is in its actual
appearance the product of the self-procreation of the ultimate, pantheistic Being:
"It bethought itself, 'Would that I were many! Let me procreate myself!' It
emitted heat. That heat bethought itself: 'Would that I were many! Let me procreate myself.' It emitted water."[17] The cosmology of the Upanishads is, indeed,
articulated in terms of the self-unfolding of Brahma who goes through successive
phases of self-differentiation. The *Maitri Upanishad* describes Prajapati, the Lord of
Creation, as creating numerous offspring by meditating upon himself and making
them animate by entering into them after turning himself into wind.[18]

This pantheistic self-differentiating process of the Lord of Creation is often given
in the Upanishads in rather crude, organismic details. The opening section of the
Aitareya Upanishad describes how Atman, the Soul of the Universe, broods "in
the beginning" upon himself. The result is that as his mouth, nostrils, eyes, and ears
become distinct, fire, winds, light, and the heavens evolve as separate entities. With
the appearance of Atman's skin and hair come forth the plants and trees while the
emergence of his heart (mind) gives rise to the moon. Finally, the semen from his

virile member turns into water, and from his navel corruption begins to exude.[19]

It is through this last theme that the *Maitri Upanishad* comes face to face, right at its very beginning, with the problem of universal decay as the question is raised: "In this sort of cycle of existence (*samsara*) what is the good of enjoyment of desires, when after a man has fed on them there is seen repeatedly his return here to earth?"[20] Obviously, the ubiquitous evidences of birth-death cycles may easily lead to an extrapolation on a cosmic scale. With a single-minded resolve this trend is erected into a supreme metaphysics in the Upanishads. The "world-soul," the "overlord of all things," "the king of all things," acts itself out as a cosmic wheel: "As all the spokes are held together in the hub and felly of a wheel, just so in this Soul all things, all gods, all worlds, all breathing things, all these selves are held together."[21] The *Svetasvatara Upanishad* insists in turn that one must understand the world soul as a wheel, but there one also finds in full evidence the morbid engrossment with the symbolism of the wheel:

> We understand him [as a wheel] with one felly, with a triple tire,
> With sixteen end-parts, fifty spokes, twenty counter-spokes,
> With six sets of eights, whose one rope is manifold.
> Which has three different paths, whose one illusion (*moha*) has two conditioning
>   causes.
> . . . .
>
> In this which vitalizes all things, which appears in all things, the Great –
> In this Brahma-wheel the soul (*hamsa*) flutters about,
> Thinking that itself (*atamanam*) and the Actuator are different.
> When favoured by Him, it attains immortality.[22]

It should not be surprising that king Brihadratha exclaims as he surveys with his mind's eyes the prospect of an endless cycle of births and decays: "In the cycle of existence I am like a frog in a waterless well."[23] The king represents, of course, the individual caught up in the usual activities of daily life, while Sakayanya, to whom he turns for information about an escape route from the hopeless predicament, is one of those who have already found the solution in a resolute detachment from the world. To achieve this detachment one had first to convince himself about the illusory nature of every human enterprise, however noble and productive. Clearly, this could hardly be done more effectively than by fixing one's mind on the endless recurrence of the same hopes, same frustrations, same triumphs, and same defeats. As the imagination of each generation of Hindu sages kept adding its contribution, the final picture became gigantically mesmerizing.

To study that picture one has to turn to the literature known as the *Puranas*, which represents the popular version of Hindu thought, recorded during the middle part of the first millenium, or about a thousand years after the last Upanishads had been composed. One of the classic examples of the huge instructional pictures about the eternal returns is in the *Brahmavaivarta Purana*.[24] It depicts the god Indra, who, following his victory over the dragon Vrta, engages the services of the divine architect Vishvakarman to provide the gods with the finest palace ever built. Before long, Indra's drive for perfection compels Vishvakarman to complain to Brahma, who in turn convinces Vishnu, the Supreme Being, about the need to teach Indra a harsh lesson.

It is Vishnu himself who takes upon himself the task, and he shows up one day as a ten-year-old pilgrim boy at the gate of Indra's palace. The attractive and holy-looking boy gains immediate admittance, and with his very first words he strikes at the heart of the matter. His question to Indra about the number of years required to complete the palace is followed by the portentous remark: "No Indra before you has ever succeeded in completing such a palace as yours is to be." The cue is eagerly taken up by Indra in his question to the boy: "Tell me, Child! Are they then so very many, the Indras and Vishvakarmans whom you have seen – or at least, whom you have heard of?"

From that moment on the respective roles change. It is now the boy who addresses Indra, "my dear child," as he calmly points out that he knows Indra's father, Kashyapa, the progenitor of all creatures, his grandfather, Marichi, the beam of celestial light, and his great-grandfather as well, or Brahma, who was brought forth by Vishnu from the lotus stemming from the navel of Vishnu himself. As Brahma and Vishnu represented the totality of beings it was most natural for the young boy to turn to the successive generations of the universe, or rather universes: "I have known the dreadful dissolution of the universe. I have seen all perish, again and again, at the end of every cycle. At that terrible time, every single atom dissolves into the primal, pure water of eternity, whence originally all arose. Everything then goes back into the fathomless, wild infinity of the ocean, which is covered with utter darkness and is empty of every sign of animate being. Ah, who will count the universes that have passed away, or the creations that have risen afresh, again and again, from the formless abyss of the vast waters? Who will number the passing ages of the world, as they follow each other endlessly? And who will search through the wide infinities of space to count the universes side by side, each containing its Brahma, its Vishnu, and its Shiva? Who will count the Indras in them all – those Indras side by side, who reign at once in all the innumerable worlds?" Obviously, the answer is in the negative. The patently impossible task of counting all the grains of sand in the world should appear feasible, the boy notes, when compared with the impossibility of counting all the Indras.

The education of Indra is not over yet. He is still to be stunned by the pre-cariousness which characterizes the existence of humans, of gods, and of the whole cosmos. The lifetime of Indra stretching over seventy-one eons may appear immensely long, yet it is a fleeting moment in comparison with one "day" in the life of Brahma. But even the one hundred and eight (divine) "years" of one Brahma are not exempt from the remorseless turn of the wheel of cycles: "Brahma follows Brahma; one sinks, the next arises; the endless series cannot be told. There is no end to the number of those Brahmas – to say nothing of Indras."

The intimation of the true psychological impact of all this is done by the author of the tale in two ways. The first consists in a classic, concise portrayal of the whole universe through the perspectives of a pantheistic, animistic, perennial treadmill. The universes come and go; they are innumerable at every moment, each harbour-ing its Brahma and Indra. The situation is described in an unforgettable simile: "like delicate boats they [the universes] float on the fathomless, pure waters that form the body of Vishnu. Out of every hair-pore of that body a universe bubbles and breaks."

If this is already demoralizing from the viewpoint of ever achieving a rational understanding of the world, the death-knell is spelt on such hopes by the concluding part of the "cosmology" of the *Brahmavaivarta Purana*. As the boy prophet speaks, a dark swarm of ants appears at the entrance of the hall. The next moment the ants are already moving across the hall at a steady, silent pace in a column four yards wide, row upon row, relentlessly. Indra is seized with terror, and as he asks with dry lips and broken voice for an explanation, the boy unfolds the true meaning of the endless parade of ants: "I saw the ants, O Indra, filing in long parade. Each was once an Indra. Like you, each by virtue of pious deeds once ascended to the rank of a king of gods. But now, through many rebirths, each has become again an ant. This army is an army of former Indras."

As an imagery, the parade of countless ants is meant to convey not only the prospect of endless returns. It also expresses the Hindu belief in the moral retribution through reincarnation in lower or higher forms. Because of one's wicked actions one can reappear in the next cycle of life as a wild animal or as an insignificant insect, whereas good actions can elevate one to the rank of the supreme gods. Such a rule may in itself act as a most powerful prompting to lead a virtuous and industrious life, but the vision of eternal returns overpowers such consideration with the mesmerizing impact of a vicious treadmill. This is stated unmistakably in the last words of the boy to Indra: "Perishable as bubbles are the good and the evil of the beings of the dream. In unending cycles the good and evil alternate. Hence, the wise are attached to neither, neither the evil nor the good. The wise are not attached to anything at all."

This attitude of total detachment from the categories of good and evil, or real and unreal, was not an exclusive and universal attitude among the ancient Hindus. Theory and practice hardly ever match one another and human nature can reassert itself against theories not sufficiently consonant with its basic aspirations. But the potentialities of human nature can at the same time be effectively thwarted by such pervasive ideologies as the ancient Hindu doctrine of eternal recurrence. Its hold on the Hindu mind was strong enough to prevent the emergence of a basically positive and confident outlook on nature and on the value of man's activities concerning nature and society.

There were some notable attempts aimed at justifying a truly productive way of life. The classic case is in the *Mahabharata*, the great epic poem dating from the period that separates the age of Upanishads from the age of the Puranas. The finest part of the poem, the *Bhagavad-gita* ("Song of the Blessed One"), is probably the most influential sacred writing in the Hindu context. In essence, it is an exhortation by the god Krishna to Prince Arjuna who suddenly comes to the conclusion that there can be no justification for a fratricidal war, grave as the grievances of one side may be. In emphasizing that even a civil war can be a duty Krishna obviously tries to vindicate the notion of duty in general: if a most repugnant course of action, like civil war, can be a duty, more agreeable actions can all the more be considered such. Although Krishna's pleadings are admirable, some of his arguments are patently self-defeating. The vision of an ultimate escape from the weariness and conflicts of life is hardly buttressed by the frank admission that only the middle part of existence is clear; the beginnings and endings are utterly obscure in the inevitable process of births, deaths, and rebirths.

The desire to reassert the positive is also evident in the *Mahabharata's* lengthy portrayal of the golden age. The *Mahabharata* dwells on the happiness and perfection of every individual during the golden age, while it sums up in one phrase the calamities of the fourth age or Kaliyuga: "then excessive rain, drought, rats, locusts, diseases, lassitude, anger, deformities . . . anguish and fear of famine take possession of the world."[25] Such emphasis on the positive was largely a rearguard action. The psychological momentum generated by the doctrine of cycles, cosmic and human, could have only one main direction which pointed to the gloomy atmosphere of the Puranas. They are full of lengthy and morbid details about the physical and moral horrors of four hundred thousand years of decadence. The sixth, or concluding book of the *Vishnu Purana*[26] paints a most depressing picture of what is in store for mankind. The Kali-age would witness the spread of religious unbelief, the dissolution of family life, the loss of respect for private property, and a physical deterioration of man as well. Infant mortality would rise sharply, and life expectancy would be drastically reduced. "Women will bear children at the age of five, six, or seven years; and men beget them when they are eight, nine, or ten. A man will be grey when he is twelve; and no one will exceed twenty years of life. Men will possess little sense, vigour, or virtue, and will therefore perish in a very brief period."[27]

Parallel with the physical rundown of mankind another transformation, the steady rise of materialistic interests in man, also draws severe strictures from the author of the *Vishnu Purana*. A preoccupation with things material might entail more concerted investigation of the phenomena of nature in general, and of the properties of various substances in particular. Coupled with some progress in the various branches of production, the "materialistic" trend could be conducive to the emergence of some scientific activity. One would, however, look in vain for as much as a rudimentary concept of science, of scientific enterprise, or of scientific spirit, in the rather depressive pages of the *Vishnu Purana*. Its main purpose is rather to teach man how to escape from the clutches of the sensual, tangible world. Concern for externals, for knowledge other than ascetical, is frowned upon by Parasara, the revered teacher. He instructs his disciple, Maitreya, to look upon "mountains, oceans, and all the diversities of the earth" as "illusions of the apprehension."[28] The pure knowledge one should strive for is independent of the varieties of things and of changes. Change means the absence of reality, and the teacher minces no words about what he has in mind as he instructs the disciple on this crucial issue: "Earth is fabricated into a jar; the jar is divided into two halves; the halves are broken to pieces; the pieces become dust; the dust becomes atoms. Say, is this reality? though it be so understood by man, whose self-knowledge is impeded by his own acts."[29]

It is in that self-knowledge that man attains the simple, universal, and eternal truth that is diametrically opposed to the uncounted pieces of information about the external world. These latter form an endless series of traps, and keep one from entering into the deity, the unchanging, eternal, and universal form. For as the teacher's momentous warning reads: "In that universe which I have described, he forever migrates who is subject to the influence of works."[30] As these "works" included practically all preoccupation with the external world, cultural and scientific endeavour could hardly find any encouragement in the idea of the eternal return of all. On the contrary, the fearful prospect of never-ending cycles filled one with distrust and hostility toward the world of senses. This is not to say that everybody in ancient India led an ascetical life removed in every respect from worldly interests, intellectual or material. Yet, the notion of eternal returns certainly created a climate of opinion, the influence of which nobody in ancient India could escape.

The two great Hindu heterodoxies, Buddhism and Jainism, implied no departure from the vision of immensely long and forever recurring world periods. Characteristically enough, Buddha's first speech in the Deer Park at Benares was remembered as an act that set "the wheel of the most excellent law" a-rolling. The spokes of that wheel represented the rules of pure conduct; the uniformity of their length symbolized justice. Wisdom corresponded to the rim of the wheel; thoughtfulness and modesty to its hub, in which the immovable axle of truth was fixed. This emphasis on the immovability of truth could have served as a radically new departure in thinking in ancient India. The saints and all the good spirits of all departed generations shouted, indeed, their approval of Buddha's speech in words that seemed to convey the very same expectation: "Truly, the Blessed One has founded the kingdom of righteousness. The Blessed One has moved the earth; he has set the wheel of Truth rolling, which by no one in the universe, be he god or man, can ever be turned back. The kingdom of Truth will be preached upon earth; it will spread; and righteousness, good-will, and peace will reign among mankind."[31]

Good intentions and sound aspirations are not, however, necessarily stronger than the logic built into some basic standpoints. What ultimately powered the Buddhist movement was an escapist attitude, which fixed its sight on reaching Nirvana. The road there led through practically endless reincarnations. Consequently, Buddhist thought readily assimilated from orthodox Hinduism the idea of reincarnation for the cosmos as a whole, although its original aim largely concerned the perfection and deliverance of the individual. The great champion of Buddhism, King Asoka (c. 269 B.C.–c. 232 B.C.) had all his monuments inscribed with the swastika, the classic symbol of cyclic returns, the original form of which probably consisted of the diagram of the bodies of two intertwined snakes, symbols of evil and pessimism. What the swastikas stated concisely about the cosmos and existence, the great Buddhist ascetical classics gave in detail.

Buddhist spirituality often equated perfection with one's ability to recall as many past world cycles as possible. A typical presentation of the grades of perfection along these lines is given in the *Visuddhi-Magga* ("The Way of Purity"), the great synthesis of Buddhist doctrine compiled by the famous Buddhaghosa during the fourth century A.D. According to it those at the lowest echelon of perfection can recall only forty previous world cycles, the ordinary disciples are said to remember one hundred and one thousand such cycles, while the memory of the eighty great disciples is claimed to reach back over one hundred thousand world cycles. On the next higher step of the ladder of perfection are the two chief disciples with their ability to recall one immensity in addition to one hundred thousand world cycles. Above them are the private Buddhas who can remember two immensities and one hundred thousand world cycles. At the summit of perfection one finds the full-fledged Buddhas whose memory can go back and forth, both in the past and in the future, over an uncounted number of immensities each consisting of many times ten million world cycles.[32]

Within Jainism these cycles were largely interpreted in terms of the fate and fortunes of the sect.[33] Thus the end of *avasarpini*, or the downward motion of the "now" on the rim of the wheel of time, meant also the complete disappearance of all Jaina. This was to come in the end of the present period, called *Dusama*. This in turn was believed to have been preceded by four periods, of which the *Susama Susama*, the golden age, was the best, when every man's height was six miles, and the number of his ribs two hundred and fifty-six. The worst age, or *Dusama Dusama*, was expected to follow the present one, which was already considered thoroughly evil. Very significantly, little was said about the ascending half of the wheel or *utsarpini*. The mesmerizing impact of cyclic returns among the Jaina is also evident by the emphasis on the indefinitely large magnitude of the various phases of a cycle. One long Jaina unit of time was the *palya*, or "countless years," and the highest unit of time, or *sagaropama*, was one hundred millions of "countless years" multiplied by one hundred million. If such indefinitely large figures meant little, some Jaina practices kept the minds of the devout vividly centred on the prospect of endless ups and downs. Of these, two are particularly telling: the arrangement of rice offering in the form of a swastika, and the marking of the same symbol with one's finger on a board upon entering the temple.

It should not, therefore, come as a surprise that one of the areas where sectarian boundaries did not exist in ancient India consisted in an unbridled exercise of imagination about the details of the cyclic consumption of the cosmos by fire, water, and wind. General agreement also existed in picturing the emergence of "new" worlds as an effect of the ascendancy of evil. There was nothing sectarian in the Buddhist insistence that even the most elementary cultural endeavours, such as the first home building, had their roots in materialistic, egotistic inclinations. Thus, the best that man could do in the midst of an illusory and depraving reality was to keep himself aloof, and view everything with compassion. In the presence of perennial cosmic, historic, and social cataclysms, man was urged to adopt an attitude summed up in the classic injunction: "Cultivate friendliness, cultivate compassion, joy and indifference, wait on your mothers, wait on your fathers, and honour your elders among your kinsfolk."[34] Yet, in its essence this was not so much an active programme as a meek attitude of non-involvement. It frowned on activity as the source of the world's troubles, it aimed at leaving them untouched lest they hurt more by being treated through intervention. If the foregoing injunction still generated salutary deeds, it was because through them one's exile to earthly existence was shortened.

The idea of cyclic returns and the accompanying urge to escape its hold played havoc not only with existence on earth. It also degraded the ideal of heaven. The sacred books made no secret about the real condition of those who had reached heaven. Within the framework of eternal returns the heaven could not be a final resting place, but only a transitional abode of respite. "Its temporary inhabitant," as stated in the *Vishnu Purana*, "is ever tormented with the prospect of descending again to earth."[35] Once back on earth, man was again caught up in the inevitability of external acts and constructive work, of which even the best results were stated to be more germane to misery than to happiness. Tied to the world of senses, man was truly a captive, or as the *Vishnu Purana* graphically put it, he was "like the seed of the cotton amidst the down that is to be spun into thread."[36] A thin thread, to be sure, that is quickly worn off into minute fragments of lint. Or to recall another simile of the *Vishnu Purana*, man, while living in a body, is scorched by the fires of the sun of this world, and the only shade that is available for him is the one cast by the tree of emancipation.[37] Obviously, if even the basic earthly commodities such as family, home, and land are but subtle means to impede man in his quest for the final emancipation, he is much less likely to engage in a systematic cultivation and investigation of external things. He will not even let himself be trapped in a purely aesthetic enjoyment of nature. Such is at least the advice of the teacher in the *Vishnu Purana*, who illustrates through a lengthy case history the sad consequences of being captivated by the beauty of the external world.[38]

In such an outlook on life, that type of knowledge and experience was mainly encouraged which contributed to man's escape from bodily existence repeating itself in cycles. While Hinduism spoke about a final and absolute deliverance from existence, its date was left undefined. It was indefinitely removed in time for all practical purposes. One need only recall that after each 4.3 million years the world was believed to undergo a thorough transformation and rebirth. An even more cataclysmic transformation was believed to occur at the end of a thousand such cycles, that is after each 4.3 billion years. Long as such a period of time is, it represented in the life of Brahma only one day, which was to repeat itself until after the "hundred years" of Brahma's life, or some 300 trillion years, a most fundamental dissolution of the universe had set in. The texts, however, do not specify how many times this fundamental dissolution was to repeat itself before the onset of the *absolute* or final deliverance from all forms of existence. At any rate, even between two fundamental dissolutions there was enough time for a billion recurrences of the Kali-age. This depressingly large number became even more so when multiplied by the number of reincarnations in store for each individual.

In this grim spectacle there was one cheerful aspect, which is noticeable only in retrospect. As the teacher in the *Vishnu Purana* tells his disciple, the most fundamental dissolutions occur at the end of periods which in terms of years correspond to a number that "occurs in the eighteenth place of figures enumerated according to the rule of decimal notation."[39] The occurrence of the expression "decimal notation" in a jungle of morbid world view is of no small relevance. It evokes one of the most portentous questions that can be raised about intellectual history – the question of why science was born in the West and not in the East. This question can be phrased in a number of ways, and one of them would refer to the inability of the great cultures of the East to develop brilliant insights, discoveries, and inventions into a sustained systematic pursuit of scientific investigation.

The decimal system and notation invented in ancient India is a case in point. It undoubtedly represents the most noteworthy single contribution of ancient India to science and its importance cannot be overstated. Had the late Middle Ages been familiar with only the cumbersome Roman numerals, the vigorous development of quantitative methods would have been delayed and with it the emergence of the new science in the seventeenth century. (Yet, the decimal system needed a Stevin and a Vieta in the sixteenth century to acquire the conceptual precision demanded by systematic scientific work.) While the detailed steps of the emergence of the decimal system in India between the fourth and seventh centuries are still unclear,[40] the decimal way of reckoning is indicative of a mental attitude much different from that which pervades the Puranas and other sacred books of India.

Nothing would, indeed, be more mistaken than to picture ancient India as completely submerged in the escapist attitude generated by the notion of cyclic returns. Excavations show that already in the pre-Aryanic India there were cities with houses built of burnt bricks and provided with adequate drainage facilities. There are also evidences of the very early use of copper and bronze. Following the invasion by Vedic Aryans, India achieved an unsurpassed reputation in the art of glass-making. The monolithic pillars of Asoka, dating from the third century B.C., attest a considerable skill in stone-cutting and transportation engineering. The advanced stage of metallurgy in ancient India is evidenced by the Iron Pillar of Delhi and by the Sultanganj copper colossus of Buddha, to mention only some world-famous monuments. Behind these and countless smaller feats there lay a lively interest in industrial arts. The main literary evidence of its extent is Kantilya's *Arthasastra*,[41] a book devoted to the manifold aspects of the business of government. Some of its chapters deal with legislation relating to business, measures, construction of buildings, ships, and roads, to husbandry, agriculture, land surveying, mining, and medicine.

Practicality, craftsmanship, and organizational talent do not, however, qualify as science. The very soul of science consists in theoretical generalization leading to the formulation of quantitative laws and systems of laws. It is rather doubtful whether science did exist in that sense in India prior to its invasion by Alexander the Great. The invasion itself, which rather understandably was treated with contemptuous if not chauvinistic silence in subsequent Hindu historical documents, was not followed by any sudden rise of interest in Greek or Hellenistic science. But the lure of Greek scientific attainments proved insuperable in the long run, though reluctance to give credit where it belonged also prevailed. Thus, as late as the beginning of the sixth century A.D., the astronomer, Varahamihira, gave only passing reference to Greek and (Roman) science in his *Panca Siddhantika* ("On the Five Solutions"), a critical appraisal of the previous astronomical tradition in India.[42]

The claim about the originality of science in ancient India is seriously handicapped by the fact that critical editions of early Hindu writings with some scientific content are still largely wanting. In this respect modern Indian scholars are faced with a monumental task that is aggravated not only by the vastness of the source material, but also by what O. Spengler called the ahistorical structure of the world-consciousness of the Hindu man.[43] This aversion to history was a natural product of the ubiquitous doctrine of perennial returns that debased any phase of history to one of countless repetitions of the same cyclic pattern. Even when this slighting of history did not go to the extreme of an ascetical rejection of social and cultural life, it certainly prevented the development of a proper sensitivity for the historical dimension of individual and social existence. A telling and portentous manifestation of this is the chronic neglect on the part of ancient Hindu authors and scholars to treat, as Spengler put it, "the appearance of a book written by a single author as an event determinate in time."[44] To understand this it is enough to recall that not until the end of the fifth century A.D., did Indian science produce a treatise, the *Aryabhatiya* of Aryabhata, that contains both the name of its author and its date of composition (496). Prior to that and even to some extent after it, the written record of ancient Hindu philosophy and science is not a series of compositions assigned to specific persons but one lengthy discourse into which subsequent thinkers and copyists inserted their own ideas and felt no compunction about altering what was already recorded.

Unflattering as such an appraisal may be, it is anything but new. The famous Persian man of science, al-Biruni, noted it almost a thousand years ago in his classic work on India,[45] in which he deplored the indifference of Hindu scribes to produce reliable copies. Al-Biruni, whom G. Sarton once called one of the most enlightened and tolerant minds of all times,[46] offered his strictures in a tone that sharply contrasts with the usual evenness of his statements. "In consequence," reads al-Biruni's comment, "the highest results of the author's mental development are lost by their [the scribes'] negligence, and his book becomes already in the first or second copy so full of faults that the text appears as something entirely new, which neither a scholar nor one familiar with the subject, whether Hindu or Muslim, could any longer understand."[47]

In speaking so, al-Biruni may have been somewhat carried away, yet by and large the picture he painted was not unduly dark. Modern scholars trying to reconstruct the true development and attainments of ancient Indian science voiced time and again their utter frustration when faced with the problem of dating their sources. In the late thirties W. E. Clark noted the unhistorical character of Sanskrit literature, the impossibility of assigning dates to particular passages, and the large amount of spurious accretions even in works ascribed to individual authors.[48] The dating of some of the most pivotal documents of early Hindu science is still largely conjectural, as can be seen from the widely differing evaluations of the true age of the Bakshali manuscript.[49] If its date of composition could definitively be assigned to around A.D. 200, or even earlier, as some modern Indian scholars argue,[50] doubters of Hindu originality in science would effectively be answered. For the time being the lack of a firm consensus on the true age of the manuscript considerably weakens the documentary value of its impressive contents.[51]

Such and similar realities about the true status of scientific origins in India would hardly delight scholarly admirers of the Eastern mind. It is, therefore, especially gratifying to see some of them subordinate their obvious preferences to the dictates of objectivity. Joseph Needham, a leading Western student and admirer of Oriental science, felt impelled to warn that a reliable dating of the ancient records was an indispensable prerequisite for an objective appraisal of ancient Indian science. His warning came in the wake of a conference of Indian and South Asian scholars gathered in Delhi, in November 1950, for a discussion of matters relating to the history of science in India and Southeast Asia. Though the Conference strengthened Needham's belief in the eventual success of proving the equality of early Eastern and Western science, it also reminded him of "the extreme uncertainties in the dating of the most important texts and even of actual objects which have survived."[52] While the next ten years did not pass without some valuable critical investigation of ancient Indian documents, the task is evidently too enormous to be accomplished even with concerted efforts within a few years or decades. No wonder that the results of the Symposium on the History of Sciences in Ancient and Medieval India, held in Calcutta in 1961, turned out to be disappointing for precisely the same reason. The chairman of the Symposium, A. C. Ukil, rightly called attention to the fact that much of the source material was still scattered in manuscript form, and that the use of what was available was very often hampered by "our lack of knowledge regarding chronology."[53]

Without a definitive chronology, Indian scholars proud of their cultural heritage shall not prevail in their efforts to dissipate the persisting suspicion about the Greek origin of most of what is best in ancient Hindu science. They may feel incensed by the trend whose principal originator, G. R. Kaye, concluded to the complete dependence of early Hindu science on the Greek. According to Kaye's general thesis, which he argued in numerous publications,[54] all the noteworthy contributions by the Hindus of old consisted in embellishments to various achievements of Greek mathematics. Undoubtedly, the invasion of India by Alexander the Great served as a powerful starting point for a cultural flow from the West to the East, but this cannot in itself prove the lack of originality of the whole of ancient Indian mathematics.[55] Nevertheless, the proof of originality of any outstanding scientific contribution largely depends on its proper dating. And therein lies the real crux of much of ancient Indian science.

The lack of accurate dating of records is not the only factor that keeps the modern scholar frustrated about science in ancient India. There is also the intermingling of instructive and foolish elements in ancient Hindu writings relevant to matters scientific. The extent of that mixture is at times shocking. It prompted al-Biruni to describe the mathematical and astronomical lore of his Hindu hosts as "a mixture of pearl shells and sour dates, or of pearls and dung, or of costly crystals and common pebbles."[56] He accused them of having no love for truth, for being unable to overcome the many absurd notions about the physical world that infested their religious literature. In this latter connection al-Biruni pointed to their infatuation with very large numbers and very long epochs of great variety. One can clearly see here the sad inhibitory influence of the doctrine of eternal cycles on ancient Hindu thought regarding the adoption of sober scientific thinking. "They cannot raise themselves," wrote al-Biruni, "to the methods of strictly scientific deductions."[57]

Severe as such a stricture may appear, one would look in vain in ancient Hindu scientific literature for anything comparable to the geometry of Euclid, to the treatises of Archimedes, or to the account of planetary motions by Ptolemy. All these and a number of smaller products of ancient Greek science testify a love for scientific systematization and reveal a mind that, when in pursuit of science, can leave behind the murky realm of legends and magic. The classic illustration of this is the difference between the two major works of Ptolemy, the *Almagest* and the *Tetrabiblos*. The latter shows a thinker completely lost in the labyrinths of astrology, while in the former no magic or obscurantism taints the long, sustained geometric account of the motion of planets. This purity of scientific air does not constitute the exclusive atmosphere of any notable document of ancient Indian science with the logical exception of some writings on algebra. As to geometry, it would have offered a splendid opportunity for the speculative and philosophical bent of the Indian mind to exercise itself unhampered by nonsensical and superstitious notions. Geometry is, however, the area where even the most extreme modern advocates of the grandeur of ancient Indian science restrain their extravagant claims.[58]

The modern reader can only be dazed as his reading of ancient Indian scientific writings leads him through a continual alternation between sensible arguments and repulsive fantasies. A rare exception is that little gem of ancient Hindu astronomy, the *Aryabhatiya* of Aryabhata.[59] A metric composition of some 120 lines, it was intended to serve as a capsule formula for the principal results of astronomical knowledge toward the end of the fifth century A.D. Short as it is, the work reveals the keen, progressive mind of its author who seems to have recognized the rotation of the earth on its axis. At the same time the work also shows that the cyclic ages and the calculations of their lengths in terms of Brahma's life-span were of fundamental importance even to Aryabhata, the most scientifically inclined figure of early Hindu astronomy. This is evident by the corrections that Aryabhata made concerning the respective lengths of the four yugas, which he claimed to be equal. He also took pains to fix the date of the onset of the Kaliyuga. According to him it did not take place around the middle of the second millenium B.C., but rather on the Thursday of the great Bharata battle in 3102 B.C.[60]

Needless to say, such precision reflects not scientific accuracy but preoccupation with mythological cosmology. Nothing reveals this more than the prominence given by Aryabhata to the concept of yuga. Immediately following the two introductory phrases of his little treatise, he gave the number of revolutions for each planet during the four yugas, or 4,320,000 years.[61] In a later section he spelled out the various steps that lead from the solar year through the years of gods to the day of Brahma[62] in essentially the same way in which this was done in religious literature. The grip of a world view based on perennial cycles is evident in Aryabhata's definition of the day of Brahma as consisting of a thousand and eight yugas of the planets, and also in his contention that time has neither beginning nor end. The fact that he made room for such details in an extremely short compendium of astronomy should speak for itself.

If an Aryabhata was unable to discourse about astronomy without paying homage to the day of Brahma and to the innumerable cycles it entailed, much less could secondary figures of early Hindu astronomy avoid the mental snare of those topics. Of the many examples let it suffice to recall the *Surya Siddhanta* ("Solution by the Sun"), possibly the most widely diffused ancient Hindu text on astronomical matters.[63] A part of its first chapter is devoted to the specification of the length of time that had elapsed since the onset of the present Kali-age. More importantly it aims at determining the length of time between that date and the beginning of the most recent of Brahma's days, which also marks the latest in an

c

endless series of general creations. The length of time is given as 155,521,970,784,000 years.[64]

The significance of the invariable presence of statements implying such large figures in Hindu astronomical literature cannot be emphasized enough. It indicates forcefully the extent to which the cyclic concept dominated even the more sober segment of ancient Hindu discourse about the physical universe. Partly because of that concept, ancient Hindu astronomy remained pervaded by magical and astrological notions that reveal themselves in countless small details. The *Surya Siddhanta* carefully warns of the malignant aspects of the sun and moon when in equal declination: "Owing to the mingling of the nets of their signal rays, the fire arising from the wrathfulness of their gaze . . . is originated unto the calamity of mortals . . . Being black, of frightful shape, bloody-eyed, big-bellied, the source of misfortune to all, it is produced again and again."[65] The ominous ring of these last words should be no surprise. It merely forms a small echo of that dispiriting tonality which the concept of perennial returns was bound to produce. Its manifestation in the *Surya Siddhanta* is restrained. Other astronomical writings seem to have fallen completely under its sway, and as a result they readily mix reason and observation with gross superstition. It can provoke but dismay and frustration when the correct explanation of eclipses is followed by a section such as, "If at the time of an eclipse a violent wind blows, the next eclipse, will be six months later. If a star falls down, the next eclipse will be twelve months later. If the air is dusty, it will be eighteen months later. If there is an earthquake, it will be twenty-four months later. If the air is dark, it will be thirty months later. If hail falls, it will be thirty-six months later."[66]

Such passages, the number of which is almost countless, are consistently overlooked by the chauvinistic wing of twentieth-century Indian historians of early Hindu science. Hand in hand with their readiness to gloss over the obvious goes their puzzling insensitivity to chronology. The results are publications which constitute an overt affront to the intellect of the average reader, to say nothing of the historian of science.[67] But even when the complexities of early Hindu science are tackled in a scholarly manner, a yielding to chauvinistic inclinations merely produces works that at best distinguish themselves by their vague generalities.[68] It is, of course, not difficult to understand the uneasiness of the present-day Indian historian on being faced with the indisputable prominence of science in modern life and of its Western origins and developments. A spectacular achievement, such as the creation of the largest democratic community on earth, the modern Indian state, naturally wants to boast of ancient roots. Among these roots the one relating to science has a most special appeal as no modern state is viable without full reliance on scientific facilities, research, and education.

In trying to put the scientific root of Indian culture in evidence the modern Indian scholar cannot help being frustrated by the uncertainty of the datings of what is best in ancient Hindu scientific heritage. On the other hand, in the absence of a strict chronology, a scholar proud of his own national background can easily be swayed by what Needham did not hesitate to call "a chauvinistic tendency, an effort to minimize foreign influences on Indian science and to emphasize all outward transmissions."[69] Another Western scholar and admirer of Indian culture,

J. Filliozat, also chided his Indian colleagues, who in their excessive national pride are "prone to believe that their sciences in high antiquity surpassed even those of today."[70] It seems that conditions in this regard did not improve considerably from the time al-Biruni engaged in scholarly exchange with the Hindus of his day. Only rarely is the voice of sincere admission heard among modern Indian scholars concerning the negative sides of the scientific past in India. To most Indians it should be harsh to hear a compatriot of theirs, N. R. Dhar, admit that there was no serious experimentation and no acceptance of the experimental method in India after the eighth century. Moreover, he blamed on this fact the backwardness and the scant measure of honesty that are often in evidence in various Indian efforts and in everyday life.[71]

The existence of a systematic scientific experimentation in India prior to the eighth century still demands convincing proof though the cause may not be completely hopeless. It may also be hoped that the Indian mind will soon shake off the shackles of that incomplete measure of honesty, or insufficient respect of reality. Its hold is not the result of the neglect of experimental method. More likely it is the other way around. The spirit of experimental method simply could not assert itself in a cultural ambience in which the urge to escape from reality constituted a pervasive pattern. With the slighting of reality there came a weakening of the search for truth about the external world. Science, however, cannot arise, let alone gain sustained momentum, without an articulate longing for truth which in turn presupposes a confident approach to reality. These elementary verities should be kept in mind by modern Indian scholars who feel dejected because of the absence of a robust, genuine scientific movement in their ancient and more recent history, to say nothing of those who try to talk away the factual record. The outstanding feature of that record is the atrophying impact that can be exercised by the belief in perpetual cosmic and historic cycles on a culture, however vast and potentially great.

The wheel of Asoka, now adorning the Indian national flag, is as telling a symbol to the perceptual historian as are the statues of Shiva to the general onlooker. The six limbs, four arms, and two legs of the Lord of the Dance follow one another in a frenzied rotation, with the same inevitability as do the spokes of a spinning wheel. Attachment to that symbolism may appear a trifle, but symbols mould minds and attitudes with an often unsuspected effectiveness. Whether the impact of science and scientific education on the modern Indian mind can turn the emblem of the wheel into an innocuous diagram, remains to be seen.[72]

While the future is always obscure, the past, in this respect at least, is remarkably clear. The case of India shows that infatuation with a cyclo-animistic and pantheistic concept of the world put a strait jacket on thought and will alike. Contentions about psychological[73] and instructional[74] benefits of the Indian preoccupation with the wheel of cycles can hardly conceal the fact that the wheel kept in rotation an ominous and debilitating treadmill. Escape from it was well-nigh impossible either emotionally or conceptually. Only confusion in logic could be generated by a pantheistic description of the cosmos in which the supreme deity was defined as being cause and effect simultaneously.[75] Encouragement of contradictions was readily forthcoming from the belief which pictured the world as a huge egg resting in the womb of a deity endowed with bi-sexual capability.[76] Irrational proclivities were inevitably strengthened by an imagery which saw the production of serpents from Brahma's hair as it shrivelled in disgust.[77]

That Brahma itself was often depicted as reclining on a bed of convoluting snakes served as a powerful reminder that in the universe everything was the prey of blind, capricious convolutions or cycles. The laws of those cycles permitted no rational explanation, as it was a patently absurd task to make a critical analysis of the breathing of Brahma, which allegedly regulated the universe. If man was a tiny part of a huge cosmic animal, there remained little if any psychological possibility that he could ever achieve a conceptual stance which would put him outside the whole for a critical look at it. Overpowered by the illusion that he was the senseless product of an all-pervading biological rhythm, man had no choice but to capitulate to a perennial rise and fall of the cosmic waters of existence, whose murky depths exuded no sense of purpose.

NOTES

1. *Le système du monde: histoire des doctrines cosmologiques de Platon à Copernic* (Paris: Hermann, 1913), vol. I, p. 5.
2. For a recent and highly acclaimed discussion of this point and of the topics of the next few paragraphs, see M. Eliade, *The Myth of the Eternal Return*, translated from the French by W. R. Trask (New York: Pantheon Books, 1954), pp. 19–21.
3. *The Vishnu Purana: A System of Hindu Mythology and Tradition*, translated from the original Sanskrit and illustrated by notes by H. H. Wilson (London: John Murray, 1840), Book I, chap. iii, pp. 22–25.
4. For further details, see H. Zimmer, *Myths and Symbols in Indian Art and Civilization*, edited by J. Campbell (New York: Pantheon Books, 1946), pp. 13–14.
5. Wilson's still instructive interpretation of the Puranas identifies this date with the war of the Mahabharata in the fourteenth century B.C.; see *The Vishnu Purana*, p. 485 note.
6. Such is the case in the great epic poem, *The Mahabharata*, a prose English translation, edited and published by Manmatha Nath Dutt (Calcutta: H. C. Dass, Elysium Press, 1896), vol. III, p. 275; Vana Parva, chap. 188.
7. On this explanation of the change of time-scale, see V. G. Aiyer, *The Chronology of Ancient India* (Madras: G. A. Natesan, 1901), pp. 112–16.
8. See Book X, Hymn 129, in *The Hymns of the Rgveda*, translated with a popular commentary by R. T. H. Griffith (Varanasi-1, India: The Chwokhamba Sanskrit Series Office, 1963), p. 129.
9. Book X, Hymn 91, vv. 12–14; *ibid.*, pp. 519.
10. Book XI, Hymn 8, v. 7; in *The Hymns of the Atharva-Veda*, translated with a popular commentary by R. T. H. Griffith (2d ed.; Benares, India: E. J. Lazarus, 1917), vol. II, p. 80.
11. Book X, Hymn 8, v. 39; *ibid.*, p. 41.
12. Book XIX, Hymn 53, v. 309; *ibid.*, p. 309.
13. Subsequent references are to the *Thirteen Principal Upanishads*, translated from the Sanskrit by R. E. Hume (2d rev. ed.; London: Oxford University Press, 1934).
14. Third Prapathaka, fourteenth Khanda; *ibid.*, p. 209.
15. Second Mundaka, second Khanda; *ibid.*, p. 373.
16. Sixth Prapathaka, second Khanda; *ibid.*, p. 241.
17. *Ibid.*

18. Second Prapathaka; *ibid.*, pp. 415–16.
19. First Adhyaya, first Khanda; *ibid.*, pp. 294–95.
20. First Prapathaka; *ibid.*, p. 413.
21. *Brihad-Aranyaka Upanishad*, second Adhyaya, first Brahmana; *ibid.*, p. 104.
22. First Adhyaya; *ibid.*, pp. 394–95.
23. *Maitri Upanishad*, first Prapathaka; *ibid.*, p. 414.
24. For further details of the following story, see H. Zimmer, *Myths and Symbols in Indian Art and Civilization*, pp. 3–11.
25. *The Mahabharata*, Vana Parva, chap. 149; *ed. cit.*, vol. III, p. 218.
26. See especially chap. 1; *ed. cit.*, pp. 621–26.
27. *Ibid.*, p. 624.
28. Book II, chap. 12; *ibid.*, p. 241.
29. *Ibid.*, p. 242.
30. *Ibid.*
31. *The Gospel of Buddha according to Old Records*, by P. Carus (6th ed.; Chicago: The Open Court Publishing Co., 1898), p. 43.
32. See chap. xiii, translated in H. C. Warren, *Buddhism in Translations* (Cambridge, Mass.: Harvard University Press, 1896), pp. 315–30. Another Buddhist classic of spirituality, the *Samyutta-Nikaya*, speaks of a mountain, one league in height, width, and depth, which is gently rubbed by a man once in every hundred years with a silk kerchief. The mountain, it is claimed there, would sooner wear away than a world cycle would come to an end. In the same source one also reads the very topical statement that not only many hundreds or thousands, but many hundreds of thousands of such cycles had already rolled by. There one also finds perfection equated with one's ability to recall as many past world cycles as possible. For the lengthy passage in question, see *Buddhism in Translations*, pp. 315–16 note. The same work abounds in references concerning the prevalence of cyclic world view in Buddhist thought.
33. For further details, see *The Heart of Jainism*, by Mrs. Sinclair Stevenson (London: Oxford University Press, 1915), pp. 272–78, and 251–53.
34. *Visuddhi-Magga*, chap. xiii; *Buddhism in Translations*, p. 322.
35. Book VI, chap. v, p. 641.
36. *Ibid.*
37. *Ibid.*
38. The story is that of a king who saves a fawn from drowning in a river, and takes a special liking to it. As a punishment he is born again as a deer, has to subsist on dry grass until he atones and is pure enough to be reborn as a Brahman. See Book II, chap. xiii, pp. 244–245.
39. Book VI, chap. iii, p. 630. This is, of course, in disagreement with the fifteen-digit figure given in Book I, chap. iii. See note 3 above.
40. In addition to problems of Hindu documentary evidence, there is the possibility of a dependence on Western influence, which exerted itself very strongly in astronomical matters.
41. *Kantilya's Arthasastra*, translated by R. Shamasastry, with an introductory note by Dr. J. F. Fleet (4th ed.; Mysore, India: Sri Raghuveer Printing Press, 1951). Its sole extant manuscript copy was discovered around the turn of the century. The core of the work seems to have been written as early as the third century B.C., and probably much of it is of no later date than the fourth century A.D. It consists of fifteen sections of unequal lengths, most of which deal with the art of warfare, with the running of the king's household, and with the distribution of justice for various offences.
42. Varahamihira's reticence was all the more revealing as two "Solutions" (of the original five now only one, the *Surya Siddhanta*, is extant) were known as the *Romaka Siddhanta*, or "Roman Solution," and the *Paulisa Siddhanta*, or "The Solution according to Paul of Alexandria."
43. *The Decline of the West*, translated by Charles Francis Atkinson (New York: Alfred A. Knopf, 1947), vol. I, p. 12.

44. *Ibid.*

45. *Alberuni's India: An Account of the Religion, Philosophy, Literature, Geography, Chronology, Astronomy, Customs, Laws and Astrology of India about A.D. 1030.* An English edition with notes and indices by Edward C. Sachau (London: Kegan Paul, Trench, Trübner & Co., 1910).

46. *Introduction to the History of Science* (Baltimore: The Williams and Wilkins Company, 1927), vol. I, p. 707.

47. *Alberuni's India*, vol. I, p. 18.

48. In his essay, "Science," in G. T. Garratt (ed.), *The Legacy of India* (Oxford: Clarendon Press, 1937), p. 335.

49. The name refers to the village of Bakshali, near Mardan, on the northwest frontier of India. There, in 1881, a mathematical script written on birch-bark was unearthed in a ruined stone enclosure. The script consists of some 70 leaves, of which several are small scraps, while the largest measures 14 by 9 centimetres. A complete photographic edition with commentaries was given by G. R. Kaye, *The Bakshali Manuscript: A Study in Medieval Mathematics* (Archaeological Survey of India. New Imperial Series. Vol. XLIII. Parts I & II. – Calcutta: Government of India Central Publication Branch, 1927). According to Kaye the manuscripts date from the twelfth century (p. 75), but R. Hoernle, who started Kaye's critical edition, preferred the eighth century, although not without some reservations.

50. Some of them do it with no apparent concern for contrary evidence and opinion, as can be seen in the work by B. Datta and A. N. Singh, *History of Hindu Mathematics: A Source Book* (new ed.; Bombay: Asia Publishing House, 1962). The first part of this work was originally published in 1935 and the second in 1938. In the Preface the authors state that their aim is to provide a reliable authentic history of Hindu mathematics "from the earliest known times down to the seventeenth century of the Christian era." Obviously, in such a work one would expect some discussion of questions relating to the dates of composition of individual works. This is done to some extent, but not, for instance, in connection with the Bakshali manuscript. While the authors assign the manuscript to the eighth century, they simply claim that its original form goes back to the third century A.D., or earlier (p. 81). Such is a curious procedure, though not the only one of its type in the book. In stating the very early origin of Hindu numerals and place value notation the authors add: "[this] is further strengthened by Indian tradition, Hindu, Jaina as well as Buddhist, which ascribes the invention of the numerical notation to Brahma, the Creator, and thereby claims it as a national invention of the remotest antiquity" (pp. 36–37). Such passages speak for themselves.

51. This includes problems involving systems of linear equations, indeterminate equations of the second degree, arithmetical progressions, quadratic equations, approximate evaluations of square roots, complex series, practical calculations and measurements.

52. See his comments in his note, "History of Science and Technology in India and South-East Asia," in *Nature*, 168 (1951): 64–65.

53. *Proceedings of the Symposium on the History of Sciences in India held at Calcutta on August 4 and 5, 1961,* in *Bulletin of the National Institute of Science of India*, No. 21, 1963. (New Delhi: National Institute of Sciences of India, 1963), p. v.

54. Of these the most important are his *Indian Mathematics* (Calcutta & Simla: Thacker, Spink & Co., 1915), his edition of the Bakshali manuscript quoted above, and his studies on Aryabhata, the great sixth-century Indian astronomer and mathematician. S. Ganguli in his "Notes on Indian Mathematics: A Criticism of George Rusby Kaye's Interpretation" (*Isis*, 12 [1929]: 132–45), failed to bring forth any argument that would impose a substantial revision of Kaye's principal conclusions.

55. Swayed by the magnificent spectacle of Greek science, even such an observant scholar as Paul Tannery refused to recognize that the substitution of semi-chords or sines for chords in trigonometry was not a Greek but an Indian achievement. "For Tannery, the fact that the Indians knew of sines was sufficient proof that they must have heard about them from the Greeks," as was pointedly noted by J. Filliozat in his *La doctrine classique de la médecine indienne* (Paris: CNRS and Geuthner, 1949), preface.

56. *Alberuni's India*, vol. I, p. 25.

57. *Ibid.*

58. Thus S. Ganguli (*art. cit.*, see note 54 above) stated on the one hand that Hindu geometry was "mainly experimental and intuitional," and claimed on the other hand that the demonstrative character of Greek geometry would have appealed to the speculative Hindus, had they become familiar with it at a much earlier date, that is prior to Brahmagupta, or about the end of the sixth century A.D. (p. 144).

59. An English translation of it with extensive notes was made by W. E. Clark under the title, *The Aryabhatiya of Aryabhata: An Ancient Indian Work on Mathematics and Astronomy* (Chicago: University of Chicago Press, 1930). The work is also the earliest known Indian mathematical and astronomical text bearing the name of an individual author.

60. *Ibid.*, p. 12. Friday, February 18, 3102 B.C., according to H. Zimmer, *Myths and Symbols in Indian Art and Civilization*, p. 15.

61. *Ibid.*, p. 9.

62. *Ibid.*, p. 53.

63. In its present form it dates from the tenth century, though its substance goes back probably to c. A.D. 400. As was already noted (see note 42 above) it was preceded by other *Siddhantas*, all of which show the Vedic and Western influences. Subsequent references are to the "Translation of the Surya-Siddhanta, A Text-Book of Hindu Astronomy; with Notes and an Appendix," by Rev. Ebenezer Burgess, in *Journal of the American Oriental Society*, 6 (1860): 141–498.

64. Chap. I, vv. 22–23, p. 155.

65. Chap. XI, vv. 3, 5.

66. Al-Biruni, upon quoting this passage from the *Samhita* of Varahamihira, could only add: "To such things only silence is the proper answer." *Alberuni's India*, vol. II, p. 114. This overbearing presence of wholly useless details in much of the ancient Hindu astronomical literature is usually minimized in modern monographs on the subject, such as *Hindu Astronomy*, by W. Brennand (London: Chas. Straker & Sons, Ltd, 1896) and *History of Astronomy during the Vedic and Vedanga Periods*, by Sankar Balakrishan Dikshit (Calcutta: Government of India Press, 1969).

67. As examples of this, mention should be made of at least two works: *The Positive Sciences of the Ancient Hindus*, by Brajendranath Seal (London: Longmans, Green and Co., 1915), and *Hindu Achievements in Exact Sciences: A Study in the History of Scientific Development*, by Benoy Kumar Sarkar (New York: Longmans, Green and Co., 1918). The quality of reasoning present in these works can be gauged from a comment of Seal on the ignorance of Hindus of Archimedes' Principle: "I think it may be regarded as fairly certain that the Hindus were ignorant of Archimedes' discovery, an ignorance which, at any rate, they could not have well borrowed from the Greeks, no more than they could have thus borrowed their knowledge of things unknown to the Greeks themselves" (p. 249).

68. As typical examples, one may refer to *Ancient Indian History and Culture*, by Chidambra Kulkarni (Bombay: Karnatak Publishing House, 1966), see especially p. 298; *A History of Hindu Chemistry from the Earliest Times to the Middle of the Sixteenth Century A.D.* (London: Williams and Norgate, 1902–09) by Praphulla Chandra Ray; "Scientific Spirit in Ancient India," by R. C. Majumdar, in *Journal of World History*, 6 (1960–61): 265–73.

69. "History of Science and Technology in India and South-East Asia," p. 64. As illustrations of this one may mention "Some Notes on Indian Astronomy," by S. R. Das, in *Isis*, 14 (1930): 338–402; "On Indian Atomism," by B. V. Subbarayappa in *Proceedings* (see note 53 above), pp. 118–29; "The Nature of the Physical World," by Umesh Mishra in Haridas Bhattacharyya (ed.), *The Cultural Heritage of India*, Vol. III. *The Philosophies* (2d rev. ed.; Calcutta: The Ramakrishna Mission, 1953), pp. 494–506.

70. *La doctrine classique de la médecine indienne*, preface.

71. "Progress of Science and the Experimental Method," in *Proceedings* (see note 53 above), pp. 31–35. Its conclusion reads: "Instead of following the path of truth, progress and science, we succumbed to moral and mental slavery."

72. For a most recent, articulate and outspoken statement of this point, see the monumental

study by Gunnar Myrdal, *Asian Drama: An Inquiry into the Poverty of Nations* (New York: The Twentieth Century Fund, 1968), vol. I, pp. 93–112.

73. Thus, for instance, Mircea Eliade in his most perceptive study, "Time and Eternity in Indian Thought," in *Man and Time: Papers from the Eranos Yearbooks* (New York: Pantheon Books, 1957), pp. 173–200.

74. See, for instance, the statements of J. Filliozat in his "India and Scientific Exchanges in Antiquity," in G. S. Métraux and F. Crouzet (eds.) *The Evolution of Science* (New York: The New American Library, 1963), pp. 94–96.

75. *The Vishnu Purana*, Book I, chaps. ii and iv, pp. 8 and 29.

76. *Ibid.*, Book I, chap. ii, p. 19.

77. *Ibid.*, Book I, chap. vi, p. 46.

CHAPTER TWO

The Lull of Yin and Yang

"China has no science, because according to her own standard of value she does not need any . . . China has not discovered the scientific method, because Chinese thought started from mind, and from one's own mind."[1] So wrote a distinguished Chinese scholar, Yu-Lan Fung, about half a century ago. His primary aim was to explain the backwardness of China which he traced to the fact that China had neither developed nor adopted a vigorous cultivation of science. Acknowledgment of this fact hardly represented a novel insight. What made his words significant was rather the tone of subtle defiance which should convey something very fundamental about the great drama of Chinese civilization.[2]

The claim that China needed no science is arguable to say the least. But no exception can be taken to the suggestion that it was a state of mind that prevented science from taking roots in Chinese soil up to very recent times. By strange coincidence, that state of mind achieved self-consciousness at about the same time when the Greeks of old began to display their highest form of creativity in pre-Socratic Athens. Chinese and Greeks were alike at grips with the problem of how to achieve a stable course for the social and political enterprise of man in the midst of great turmoils and continual warfare. In China the main solutions regarding social stability came from the main trends of thought represented by the Mohists, the Confucians, and the Taoists. Of these, the Mohists, or the followers of Mo Ti (c. 500 B.C.–c. 425 B.C.), made the least mark on Chinese cultural history. Their insistence on equality, brotherhood, and practical artisanship had no lasting appeal to the Chinese. This is all the more significant as the Mohist way of thinking had some unmistakable characteristics of scientific mentality. It was in a book ascribed to Mo Ti that there appeared definitions about space, duration, causality, geometric figures, and energy that have some depersonalized, abstract, and quantitative flavour.[3] In addition, the Mohists strove for practical implementation of general propositions. Mo Ti himself earned fame not only for his dicta but also for the building of catapults and other war machines in defence of security and equality. But what for the Mohists was a systematic effort to control the future on the basis of past experiences, soon became branded as cheap utilitarianism and lack of refinement.[4]

The charge came from the Confucians whose views gained an official status in Chinese society as time went on. The refinement at which Confucius (551 B.C.–479 B.C.) and his followers aimed, represented a way of thinking and a state of mind that was ultimately beyond the reach of plain logic and of criticism rooted in the observation of nature. For the Confucians man's social existence served as the principal source of information about nature as well. Such an approach to nature could hardly inspire a search for quantitatively exact laws. Social interaction had its intangible finesses, its unwritten patterns, and their largely unpredictable modifications. The laws of social living were in reality customs, imperceptibly emerging and slowly imposing themselves as "laws." In the eyes of the Confucians, the "laws" that ruled nature were something closely akin to social customs. As Hsun Tzu, a leading Confucian of the third century B.C., wrote:

> Custom [*li*] is that whereby Heaven and Earth unite, whereby the sun and moon are brilliant, whereby the four seasons are ordered, whereby the stars move in their courses, whereby rivers flow, whereby all things prosper, whereby love and hatred are tempered, whereby joy and anger keep their proper place. It causes the lower orders to obey, and the upper classes to be illustrious; through a myriad changes it prevents going astray. If one departs from it, one will be destroyed. Is not custom [*li*] the greatest of all principles?[5]

The generality of the concept *li* was well matched by its vagueness. There was something inevitable in this as can be easily seen if attention is turned to a major Confucian document, the *Shi Ming* dictionary, compiled around A.D. 100. There *li* is defined in terms of *thi*, or living body. This biological or organismic connotation of the concept *li* should be kept in mind as one reads in the same dictionary the description of the basic law of external nature as "the way in which the affairs of the society are (or should be) handled."[6] The handling of the affairs of society therefore relied heavily on the vagueness of intuitive or introspective reflections. The ways of society and of nature had to be "felt," and this implied a ready acceptance of what is an invariable component of the organismic outlook on existence, namely, that every process has a rhythmic or cyclic patterning.

Thinking in terms of cycles was a prominent feature of the Confucian concern with the correct diagnosis of the past and future course of political life. Confucius regarded his own times as the Age of Disorder, and hopefully projected its eventual replacement by the Age of Rising Peace, which in turn was to be followed by the Age of Great Peace. These three were believed to form a major cycle in the unfolding of history, and by endorsing this view Confucius greatly enhanced the stature of an already traditional idea.[7] It was natural that the doctrine of historical cycles should receive further elaborations by subsequent exponents of Confucius' ideas. Thus Meng Kho (Mencius, c. 374 B.C.–289 B.C.), the greatest disciple of Master Kung (Confucius), laid it down as a rule "that a true royal sovereign should arise in the course of five hundred years, and that during that time there should be men illustrious in their generation."[8]

Such crude systematizing was not to the liking of history. "Judging numerically," Mencius noted, "the date is past" for the expected appearance of distinguished leaders. The beginning of the Chou dynasty had taken place 700 years earlier, and as a result the 500-year cycle was running 200 years behind schedule. Belief in historic and cosmic cycles exacted once more its typical ransom from minds captivated by it. The resigned, disappointed, if not fatalistic state of mind can easily be sensed in Mencius' closing remark: "But Heaven does not yet wish that the kingdom should enjoy tranquillity and good order. If it wished this, who is there besides me to bring it about? How should I be otherwise than dissatisfied?"[9]

When Tung Chung-Shu succeeded in making Confucianism the official state doctrine in 136 B.C., the emphasis on organismic considerations and their close ties with cyclic patterns became even more evident. The classic document of this is Tung Chung-Shu's *Luxuriant Gems of the 'Spring and Autumn Annals'*, where the organismic and cyclic interpretation of individual, social, and cosmic existence is presented in elaborate detail.[10] Tung Chung-Shu described the erect posture of man as expressive of man's direct communication with the skies, and the roundness of man's head was for him a replica of the shape of heaven. He correlated man's hair with the stars and constellations, his ears and eyes with the sun and the moon, his breathing with the winds. The abdomen and the womb were, according to him, as full of things as the earth was, and he envisioned the human body from the waist down as a mirror image of the earth. From this followed the parallelism between the spread-out feet and the supposedly square shape of the earth.

Needless to say, the parallelism between man's body and the parts of the cosmos did not consist of static features alone. An organism is an entity existing in time, and Tung Chung-Shu was most eager to suggest connections between the structuring and operations of the human body on the one hand and the units of time and their recurrences on the other. It is in this light that one should take his patently absurd claim about the number of lesser joints in the human body as being exactly the same as the number of days in a year. The same holds of his "count" of twelve large joints in the body. This figure and the four limbs matched the twelve months and the four seasons, or one complete revolution of the heavens. According to Tung Chung-Shu the continual opening and closing of one's eyes corresponded to the succession of day and night. Winter and summer were in turn reflected in man's periods of weakness and strength, while the alternation of sorrow and joy was due to the oscillation of the twofold cosmic forces, the Yin and Yang.

For a true Confucian like Tung Chung-Shu, the matching of the individual and the heavens went hand in hand with the analysis of human society and history. According to him, the succession of dynasties followed the changes of heaven, or rather the threefold stages of the integration by heaven of the material force, *ch'i*. These stages were also manifested in the development of a flower from its budlike (or closed) condition through the opening of its petals to full blossoming. All this in turn was duplicated in the behaviour of the ruler's court, where dark, white, or red vestments were worn, depending on the stage of development of the political system. By acting out this symbolism the ruler demonstrated that he and his kingdom were in full agreement with the actual condition of the cosmic force. Or as Tung Chung-Shu put it: "The reason why the Three Systems are called the Three Correct Systems is because they make things operate. When the integration is extended to cover the material force of all things, they all will respond [to Heaven]. As the correct system is rectified, everything else will be rectified."[11]

The key to success and harmony was therefore one's docility and willingness to merge into the rhythm of cosmic cycles that were also the patterns of human history. Such a tenet embodied a most genuine inclination of the Chinese that asserted itself powerfully even in modern times. In 1898, as the Hundred Days Reform was initiated through the influence of Kang Yu-Wei, the leading Chinese intellectual of his time, he spoke of it as the beginning of the Age of Small Peace, or the second of the three great periods in history. Past cycles of history contained for him the exact pattern to be followed in the future, and he believed, for instance, that a careful analysis of the three political epochs, that of the Hsia (c. 2193 B.C.– c. 1752 B.C.), Shang (c. 1751 B.C.–1112 B.C.), and Chou (1111 B.C.–249 B.C.), should yield the basic pattern of changes and events in a hundred generations to come.[12] To account for smaller variations, he distinguished in each great age three rotating phases. Into each of these he assigned three shorter periods, which in turn were divided into three parts with no limits set for successive subdivisions. His view of history could then be represented by a huge wave, the broad undulation of which was modified by smaller waves which in turn were broken up into smaller and smaller ripples. In all probability, Kang Yu-Wei did not agree literally with traditional Confucianism which gave the length of each of the great ages as the age of Great Peace, that is, a hundred generations. But he foresaw its eventual replacement by another age of Great Disorder. His was, indeed, a revealing commitment to the cyclic concept of human history, which also formed the very core of the Confucian interpretation of the world.

The enduring influence of Confucianism on Chinese thought was almost equalled by Taoism. In a sense, the two could not be more different. The early Taoists lived as hermits whose withdrawal from society stood for the rejection of the Confucian method of finding the pattern of cosmic order through reflection on social life. In the eyes of the Taoists the only viable approach to the order of nature (Tao) lay in a constant communing with nature. Such a stance was diametrically opposed to that of the Confucians, and received their strictures time and again. An example of this was the peremptory statement of Hsun Chhing (fl. 250 B.C.) directed against the Taoists: "To neglect man and speculate about nature is to misunderstand the facts of the universe."[13] Yet, the charge was not without some profound irony, for in one basic respect Confucians and Taoists were much alike. Just as the Confucians tried to understand society by overemphasizing the intuitive aspects of reasoning, the Taoists too considered intuition the chief, if not exclusive, avenue to the understanding of nature.

Some of the main features of the Taoist understanding of nature are already well formulated in the *Tao Te Ching*[14] (Canon of the Virtue of the Tao), a work composed in part by Lao Tzu, the principal proponent of Taoism. It is emphasized there, that the Tao, or the order, or rather normality, that should prevail in every process, is a mysterious factor that cannot be adequately described in words. The stillness of passive contemplation can, however, make one a full sharer in the Tao: "He who devotes himself to learning (seeks) from day to day to increase (his knowledge); he who devotes himself to the Tao (seeks) from day to day to diminish (his doing). He diminishes it and again diminishes it, till he arrives at doing nothing (on purpose). Having arrived at this point of non-action, there is nothing which he does not do."[15]

This paradoxical reasoning pervades much of the Taoist statements about nature, which is conceived as the continual interplay of pairs of opposite forces and qualities. The Taoist analysis of this interplay does not follow the lines of ordinary logic. According to Lao Tzu "gravity is the root of lightness; stillness, the ruler of movement," and in much the same manner are correlated softness and hardness, weakness and strength, greatness and smallness.[16] As a fundamental truth about nature it is proposed that "the movement of the Tao by contraries proceeds; and weakness marks the course of Tao's mighty deeds."[17]

Not surprisingly, such paradoxical tenets do not issue in practical implementation. In the *Tao Te Ching* it is plainly recognized that "Everyone in the world knows that the soft overcomes the hard, and the weak the strong, but no one is able to carry it out in practice."[18] Thus, the Tao remains, as if by definition, beyond one's skill, in a striking consistency with the Taoist view, according to which the Tao never exercises mastery over anything, although everything is its manifestation: "All-pervading is the Great Tao! It may be found on the left hand and on the right. All things depend on it for their production, which it gives to them, not one refusing obedience to it . . . It clothes all things as with a garment, and makes no assumption of being their lord."[19]

The attitude that should be inspired by such considerations in the wise interpreter of nature is an imitation of the unfathomable skill of the Tao, which "in its regular course does nothing (for the sake of doing it), and so there is nothing which it does not do."[20] To such conception of a paradoxical feat clearly no words can do justice. Lao Tzu himself characterized "as nameless simplicity" the transformation achieved in a man who had fully immersed in the Tao: "Simplicity without a name – Is free from all external aim. – With no desire, at rest and still, – All things go right as of their will."[21]

The reference to an ubiquitous though undefinable will is important to note for a satisfactory understanding of the Taoist interpretation of nature. Nature for Taoism is an all-encompassing living entity animated by impersonal volitions. This is well articulated in *Chuang Tzu*, the second most important book of Taoism, put together around 300 B.C. There the whole of nature, stars, spirits, earth, and divinity, is described as the self-unfolding of the Tao: "there is in it emotion and sincerity, but it does nothing and has no bodily form."[22] As a result Nature could not be expected to yield her secrets to analytical reasoning or to systematic research activity. Since the Tao did nothing while doing everything, "developing in the Tao" meant in turn that "men do nothing and the enjoyment of their life is secured."[23] It was in full loyalty to the teaching of the master, Lao Tzu, that the author of *Chuang Tzu* voiced the dictum of inactivity as the supreme measure of understanding: "He who practices the Tao, daily diminishes his doing. He diminishes it and again diminishes it, till he arrives at doing nothing. Having arrived at this non-inaction, there is nothing that he does not do."[24]

The precept, needless to say, was not meant to be literally implemented. But its influence was certainly not in the direction of a purposive, systematic exploration of nature. Those animated by the Tao held that "when we walk, we should not know where we are going; when we stop and rest, we should not know what to occupy ourselves with."[25] The Taoist mastery of nature consisted in the attitude of abandon. The life of the sage was a floating on the undulations of the Yin and Yang: "In his stillness his virtue is the same as that of the Yin, and in movement his diffusiveness is like that of the Yang. He does not take the initiative in producing either happiness or calamity. He responds to the influence acting on him, and moves as he feels the pressure. He rises to act only when he is obliged to do so. He discards wisdom and the memories of the past; he follows the lines of his Heaven (-given nature)."[26]

Such a recondite directive had to appear reasonable within an outlook on nature where the individual's breath was part of the one all-pervading breath: "All under the sky there is one breath of life, and therefore the sages prized that unity."[27] But the Taoist concept of man's intimate, organic unity with nature implied an unbalanced emphasis on the minuteness and impotence of the individual vis-à-vis the great nature. The passage in the *Chuang Tzu* which reminded the individual that he had received his breath from the Yin and Yang also equated mankind with a "single fine hair on the body of a horse."[28] Quantitatively speaking the simile had its merit, but it covered only a hair's width of the full gamut of attitudes that man could reasonably cultivate towards nature. Within Taoism, man was at best a ripple on the great rhythmic undulations of the Yin and Yang out of which everything emerged and into which everything dissolved again. In such a context it made little sense to ask about the beginning or end, about final purposes and goals, or about the possibility of controlling to some extent at least the forces and patterns embodied in the workings of the Yin and Yang. The great cosmological passage of the *Chuang Tzu* leaves little untold in this respect:

The Yin and Yang reflected light on each other, covered each other, and regulated each other; the four seasons gave place to one another, produced one another, and brought one another to an end. Likings and dislikings, the avoidings of this and movements towards that, then arose (in the things thus produced), in their definite distinctness; and from this came the separation and union of the male and female. Then were seen now security and now insecurity, in mutual change; misery and happiness produced each other; gentleness and urgency pressed on each other; the movements of collection and dispersion were established: – these names and processes can be examined, and, however minute, can be recorded. The rules determining the order in which they follow one another, their mutual influence now acting directly and now revolving, how, when they are exhausted, they revive, and how they end and begin again; these are the properties belonging to things. Words can describe them and knowledge can reach to them; but with this ends all that can be said of things. Men who study the Tao do not follow on when these operations end, nor try to search out how they began: – with this all discussion of them stops.[29]

The conception of nature as a cosmic cycle was eagerly reasserted on the pages of the *Chuang Tzu*, and with an uncanny wording concerning some of its principal implications. Thus, it was resolutely emphasized that in nature, or existence in general, there could be no real beginning,[30] that every process was the unfolding of a dialectical oscillation between shape and utter shapelessness. The latter was rather paradoxically also called non-existence.[31] Most significantly, the "unbroken ring" of transformations, the perennial return of all in different forms, is called in the *Chuang Tzu* the Lathe of Heaven,[32] a metaphor derived from a wheel-type instrument used by potters.

In a conceptual framework where questions about the Beginning were resolutely barred, one became engulfed in the vision of the "Great Returning,"[33] with the result that questions about "small" beginnings, about strict causal connections between events, also lost much of their appeal. The situation is graphically described in the *Chuang Tzu* in a fictive conversation between the shadow and the penumbrae who press the former for explanation of the manifold changes in its appearance. The reply of the shadow deserves to be quoted in full:

> Venerable Sirs, how do you ask me about such small matters? These things all belong to me, but I do not know how they do so. I am (like) the shell of a cicada or the cast-off skin of a snake; – like them, and yet not like them. With light and the sun I make my appearance; with darkness and the night I fade away. Am not I dependent on the substance from which I am thrown? And that substance is itself dependent on something else! When it comes, I come with it; when it goes, I go with it. When it comes under the influence of the strong Yang, I come under the same. Since we are both produced by that strong Yang, what occasion is there for you to question me?[34]

Clearly, within a world view dominated by the Yin and Yang, detailed, exact questions about the whole or parts of nature and of its processes would not be encouraged. Behind the unfathomable forces of the Yin and Yang there was nothing to look for, certainly not a Lawgiver, or a Governor of all. The author of *Chuang Tzu* could find no trace of Him in the universe,[35] and as a result the universe became viewed to be intrinsically untraceable. Distinct lines or patterns inevitably faded away in the monotonous lull, as the wheel (the Lathe of Heaven) kept going through its perennial revolutions. Endless returns left no room either for identifying definite starting and end points. The author of *Chuang Tzu* unabashedly admitted this as he tried to explain that there was no beginning which was not in itself an end: "The change – rise and dissolution – of all things (continually) goes on, but we do not know who it is that maintains and continues the process. How do we know when any one begins? How do we know when he will end? We have simply to wait for it, and nothing more."[36]

This detailed portrayal of the Taoist interpretation of nature should help one understand a tantalizing problem of cultural history, the failure of the Chinese to develop isolated rudiments of scientific feats into a self-sustaining practice of the scientific method. It seems, indeed, that the Chinese were not inclined to break out of the pattern in which periods distinguished by interesting though largely practical scientific achievements alternated with culturally dormant periods. The real start for science never came in China, and certainly not because of the lack of long centuries of relative peace, material prosperity, active social interplay, creativity of mind, and possibility of contacts with other cultures. In addition, China was a land where cultural accomplishments were carefully handed down throughout many centuries. Thus, the *Chou Pei Suan Ching* (Arithmetic in Nine Sections), originally compiled around the ninth century B.C., or in the early phase of the Chou dynasty (1111 B.C.–480 B.C.), was re-edited around 200 B.C., or the beginning of the Han period (202 B.C.–A.D. 220), culturally one of the most productive in Chinese history.

The era known as the age of the Warring States (480 B.C.–220 B.C.) was not, therefore, an unqualified turmoil causing the complete loss of a precious cultural heritage. It was around 350 B.C. that the astronomer Shih Shen drew up his catalogue of some 800 stars, and that the central storing of manuscripts got under way in the Imperial Library. In the field of technology the same age also witnessed further improvements in water works and the extension of the Great Wall. Again, cultural activity did not come to a standstill, nor did it become wiped out, during the three and a half centuries known again as the age of the Warring States (220–581). During those centuries the Chinese invented the vertical waterwheel and the wheelbarrow, devices highly indicative of continued technological interest. Around the middle of the fourth century the astronomer Hu Hsi concluded from his observations to the precession of equinoxes, probably without knowing anything of the discovery made by Hipparchus centuries earlier.

Clearly then, the peak periods of Chinese culture, the Han (220 B.C.–A.D. 220) and the Sung (960–1279), and the culturally less productive Thang (618–906), Yuan (1279–1368) and Ming (1368–1644), represented a length of time during which scientific endeavour could have more than once received a decisive spark. For the triggering of a breakthrough there were technological feats of which the Chinese were the proud inventors and sole possessors for a number of centuries. Long before the Han period the Chinese made far more effective use of horses than other civilizations. The foot-stirrup originated in China and so did the breast-strap harness, which the Chinese modified in medieval times into a collar harness. During the Han period there came the discovery of the directional behaviour of pieces of magnetic ore and the invention of paper making. The Thang period witnessed the production of the first printed books, the processing of gunpowder, major improvements in porcelain craftsmanship, and the development of water-driven mechanical clocks. Among the chief technological achievements of the Sung period were the use of magnets in travel and the development of movable clay types for printing. The high skill of Chinese iron mongers during that time is evidenced by the oldest extant cast-iron pagoda (A.D. 1061), at Yü-ch'üan ssu, in Hupei province. The same period closed with a new edition of the Arithmetic in Nine Sections. It showed Chinese algebra at a level comparable with the best that was known in Europe around 1250.

Printing, gunpowder, and magnets, these were the factors which in Francis Bacon's estimate did more than anything else to usher in the age of science.[37] It might have shocked him to learn that for all their gunpowder, magnets, and printing skill the Chinese remained hopelessly removed from the stage of sustained, systematic scientific research. They had rockets for centuries, but failed to investigate their trajectories, or to probe into the regularities of free fall. Unlike in the West, bookprinting did not lead in China to a major intellectual ferment. Although magnets were installed on Chinese ships, which formed the best navy in the world during the fourteenth and fifteenth centuries, their captains never had the urge of a Vasco de Gama, Columbus, and others to circumnavigate the globe.

The reason for this strange reluctance lies in what is probably the most revealing aspect of the status of science in China prior to the seventeenth century. The recognition by the Chinese of the sphericity of the earth was fainthearted at best. One would search in vain in the writings of ancient and medieval Chinese astronomers and cartographers for numerical estimates of the radius and circumference of the earth, which the Greeks calculated with stunning accuracy. Familiar only with the very rudiments of geometry, the Chinese formed a concept of the earth and the universe which oscillated between the image of a disc shaped earth and a vague geocentric arrangement of the planets surrounded by other "heavens." Values given by Chinese astronomers about the height of the "heavens" were guesses that could not be put to the test of measurements within the system itself. References to the infinity of space made occasionally by Chinese astronomers from the first century on gave Chinese cosmology nothing more than the veneer of science. Their cosmology remained estranged from the massive body of Chinese astronomical observations among which was the recording of each return of Halley's comet for over two thousand years. Chinese watchers of the sky could also boast of their sighting in 1054 of what modern astronomy classifies as a supernova. But Hu Hsi, the astute Chinese proponent of the precession of the equinoxes, also spoke of the periods of the sun, moon, and planets as something analogous to the behaviour of animals that go in and out of their hiding places.

The prominence and persistence of organismic analogies was well matched in Chinese astronomy by the eagerness to establish celestial cycles of different lengths.[38] In this respect the penetration of Buddhism into China in the second century added further support to an already strong preoccupation with cycles dominating the cosmos. In *Chi Ni Tzu* (The Book of Master Chi Ni), which transmits much of the pre-Han tradition, there is a detailed description of cyclic succession of abundant harvests, floods, prosperity, and drought. The length of each period is 12 years, the time during which Jupiter very nearly completes its sidereal period. But it is not Jupiter itself but the revolution of a mythical "counter-Jupiter," called Thai Yin, or Great Yin, that triggers the periodic changes on earth. This particular occurrence of the word Yin should be suggestive enough of the intimate connection of the cyclic and organismic elements which can also be seen in some Chinese explanation of eclipses. Thus, the recurrence of eclipses was ascribed by Wang-Chhung (fl. A.D. 80) to periodic changes in the "life-strength" of the moon and sun, and to the consequent rhythmic variations in their intrinsic brightness.

The metonic cycle, consisting of 19 years, or 235 lunations, was known in China as the *chang*, while its fourfold multiple, 76 years, was called *pu*. There was also a special name, *hui*, for 27 *chang*, 513 years, or the almost exact equivalent of 47 lunar-eclipse periods. Three *hui* (1539 years) constituted one *thung*, while 20 *pu*, or 1520 years, were called *chi*. Three *chi* formed a *shou*, seven of which, or 31,920 years, gave the length of the Grand Period. Its completion supposedly marked the return of every process to its original form and condition. This opinion is stated in *Chou Pei Suan Ching*, the great mathematical classic of the Han period.

Calculations of planetary conjunctions gave Han astronomers even bolder vistas concerning cosmic cycles. They combined 3 *thung* periods, or 3×4617 years, with 138,240 years during which all the planets were believed to repeat exactly their combined motion. The result was the "world cycle," or 23,639,040 years, the beginning of which was regarded as the Supreme Ultimate Grand Origin. The idea of such large periods of time obviously remained familiar in scholarly circles at least, as can be seen in the writings of the eighth-century mathematician, I-Hsing,

D

who claimed that prior to the 12th year (A.D. 724) of the Khai-Yuan period, 96,961,740 years had already passed since the onset of the latest Grand Period.

As the scholarly tradition among Chinese astronomers and mathematicians was attracted to the idea of cosmic cycles, so was the widely shared organismic, or animistic, view of nature suggestive of rhythmic periodicities in the life of the universe. Thus, the ground was well prepared for the reception of Hindu and Buddhist speculations about cyclic repetitions of cosmic ages. They were explicitly entertained and assimilated by the Neo-Confucians from the eleventh century on, and because of this, Hindu conjectures about the length of yugas and kalpas became a staple feature of Chinese discussions of cosmological questions. The first steps in this direction seem to have been taken by Shao Yung (1011–1077) who, relying on the analogy of the clock, divided into twelve parts each cosmic cycle of destruction and reconstruction. A hundred years later Chu Shi (1131–1200), an outstanding figure of Neo-Confucianism, supported his contention of the eternity of the world by referring to the never-ending series of cosmic periods.[39] He even quoted with full approval the figure of 129,600 years as the length of each period. This figure is the product of $30 \times 360 \times 12$ and it reflects the role of both the clockwork analogy (12) and the Hindu notion that cosmic or divine units of time (days) are 360 times longer than the "human" day. The number 30 stands for the number of years in one generation.

The Chinese name for the cosmic period was *yuan* and its twelve subdivisions, each of 10,800 years, were called *hui*. With the onset of the eleventh *hui* the extinction of all living beings was believed to come. The middle of the twelve periods marked the dissolution of all heavy matter into a dust cloud so refined as to make possible its fusion with the ethereal matter composing the heavens. Once this fusion had been completed the whole mass was believed to rotate at an increasingly rapid rate until everything was enveloped in total darkness. At this point the *yuan*, or world-period, reached its end to start all over again. To the first half of the first *hui*, also called *tzu*, was ascribed the gradual emergence of light. During the second half of *tzu* the lightest kind of matter began to separate from the rest to produce the sun, the moon, the planets, and the fixed stars. The formation of the earth, and its separation from the elements of water and fire, were assigned to the second *hui* or *chhou*. The third *hui* or *yin* was the phase that witnessed the spontaneous appearance of man. The rest of *huis*, from the fourth to the eleventh, were less active cosmologically: they provided the time needed for the unfolding of human history.

The interpretation of cosmic and human history in terms of cycles was far more than intellectual entertainment for a few scholars. Rather, it acquired at a very early age and enjoyed until very recently a semi-official status. Chu Hsi's work on cycles received a number of official or imperial sanctionings, of which the latest came on July 17, 1894.[40] Therefore, it should not be surprising that analysis of nature and history in terms of cycles moulded Chinese thought with some very far reaching consequences for the fate and fortune of science in China. As the great French sinologist, M. Granet, pointed out, "the conviction that the All and everything composing it, have a cyclic nature" drastically stymied the Chinese awareness of causal connection between events.[41] A telling evidence of this is the fact that the Chinese saw nothing inordinate in attributing the political failure of a certain prince to the sacrificing of humans at his burial. As both political impotence and cruelty evidenced the absence of the same virtue, one could replace the other as explanation regardless of their sequence. What the Chinese preferred to register were not, in Granet's words, "causes and effects, but manifestations, whose order mattered little, conceived as they were separate, but grafted nevertheless on the same root. *Equally expressive, they appeared interchangeable.*"[42] If at a particular time, a mountain collapsed, a river ran dry, a man allegedly changed into a woman, and a dynasty came to an end, the Chinese sage took all these as equally significant indications of a "change of order" both in cosmos and in history, without feeling any urge to search into a causal relationship among them.

In such an outlook, measurable, quantitative aspects of events occurring closely in time could have no particular significance. Their frequency or order of magnitude commanded no special interest, nor did the normal sequence of events. The Chinese cataloguing of events and facts stood poles apart from the collections of facts or "Natural Histories" proposed by Francis Bacon. This is not to suggest that his precepts represented the best in scientific method. But they pointed at least in the right direction. The Chinese, bent on seeking the poetical, empathic, and organismic solidarity among facts, had no interest in their regular sequence. In their eyes, it was cyclic anyway, bringing along much the same situation after the completion of each period. Or as Granet remarked: "Instead of considering the course of things as a succession of phenomena, susceptible to measurement, and to subsequent coordination, the Chinese see in the sensible realities only a mass of concrete signs. The task of making a repertory of them imposes itself not on the physicist but on the chronicler: History holds the place of Physics."[43]

The study of physics remained atrophied in ancient and medieval China. But even in another branch of science where the Chinese made some stunning anticipations of modern scientific insights, the promising starts failed to be followed up. A good illustration of this is the reflection of Chu Hsi in the twelfth century on the significance of fossil oyster shells found on high mountains. He argued correctly that only a great vertical displacement of the sea bed could account for such findings. He also expressed hopes that careful considerations of such phenomena would lead to "far-reaching conclusions." These failed to be drawn by Chinese men of science. Their curiosity about nature never turned into a sustained resolve to exploit fully the meaning of valuable observations and inventions.

Despondency about man's ability to decipher the exact patterns of nature made itself felt time and again before China's long isolation from the Western World came dramatically to an end around 1600. Wang Yang-Ming, the most notable Chinese thinker of the early sixteenth century, spoke in a tone of resignation about the futility of trying to find out anything or all about nature, and about the headaches which went along with such efforts. A friend of his, he noted, tried to discover the principles embodied in the structure of bamboos. For three days running he tried, but only to become completely exhausted mentally. Then it was the turn of Wang Yang-Ming who devoted seven days to the task, but to no avail. He too became ill from being unduly burdened with thoughts. The conclusion reached by the two friends is worth being quoted in full: "Thus we both sighed and concluded that we could not be either sages or men of virtue, lacking the great strength required for carrying on the investigation of things." To this Wang added: "When, while living among the savage tribes for three years, I clearly saw through this idea [the problem of the investigation of things], I knew that there was really no one who could investigate the things under heaven. The task of investigating things can only be carried out in and with reference to one's body and mind."[44]

This introspective self-centredness with respect to external nature was not merely the attitude of select individuals. It was also China's stance toward cultures surrounding her. On Chinese maps China always formed the centre of inhabited land, in clear indication of the subconscious attachment to the primitive notion of a flat earth. Areas outside China represented the periphery, which could never match the excellence of the centre. So deeply ingrained was this outlook in the Chinese that the centuries-long presence of Muslim astronomers in medieval China made no dent on it. This was in sharp contrast with the excitement and ferment created in medieval Europe by news about Arabic science. But the smug feeling of superiority among the Chinese remained largely unshaken even after Chinese cultural history experienced its sharpest jolt, following the arrival of Father Matteo Ricci.

Ricci's first and largely enthusiastic report on Chinese science was written on September 13, 1584, a year after he had settled on the Chinese mainland. He marvelled at the success of the Chinese to predict two eclipses of the moon without apparently any knowledge of Ptolemaic astronomy. But as the years went by he found out that his own relatively modest expertise gave him an extraordinary stature in the eyes of his hosts. "In truth," reads his long report of November 4, 1595, "if China was the entire world, I could undoubtedly call myself the principal mathematician and philosopher of nature, because it is ridiculously and astonishingly little what they know; they all are preoccupied with moral philosophy, and with elegance of discourse, or to say more properly, of style."[45]

A dozen or so years provided enough eclipses to show the enormous superiority of Western, that is, Ptolemaic astronomy over that of the Chinese. Following the eclipse of the sun on September 22, 1596, Ricci could point out that a particular solar eclipse has not the same duration and extent at every point on earth. Ricci also grew aware of the fact that his hosts had no idea of the physical cause of the moon's eclipse. The more he saw, the clearer the situation became to him, as shown in his letter of September 9, 1597: "About the learned among the Chinese, let me say this: the Chinese have no [physical] science at all; one may say that only mathematics is cultivated, and the little they know of it is without foundation; they stole it from the Saracens. Only the King's mathematicians teach it to their sons. They just manage to predict eclipses, and even in that they make many mistakes. All are addicted to the art of divinations, which is most unreliable and also completely false. Physics and metaphysics, including logic, is unknown among them . . . Their literature consists wholly in beautiful and stylish compositions all of which correspond to our humanities and rhetoric."[46]

The numerous references in Ricci's letters to the addiction of the Chinese to astrology indicate his awareness of the deepest reason for the backwardness of Chinese science. The ten years (1601–1610) he spent at the Imperial Court in Peking brought him face to face with the variegated and overbearing implementation of the Chinese conviction that Earth and Heaven were answerable to each other. Misconduct on the part of the Emperor, or of his officials, was believed to have a disturbing effect on celestial motions with further deterioration in terrestrial affairs. Ricci knew all too well what considerations lay behind the all-consuming preoccupation of the Chinese to have their calendar meticulously co-ordinated with the revolution of celestial bodies. He also saw how ill-equipped were Chinese astronomers for the task: "The King employs," Ricci wrote on May 12, 1605, "more than two hundred people, I believe, at great expense to have every year the ephemerides of that year computed; in addition there are two colleges, one known as the College of the Chinese Method, it also enjoys the higher reputation; . . . and the other, of lower rank, known as the College of the Moorish Method, which predicts the eclipses more exactly; . . . but with all this they know nothing more than to make computations, without any insight into the rules, and when the result does not come out right, all they say is that they kept to the rules of their forebears."[47]

In the same letter Ricci also urged the dispatching to Peking of someone fully trained in astronomy. Ricci, the philosopher and theologian, and a skilled clock-maker and architect, felt that he knew enough of geometry, sun-dials, and astrolabes. But it was the position and trajectory of planets, the calculation of eclipses, and the art of drawing up ephemerides that could best impress the Chinese whose political and social life so heavily depended on calendar making. Meanwhile, Ricci kept supervising the first translation of Euclid into Chinese. Its printing was completed in August 1608. Ricci's choice of Euclid as the first scientific treatise to be translated into Chinese had a motivation worth recalling. He wanted to expose the Chinese to the paragon of clear, stringent, demonstrative procedure in science with which they were wholly unfamiliar. The translation, as Ricci noted, "was much more admired than understood. It, however, served well the purpose to humble the Chinese pride; it forced their best scholars to admit that they had seen a book which, though printed in their own language, they could not understand even after studying it with great attention; such was indeed a fact that may not have happened to them before."[48]

The reference to the pride of the Chinese was most appropriate. It impeded Ricci's efforts as much as it prevented the seeds of the "new" science from taking deep roots in Chinese soil. The resentments caused by Ricci's world map are a case in point. The sphericity and true dimensions of the earth posed an overwhelming challenge to the traditional Chinese belief in the central and dominant position of China in the world. The following remarks of a younger contemporary of Ricci, Wei Chün, contain the whole drama in a nutshell:

Lately Matteo Ricci utilized some false teachings to fool people, and scholars unanimously believed him . . . The map of the world which he made contains elements of the fabulous and mysterious, and is a downright attempt to deceive people on things which they personally can not go to verify for themselves. It is really like the trick of a painter who draws ghosts in his pictures. We need not discuss other points, but just take for example the position of China on the map. He puts it not in the center but slightly to the west and inclined to the north. This is altogether far from the truth, for China should be in the center of the world, which we can prove by the single fact that we can see the North Star resting at the zenith of the heaven at midnight. How can China be treated like a small unimportant country, and placed slightly to the north as on this map? This really shows how dogmatic his ideas are. Those who trust him say that the people in his country are fond of travelling afar, but such an error as this would certainly not be made by a widely-travelled man.[49]

The ultimate bone of contention concerned, of course, neither science, nor national pride, but the question whether the world was made by God or was itself god. It was inevitable that Ricci himself should come to grips with that crucial issue. After all, he went to China to lead souls to God. He found no saving grace in the overt pantheism of Chinese Buddhism. But he felt that Confucianism, in its ancient form at least, represented a noble moral ideology, which unfortunately had taken on an increasingly atheistic interpretation during the previous five hundred years. In the methodical exclusion by the Taoists of a First Cause and of a Beginning, Ricci saw a most detrimental phenomenon of Chinese thought. The second chapter of his catechism was devoted to the refutation of the Taoist claim that the "nothing" can for ever give rise to "everything." "Such a doctrine," he wrote, "not only fails to enlighten the world, but is in fact generating a great number of confusing doubts."[50] Obviously, the stricture also bore on the vagueness and ambivalence of the Taoist statements about nature. The clarity and cogency of the laws of Western science, especially astronomy, could therefore suggest in a most powerful manner that the laws of nature had an exactness that pointed to a rational Lawgiver. True, Ricci taught to the Chinese the Ptolemaic world-view and astronomy. But the latter implied an impressive body of incontrovertible scientific results which stood in irreconcilable conflict with the Chinese (Taoist) interpretation of the world, and especially with its morass of organismic, astrological, and fatalistic notions.

Being a missionary, and an extremely tactful one at that, Ricci did his utmost to see the culture of the Chinese in the best possible light. The approach taken by him set the tone of the reports on China written by his successors in the missions. Those reports launched in Europe a wave of enthusiasm for everything Chinese, porcelain as well as Confucian wisdom, silk as well as Chinese social organization, screen paintings as well as Chinese natural piety. The appreciation of the Chinese for the imposing edifice of Ptolemaic astronomy, for telescopes,[51] for tables of logarithms, and for Europe in general, was spotty at best. There were some eager learners among the Chinese, especially in mathematics, though they failed to add anything notable to what they had learned. Never in history did resentment and frustrated feeling of superiority leave unexploited a greater cultural and scientific opportunity.

To be sure, the Jesuits kept the Copernican system under cover to the end. The still heavily Aristotelian Western science was to serve in their hands as proof of the superiority of the Christian West and by implication of Christian revelation. Cardinal points of the latter, such as the redemptive death of the Son of God, were deeply repugnant to Chinese mentality raised on Confucianism. The four Aristotelian elements advocated by Ricci and his successors represented no great advancement over the five elements of the Chinese. But there could be no doubt about the ability and sincerity of Fathers Ricci, Schall von Bell, Verbiest, and a host of others to teach science and to meet the Chinese more than half-way. Yet, while medieval Christendom had the mental strength to extricate Greek scientific inheritance from its pantheistic matrix, China did not seem to have either the strength or the inclination to see that the science brought from the West was not, as far as its contents went, either Western or Christian.

Almost childish reaffirmations of the superiority of Chinese scientific tradition over anything coming from the West constituted a recurring theme as Chinese scholars began to write their accounts of Chinese science from the new vantage point forced on them by the "wise men" from the West. A classic illustration of this is Juan Yuan's famous *Chhou Jen Chuan* written in 1799, a work which gives the biographies and accomplishments of some hundred Chinese and of some two dozen Western mathematicians.[52] Needless to say, the work is far below the level of scholarship evident in histories of astronomy and mathematics published in Europe about the same time. Accounts of older Chinese mathematicians are given without exact references to their works, a procedure as unenlightening as are the ubiquitous remarks of the type: "He far surpasses any European," "Europe has added nothing new," "China knew all, but the Occidentals borrowed her science and developed it."[53]

Clearly, nothing had changed since 1645 when Father Schall von Bell was forced to change in the title of his great astronomical encyclopedia the expression "according to Western methods" to "according to the new methods."[54] But love for the "new methods" was halfhearted at most. Juan Yuan praised Chinese men of science for not falling prey to the lure of Western methods: "Our ancients sought phenomena and ignored theoretical explanation. Since the arrival of the Europeans, the question has always been concerning explanations, circular orbits, mean movements, eclipses, and squares. The foreigners think the earth revolves about a fixed sun . . . but the theory of Tycho has been modified many times during the last century and I believe that it will be again . . . Therefore I do not see upon what the Europeans base their arguments . . . and really it does not seem to me the least inconvenient to ignore the western theoretical explanations and simply to consider the facts."[55]

Such a partisan discrimination between theories and facts did justice to neither. The facts, the great inventions of science, failed to earn due respect in Chinese society up to Juan Yuan's time and even beyond. Traditional Chinese mentality posed a major barrier to a genuine assimilation of the data and method of science. At the beginning of the 19th century the dominant attitude of educated Chinese circles toward the *facts* of science was still well exemplified by the poem of Fengshen yin-te, a high ranking state dignitary, and son of a Prime Minister (Ho Shen, 1750–1799):

> With a microscope you see the surface of things.
> It magnifies them but does not show you reality.
> It makes things seem higher and wider,
> But do not suppose you are seeing the things in themselves.[56]

Nothing is easier than to blame the "feudal" and "bureaucratic" society of pre-modern China for its failure to develop or assimilate science in a meaningful manner. Continual encomiums heaped on ancient Chinese contributions to science, and endless allegations of unfavourable socio-economical factors, may easily turn into a self-defeating dialectic. Dialecticians, especially the Marxist brand, are distinguished by a sharp though one-way vision. In our case it is bafflingly insensitive to the distinctly different workings of the same socio-economical factors in feudal China and in feudal Europe. In the latter there was a burgeoning of universities and an intellectual ferment, the like of which China failed to match even remotely. Again, the more evidence, convincing or merely suggestive, is collected about the technological superiority of ancient and medieval China over the West, the more stumbling blocks are gathered for the contention that science is an inevitable result of using proper tools of production. The baffling fate of science in China should suggest that the explanation is far from being that simple. A reluctant admission of this comes from J. Needham himself, the leading Western historian of Chinese science, and an avowed Marxist. According to him,[57] it is the a-theological orientation of traditional Chinese thought that should ultimately be singled out as the decisive factor which blocked the emergence of a confident attitude toward systematic scientific investigations. All this stood in sharp contrast with the situation prevailing in Western Europe. There, according to Needham's admission, all the early cultivators of science drew courage for their pioneering efforts from their belief in a personal and rational Creator. The handiwork of such a Supreme Being had to be rational and therefore investigable in a manner satisfying the stringent demands of reason. The Chinese had no recourse to a similar belief. Well before the time of Confucius, the idea of a personal, supreme Lawgiver began to yield to a consideration of the universe as the ultimate entity, a view which was aptly called universism by the great Dutch Sinologist, J. J. M. de Groot.[58]

Among the numerous evidences of this shift in Chinese thought, the third-century commentary on *Chuang Tzu* by Hsiang Hsiu and Kuo Hsiang is particularly expressive. It contains a lengthy and sophisticated paraphrase of the assertion repeatedly made in *Chuang Tzu* that there was no absolute origin. According to the commentary, the Yin and Yang are themselves things, and therefore they cannot be said to have existed before the realm of all things. Nor can nature as such be singled out for such a role, since nature is simply the naturalness of things. The Tao itself is not an ultimate entity because it is the supreme non-being as well. "Since it is non-being, how can it be prior? Thus what can it be that is prior to things? And yet things are continuously being produced. This shows that things are spontaneously what they are. There is nothing that causes them to be such."[59]

This point is further articulated with a specific reference to the idea of a Creator and to the possibility as to whether one can infer His existence from observing the causal connection between phenomena. Among these the functional chain stretching from penumbrae to shadows, from shadows to opaque bodies, and from there to the Creator, is singled out for consideration. The reply is resolutely negative, but most significantly the denial of such a proof on behalf of a Creator generates a serious doubt about the causal role of opaque bodies in the production of shadows: "Some people say that the penumbra is dependent upon the shadow, the shadow upon bodily forms, and bodily forms upon a Creator. But I venture to ask whether the Creator is or is not? If He is not, how can He create things? If He is, then, (being one of these beings), He is incapable of creating the mass of bodily forms. Hence only after we realize that the mass of bodily forms are things of themselves can we begin to talk about the creation of things. Therefore throughout the realm of things, there is nothing within the Mystery, even the penumbra, which is not 'self-transformed.' Hence the creating of things has no Lord; everything creates itself. Everything produces itself and does not depend on anything else. This is the normal way of the universe."[60]

"The creating of things has no Lord; everything creates itself," is a most portentous statement which echoes the frequent references in Chinese philosophical tradition to a silent, voiceless heavens. While the harmony displayed by the heavens is the supreme embodiment of the "normalcy" of the universe, this is not the outcome of a superior ordinance. The Tao, or the all-embracing order, produces everything, feeds everything, but does not lord over anything. Its ordinances are wordless edicts. The sustained emphasis on the "silence" of the universe, on its being "voiceless," is more than entertaining poetry. It rather suggests that the precise constitution of the parts of the universe and of their mutual harmony is not spelled out in their behaviour. Consequently, the inquiring mind can have no hope of ever coming within hearing distance of clear-cut statements that nature and its parts may utter about laws governing them. In a universe without the voice of God there remains no persistent and compelling reason for man to search within nature for distinct voices of law and truth. The voice of nature in this perspective is at best a pleasant but diffuse hum pervading man also. To understand nature man has only one road open to him: to tune himself to that vague note not by discerning analysis but rather by an attitude of assimilation. Instead of staking out his place in the universe man must rather "feel out" his place in it, like any partial organ supposedly does, while it functions in the whole organism.

As a part of a huge organism, or nature, man should fit into the whole without asking for explicit directives from other parts of the whole. Much less should he try to influence, rule, or modify them. For nature is a large organism, the coherence of which is wholly spontaneous without the influence of superimposed rules. Such is obviously no small paradox, as it implies that the parts associate in what is effectively a non-association, and co-operate in what is for all practical purposes a non-co-operation. Yet the paradox is fully espoused in Chinese thought: "The hands and feet differ in their duties; the five internal organs differ in their functions. They never associate with one another, yet the hundred parts (of the body) are held together by them in a common unity. This is the way in which they associate through non-association. They never (deliberately) co-operate, and yet, both internally and externally, all complete one another. This is the way in which they co-operate through non-co-operation."[61]

It is undoubtedly right to emphasize, as Needham did, that this organismic view of the universe constitutes the most palpable aspect of the Chinese outlook on nature. Other Western students of Chinese thought, from Leibniz on, have stated the same on countless occasions.[62] Needham was also correct in stating that the key-word to Chinese thought is organism in which the parts react on each other "not so much by mechanical impulsion or causation as by a kind of mysterious resonance."[63] Again, he made a worthwhile contribution in pointing out that the preference of the Chinese for the organismic and for the spontaneous kept them from developing the concept of natural law in the modern scientific sense. He also recognized somewhat ruefully that it was the reluctance of the Chinese to accept the idea of a personal, supreme lawgiver (God) that ultimately blocked their vision toward scientifically formulated laws of nature. In a truly revealing phrase he declared: "It was not that there was no order in Nature for the Chinese, but rather that it was not an order ordained by a rational personal being, and hence there was no conviction that rational personal beings would be able to spell out in their lesser earthly languages the divine code of laws which he had decreed aforetime."[64]

It is rather disappointing that he left this theme undeveloped. His reticence is understandable. For Needham, a Marxist interpreter of science, it must have been a frustrating pill to acknowledge the crucial role played by faith in a personal Creator in the rise of modern science. His was indeed a sulking attitude, which forced him to adopt some strangely uneven standards of evaluation. This came through forcefully as he contrasted the respective merits of a theistic and of a pantheistic (a-theistic) concept of the cosmos. On the one hand he dwelt at length on some unsavoury outgrowths of the notion of a universe ruled by the laws of a supreme Lawgiver or Creator. He tried indeed to create the impression that it had been typical in Western Christian civilization to prosecute and destroy animals for "unnatural" acts, or in general for transgressing the laws of nature.[65] On the other hand he spared no praises for the organismic view of the universe as cultivated among the Chinese. He admired its pantheistic or a-theistic flavour and found its basic assumptions very similar to the conceptual foundations of modern physical science.

The roots of such a startling claim are not difficult to find. According to Needham there is a basic identity between the ancient Chinese concept of the world as an organism and the description by Whitehead of the modern scientific world view as an organismic notion. Oddly enough, Needham also claimed that the modern scientific world view did full justice to Marxist dialectics. No wonder then that the Chinese of old were given credit by Needham for producing the foundation of modern scientific thought: "the philosophy of organism, essential for modern science in its present and coming form, stemmed from the bureaucratic society of ancient and medieval China." Thus, the great paradox of scientific history, the unique emergence of exact physical science in the late Middle Ages, became in Needham's hands a chiefly Chinese paradox: "The gigantic historical paradox remains that although Chinese civilisation could not spontaneously produce 'modern' natural science, natural science could not perfect itself without the characteristic philosophy of Chinese civilisation."[66] In line with this astounding claim, Needham interpreted the vague statements of that organismic philosophy as anticipations of the world view of modern physics where mechanical constraints (external laws) have yielded to the "spontaneousness" of space-time curvature. To crown the comedy he found something scientifically creative even in certain unsavoury if not perverse practices of some ancient and medieval Taoists strongly committed to the organismic view of nature.[67]

It should not therefore be surprising that Needham remained blind to the psychological impact which the organismo-cyclic world view is bound to exercise upon the inquiring mind. That in the traditional Chinese interpretation of nature the organismic and cyclic concepts formed a tightly knit unity is repeatedly stated by him, though he gave to the cyclic aspect less attention than to the organismic one. In discussing the fundamental ideas of Chinese science Needham summed up in the following words the conception of the Chinese about the world: an "uncreated universal organism, whose every part, by a compulsion internal to itself and arising out of its own nature, willingly performed its functions in the cyclical recurrences of the whole."[68] Needham noted, of course, only the allegedly beneficial influence of such a concept of nature on scientific undertakings. Thus he claimed that in China the early recognition of the true nature of fossils was due to speculations on cosmic world cycles, of recurring world catastrophes and rejuvenations. While he admitted that on this point the Chinese were most likely indebted to Buddhist ideas, he also warned against dismissing the possibility of world cycles. Some of the most eminent modern astronomers are inclined to think, he advised his readers, that the basic dynamics of the universe consisted of cycles of expansion and contraction.[69]

In making this remark, Needham undoubtedly wanted to enhance the value of early Chinese speculations which he invariably put in the most favourable light. For him it was not a mockery of logic but a "timeless pattern" of thought that some Chinese of old attributed the political fiasco of that legendary prince to the fact that following his death humans were sacrificed on his behalf.[70] In that "timeless pattern" Needham saw a notion far superior to that of a world of billiard balls where every configuration is strictly determined by the parameters of the phase immediately preceding it. The billiard ball universe is, to be sure, now outdated, both in physics and in philosophy. But the experience and notion of time seems to be just as challenging and creative today as it was when science was born four to five centuries ago. Without a strong sense of the time-conditioned character of physical processes, scientific thought would have remained in the shackles of vague "timeless patterns" amounting to fantasies. That science could not surface in ancient and medieval China is largely due to that mental lull which overcomes man when his sense of the unique value and flow of time is weakened.

It should not be difficult to see the striking similarity between the mental lull generated by a belief in a universe revolving for ever in cycles, and the passivity of mind, pleasant as it may be, induced by the organismic conception of the universe. The fusion of these two can only undermine any budding intellectual enterprise along scientific lines. The organismic concept of the world (not in the Whiteheadian sense) invariably fosters a state of mind dominated by a nostalgic longing for the primitive golden age, with its idyllic settings in which everything takes place in an effortless way. In that dreamlike condition of spontaneousness men live off nature without disturbing it, and carry out their social propensities without the sense of constraint due to authorities and laws.

A classic description of what is implied psychologically in that idyllic, organismic order of things, persons, and events, is given in the writings of Pao Ching-Yen, a rather unconventional Taoist philosopher of the third century. After denouncing the various forms of political, economical, and social pressures as deviations from the true laws of heaven he evokes the perfect conditions of old, where everybody enjoyed a carefree existence. It was a golden age undisturbed by cultural efforts. The face of nature was not ruined by channels, roads, and bridges. There was universal peace as people were uninhibited, uncompetitive, and unconcerned about either honour or shame. Their life was pleasant but certainly uneventful, unfettered by ambitions, calamities, and diseases. "Having enough to eat, the people were contented, patted themselves on the belly, and wandered about for pleasure." In other words, they forgot themselves in the enjoyment of the moment eschewing anything that could come under the heading of cultural or intellectual pursuit. In doing so they complied fully with the assumed behaviour of every and all parts of that great animal, nature. For as Pao Ching-Yen summed up the state of nature when not disturbed by men committed to cultural and social advancement: "the myriad beings participated in a mysterious equality and forgot themselves in the Tao."[71]

A competent analyst of Chinese culture and thought like M. Granet was fully justified in noting the bucolic roots and colouring of the Chinese concept of order. That concept derived, according to him, from a "healthy country feeling of good understanding." As he rightly emphasized, the notion of order was conceived by the Chinese "under the aspect of a Peace which cannot be established by the abstract forms of obedience nor be imposed by the abstract forms of reasoning." What was rather needed by the Chinese, according to Granet, was a "taste for conciliation which in turn demands a sharp sense for the actual customs, for spontaneous solidarities, for free hierarchies." This, of course, determined the fundamental characteristic of Chinese logic. It is in no sense "a rigid logic of subordination, but a supple logic of hierarchies." In Chinese thinking the concept of order has always "retained everything concrete present in the images and feelings out of which it originally came forth." In every Chinese school of thought, goes Granet's memorable conclusion, there is present "the idea that *the principle of a good, universal agreement is identical with the principle of universal intelligibility.*"[72]

Unfortunately, such an intelligibility is anything but universal. It implies a marked insensitivity to clear-cut propositions and to strict, quantitative correlations concerning nature. An intelligibility of this type corresponds to a sad weakness of mind that cannot be talked away by specious distinctions between "rigid logic of subordination" and a "supple logic of hierarchies."[73] Ultimately, logic and search for truth make sense only if they exclude compromise with illusions, vagaries, and empty verbalizations. The confident expectation of obtaining exact results seems indeed to be an indispensable condition of a sustained investigation of the workings of the external world. It is this systematic research that failed to get its wings in ancient and medieval China.

Such research means far more than the compilation of encyclopedias in the traditional Chinese style. These the Chinese possessed in oppressively large numbers about every topic under the sun. But most of those encyclopedic statements about nature have not much tangible to offer. They mostly paraphrase vague and unverifiable concepts bordering often on the mystical. L. Levy-Bruhl, a keen interpreter of the frame of mind of ancient civilizations, penetrated the very core of this predicament as he wrote: "The abstract, general form which these concepts have taken on allows a twofold process of analysis and synthesis that has the semblance of being logical. This always futile though smug process is carried on endlessly. Those who are most familiar with the Chinese mentality . . . almost despair of ever seeing it emancipate itself and stop going around in circles."[74]

This conceptual merry-go-round in which the Chinese mind was trapped can be seen in almost every page written by the Chinese of old on nature. The most striking expression of this can be found in the bewildering ramifications that grew out from the primitive notion of the Yin and Yang. The latter, which originally meant bright sunlight, was subsequently identified with the principle of maleness and also with the qualities of hardness and weightlessness. The Yin which originally referred to dark clouds, became the word for the feminine, soft, and heavy. Later connotations of the Yang extended to everything hot, dry, and pure, whereas the Yin became tied to anything cold, turbid, and moist. Again, fire was spoken of as Yang, and so was everything ready to extend or to move upward. Yin, in turn, was said to be the essence of water and of downward and contracting movements. Yang produced everything round and moving, while Yin represented squareness and stillness.

Chinese scholars were not disturbed at all by the inconsistency present in their assigning the changes of weather to the stillness of Yin. Nor did they see that it was highly inconsistent to identify the allegedly square earth with the soft Yin on the one hand, and the round heaven with the hard Yang on the other. Much less did it occur to them the problem of squaring the circle, although it was evidently implied in the constant transformations into one another which the Yin and Yang were believed to undergo continually. Nothing logical or quantitatively convincing could come out from the identification of temporal and spatial sequence with the Yin and Yang, the latter always playing the role of the leader. Considerations about physical, meteorological, botanical, physiological, and chemical phenomena were stifled in a morass of verbalization. A blissful cultivation of conceptual categories with confusingly overlapping boundaries was not to develop into exact scientific thought. Concepts, like those of roundness, dryness, and weightlessness, could hardly be manipulated meaningfully as long as they were intimately correlated with peace, eating, wealth, cheerfulness, celebrity, and profit. The same holds true also of the concepts of squareness, wetness, and heaviness which had the closest affinity for the Chinese with sorrow, drinking, poverty, ignominy, and decapitation to boot.

The categories of the Yin and Yang came to dominate the theory of magnets as well as the rituals of divinations. If Chinese notions of music, directives of medicine, principles of military tactics, often show childish elements, it is largely because of the ubiquitous presence there of concepts related to the Yin and Yang. The properties of the sun, moon, and stars could only receive poetical descriptions within a conceptual framework in which the Yin and Yang ruled supreme. Speculations about the seasonal and cyclic variations of this famed pair of opposites put the crowning touch of irony on the efforts of the Chinese to understand nature, man, and cosmos. It turned out to be a patently self-defeating enterprise. A graphic description of the Yin and Yang in *Chuang Tzu* unwittingly tells the story:

"There is no robber greater than the Yin and Yang, from whom nothing can escape of all between heaven and earth. But it is not the Yin and Yang that play the robber; – it is the mind that causes them to do so."[75] The victim of the robbery was the most ancient of all living cultures, the Chinese. Lulled by the deceitful hum of the Yin and Yang, it remained deprived of a special maker of modern culture, exact physical science.

NOTES

1. "Why China has no Science – An Interpretation of the History and Consequences of Chinese Philosophy," *The International Journal of Ethics*, 32 (1922): 238, 260.
2. The work to be first mentioned concerning the fate and fortunes of science in Chinese civilization is the monumental study of J. Needham and his collaborators, *Science and Civilisation in China* (Cambridge: University Press, 1954–) of which the volumes already published contain the material on mathematics, astronomy, physics, and engineering, in addition to introductory discussions on Chinese culture in general and on the scientific elements in Chinese philosophy in particulare the. Whil work is unsurpassed for its wealth of material, its interpretative sections are heavily biased by Needham's avowed Marxism. The strength and weaknesses of Needham's account of Chinese science can be seen in a compact form in his contribution, "Science and China's Influence on the World," to *The Legacy of China*, edited by R. Dawson (Oxford: Clarendon Press, 1964), pp. 234–308. J. Needham also co-authored with A. Haudricourt the sections on ancient and medieval Chinese science in *History of Science: Ancient and Medieval Science from the Beginnings to 1450*, edited by R. Taton, translated by A. J. Pomerans (New York: Basic Books Inc., 1963), pp. 161–77 and 427–39. There is much informative material in the now dated work of the great German Sinologist, A. Forke, *The World-Conception of the Chinese: Their Astronomical, Cosmological and Physico-Philosophical Speculations* (London: Arthur Probsthain, 1925).
3. Some of these are quoted by Yu-Lan Fung, *art. cit.*, p. 250. There is something akin to operationalism or functionalism in the strongly utilitarian character of Mo Ti's thought which is stressed in Yi-Pao Mei's *Motse: The Neglected Rival of Confucius* (London: Arthur Probsthain, 1934), pp. 69–84; and in Yu-Lan Fung's *A History of Chinese Philosophy:* Vol. I, *The Period of the Philosophers*, translated by Derk Bodde (Peiping: Henri Vetch, 1937), pp. 84–87.
4. A particularly strong critic of the Mohists was the Confucian Suen Tse in the third century B.C. See on this Yu-Lan Fung, *art. cit.*, p. 249.
5. Quoted in Homer H. Dubs, *The Works of Hsün Tzu* (London: Probsthain, 1928), p. 223.
6. Quoted in Needham, *Science and Civilisation in China*, vol. II, p. 28, note *a*.
7. Although Confucius cannot be considered the author of the *Spring and Autumn Annals*, the emphasis there on the Three Ages reflects his views.
8. Book II, Part II, chap. xiii, in *The Works of Mencius*, in J. Legge, *The Chinese Classics* (3rd ed.; Hong Kong University Press, 1960), vol. II, p. 232.
9. *Ibid.*
10. For lengthy excerpts, see *A Source Book in Chinese Philosophy*, translated and compiled by Wing-Tsit Chan (Princeton: Princeton University Press, 1963), pp. 273–84.
11. *Ibid.*, p. 288.
12. *Ibid.*, pp. 723–28.
13. Quoted in *Needham, Science and Civilisation in China*, vol. II, p. 28.
14. In *The Texts of Taoism*, translated by J. Legge, with an Introduction by D. T. Suzuki (New York: Julian Press, 1959).
15. Chap. 48, p. 138.
16. Chap. 26, p. 116; see also chap. 63, p. 154.

17. Chap. 40, p. 131.
18. Chap. 78, p. 168.
19. Chap. 34, p. 124.
20. Chap. 37, p. 127.
21. *Ibid.*
22. Book VI, par. 7, in *The Texts of Taoism*, p. 291.
23. Book VI, par. 11, p. 301.
24. Book XXII, par. 1, p. 499.
25. Book XXII, par. 4, p. 502.
26. Book XV, par. 2, p. 413.
27. Book XXII, par. 1, p. 500.
28. Book XVII, par. 1, p. 424.
29. Book XXV, par. 11, pp. 568–69.
30. Book XVII, par. 6, p. 430.
31. Book XII, par. 8, pp. 363–64 and Book II, par. 6, p. 235.
32. Book XXVII, par. 1, p. 584.
33. Book XXII, par. 5, p. 505.
34. Book XXVII, par. 5, p. 587.
35. Book II, par. 2, pp. 227–28.
36. Book XX, par. 7, p. 479.
37. *The New Organon and Related Writings*, edited by Fulton H. Anderson (New York: Liberal Arts Press, 1960), p. 118 (Book I, aphorism 129).
38. For further details on what follows, see Needham, *Science and Civilisation in China*, vol. III, pp. 402–08.
39. In the eighth chapter, "Heaven and Earth," of the forty-ninth volume of the Imperial Edition of his works. Partly translated by Stanislas Le Gall, *Le philosophe Tchou Hi, sa doctrine, son influence* (2d ed.; Chang-Hai [Shanghai]: Imprimerie de la Mission Catholique, 1923), p. 120.
40. *Ibid.*, p. 22.
41. *La pensée chinoise* (Paris: La Renaissance du Livre, 1934), p. 330.
42. *Ibid.*
43. *Ibid.*, p. 331.
44. As given in the most complete English translation of Wang Yang-Ming's writings, *The Philosophy of Wang Yang-Ming* by Frederick G. Henke (Chicago: Open Court, 1916), p. 178. The radical idealism advocated by Wang Yang-Ming was a rather logical outcome of traditional Chinese pantheism.
45. *Opere storiche del P. Matteo Ricci S.J.*, edited by P. Tacchi Venturi (Macerata: F. Giorgetti, 1911–13), vol. II, p. 207. Copious quotations from Ricci's letters can also be found in the essay of H. Bernard, *Matteo Ricci's Scientific Contribution to China*, translated from the French by E. C. Werner (Peiking: H. Vetch, 1935).
46. *Opere storiche*, vol. II, p. 237.
47. *Ibid.*, pp. 284–85.
48. The passage is from Book V, chap. viii, of the original text of Ricci's own account of his missionary work in China ("Commentarj della Cina") published for the first time in *Opere storiche*, vol. I, p. 502.
49. Quoted in Kenneth Ch'en, "Matteo Ricci's Contribution to, and Influence on, Geographical Knowledge in China," *Journal of the American Oriental Society*, 59 (1939): 348.
50. Unfortunately no translation of this work, printed in Chinese in 1604, seems to be available. The phrase quoted is part of a lengthy passage, which is discussed by J. Bettray in his study of Ricci's missionary method, *Die Akkommodationsmethode des P. Matteo Ricci S.J. in China* (Rome: Gregorian University, 1955), p. 273.
51. The rapid transmission by Jesuits to China of Galileo's discoveries made by the telescope in 1610, and the contents of first treatises in Chinese on telescopes are discussed in the essay by Pasquale M. D'Elia, *Galileo in Cina: Relazioni attraverso il Collegio Romano tra Galileo e i gesuiti scienzati missionari in Cina (1610–1640)* (Rome: Gregorian University, 1947).

52. For an informative account on this work, see the article by L. van Hée, "The Ch'ou=Jen Chuan of Yüan Yüan," *Isis*, 8 (1926): 103–18.

53. *Ibid.*, pp. 117–18.

54. Reported by J. Chesneaux and J. Needham in their contribution, "Science in the Far East from the 16th to the 18th Century," in R. Taton (ed.), *History of Science: The Beginnings of Modern Science*, translated by A. J. Pomerans (New York: Basic Books, 1964), p. 589.

55. Quoted by L. van Hée, *art. cit.*, p. 116.

56. Quoted in H. Bernard, "Notes on the Introduction of the Natural Sciences into the Chinese Empire," in *The Yenching Journal of Social Studies*, 3 (Aug. 1941): 241. At about the same time Goethe displayed a similar miscomprehension about exact physical science as he passionately argued on behalf of a new study of colours unhampered by the "restrictive" features of optical instruments. See on this my "Goethe and the Physicists," *American Journal of Physics*, 37 (1969): 195–203.

57. *Science and Civilisation in China*, vol. II, pp. 580–82.

58. In his *Religion in China: Universism, a Key to the Study of Taoism and Confucianism* (New York: G. P. Putnam's Sons, 1912). Since de Groot explicitly defined universism as a philosophical worship of the universe, he might as well have called it pantheism. At this point mention should be made also of his great monograph, *Sectarianism and Religious Persecution in China* (Amsterdam: J. Müller, 1903–04; reprinted in 1940). Its extensive documentation should dispel the studiously cultivated claim of Needham and others that the organismic, pantheistic world view of the Chinese was a most effective source of harmonious and peaceful cultural development.

59. The translation of the whole passage is given in Yu-Lan Fung's *A History of Chinese Philosophy*, Vol. II, *The Period of Classical Learning*, translated by D. Bodde (Princeton: Princeton University Press, 1953), pp. 208–10.

60. *Ibid.*, p. 211.

61. In the third-century commentary by Hsiang Hsiu and Kuo Hsiang on *Chuang Tzu*, quoted in Yu-Lan Fung, *A History of Chinese Philosophy*, vol. II, p. 211.

62. See Needham, *Science and Civilisation in China*, vol. II, pp. 338–39.

63. *Ibid.*, p. 281.

64. *Ibid.*, p. 581. This all-important point is ignored in the discussion by W. Hartner of the reasons that brought about the decline of the "classic" period of early Chinese civilization. See his "Classicisme et déclin culturel dans la civilisation chinoise," in *Classicisme et déclin culturel dans l'histoire de l'Islam: Actes du Symposium international d'histoire de la civilisation musulmane* (Bordeaux 25–29 Juin 1956) organized by R. Brunschvig and G. E. von Grunebaum (Paris: Editions Besson, 1957), pp. 367–75.

65. *Science and Civilisation in China*, vol. II, pp. 574–75.

66. *Ibid.*, pp. 339–40.

67. *Ibid.*, pp. 146–51 and 425–29.

68. *Ibid.*, p. 290.

69. *Ibid.*, p. 420.

70. *Ibid.*, p. 289. Nor was Needham alarmed by the fact that the denial of the dependence of the world on a Creator resulted in the overt denial of the existence of a shadow on an opaque object blocking the passage of light. On a less philosophical, though from the Marxist viewpoint even more important level, the level of practical inventions, Needham could heap encomiums on the Chinese invention of breast-strap harness for horses, while saying nothing, in a curiously inconsistent manner, about the equally early Chinese "invention" of the exclusively man-powered sedan chair.

71. Quoted by Needham, *ibid.*, p. 435.

72. M. Granet, *La pensée chinoise*, pp. 590–91.

73. It is this type of play on words that strikes a disturbing note on almost every page in Needham's monumental work.

74. *Les fonctions mentales dans les sociétés inférieures* (Paris: Félix Alcan, 1910), p. 449.

75. Book XXIII, par. 8, in *The Texts of Taoism*, p. 524.

CHAPTER THREE

The Wheels of Defeat

Until rather recently Southeast China has been considered the most likely cradle of *Homo sapiens*. Though such distinction is now accorded to East-Central Africa, the vast expanses of China undoubtedly formed the ancient habitat of the original settlers of the American continent. The age of Confucius was still many thousand years away when they left their Asian homeland. The details of their perilous crossing of the Bering Strait and the phases of their slow advance along the coastal regions of the Pacific are known vaguely at best. So are the various aspects of their first cultural endeavours. Only tentative inferences can be made about the primordial forms of their interpretation of nature. In all likelihood, it was heavily dominated by the assumption common among primitive tribes that nature itself was a huge, living entity going through the phases of decay and renewal. Such an inference is also supported by the usual continuity between origins and full-fledged developments. The Aztec, Inca, and Maya civilizations certainly represent the highest stages of cultural development in pre-Columbian America. In various degrees of explicitness they are also the cultural embodiments of an organismic and cyclic conception of the universe.

In contrast to the Maya civilization, a product of at least a thousand years of gradual maturing, the Aztec and Inca civilizations have one particular common characteristic which is of considerable relevance for appraising their chances of going beyond some proto-scientific achievements. They both represent the rapid rise of nomadic tribes to the heights of well-organized empires that stood at their zenith in the early sixteenth century when the Spanish conquistadors reached their shores. Of the two, the Aztecs were the more advanced in most respects.[1] Unlike the Incas, the Aztecs conquered not nomadic peoples but nations that had to their credit outstanding cultural achievements. Foremost of them were the Toltecs who developed three great architectural complexes serving as religious centres, Teotihuacan, Tula, and Xochicalco. The first of these, a temple-city with an area of about eight square miles, had served as a model for Aztec city planning, architecture, and decorative arts. Having established themselves as masters of the central Mexican highland, the Aztecs relied heavily on Toltec artisans, adopted their ceremonial year of 260 days and their cyclic calendar of 52 years.

To these assets the Aztecs added some cultural contributions of their own, one of which should appear at first sight as the very denial of what culture stands for. Hardened band though they were, Cortés and his mercenaries could not help being appalled on learning about the gruesome details of human sacrifices as practised by the Aztecs. For the Aztecs, however, the smearing of their idols with the blood of still warm human hearts represented the high point of their identification with the basic cycle of nature, the setting and rising of the sun. To secure the uninterrupted recurrence of nature's cycles constituted a foremost objective for the

E

Aztecs. In fact, the chief patterns of their social, political, religious, cultural (scientific) activities were motivated by their total dedication to implementing that objective. Awareness of this should form the clue to our understanding of their elaborate ceremonies that had taken place during the five "useless" days that were attached to their 360-day calendar. Of even greater significance for them was the simultaneous start, once in every 104 years, of the sacred 260-day year, of the 365-day solar year, and of the 52-year period.[2] By offering human hearts, often in enormous quantities, on such and many similar occasions to their gods,[3] the Aztecs aimed at strengthening them, convinced that the cyclically restored vigour of their gods was the ultimate foundation of their own survival as well.

Their gods were, of course, the personifications of various forces and phenomena in nature, all of them manifesting themselves through periodic changes. At the head of the Aztec Pantheon was the sun and its various attributes. Surrounding the Sun-god were the numerous fertility gods from the Mother-earth to deities connected with plants. Added to them were the gods of rain, moisture, and fire, while other gods were associated with stars. This organismic, animistic, and cyclic conception of the wholeness of existence manifested itself also in the names given by the Aztecs to the days of their 20-day ritualistic month of which 13 made up the 260-day ritualistic year. Most of these names referred to animals and vegetation. In the same manner they associated birds with each hour of daylight. This intimate unity of nature and time's units and cycles is also in full evidence on the Aztec calendar stones, that possibly served also as altars for human sacrifices.[4]

The most famous of these, made in 1479, is a round relief-sculpture of about 13 feet in diameter.[5] In its centre is the face of the Sun-god, Tonatiuh, marked with a symbol of cyclic motion.[6] Surrounding the centre is the band of four hieroglyphs, representing in turn a jaguar, the wind, the rain, and the water. Together they convey the Aztec belief that there is a succession of worlds each of which is destined to go under in a great physical catastrophe. The four hieroglyphs actually refer to four worlds that had already been destroyed respectively by jaguars, by hurricane, by volcanic rain, and by flood. The present world was supposed to meet its end in a gigantic earthquake symbolized by a hieroglyph of motion carved beside the face of the Sun-god. Beyond the centre part of the calendar stone are three bands of carvings, the first of which is filled with the day-signs, the second illustrating the rays and feathers of the sun-eagle, while the third is reserved for the figures of two long reptiles each bent in a semicircle so as to contain the whole representation. The reptiles symbolize the universe as a living unit and so do the obsidian butterflies carved on the side-edge of the stone.

The silent message of Aztec calendar stones is corroborated by the recently begun scholarly evaluation of the written record of Aztec folklore compiled during the decades following the conquest. Needless to say, Friar Bernardino de Sahagún and his confrères, who initiated the project, were not ethnographers in the modern sense of the word. The Aztec scribes, whom the friars taught how to transcribe their native tongue, the Nahuatl, by using the Latin alphabet, could only do their primitive best. Christian notions about man and cosmos inevitably mixed in their minds with their own heritage as they put in writing the poems, songs, chronicles, and legends of their forebears. Again, they were often asked to answer questions which reflected not their thinking but that of the missionaries. The material fills hundreds of folios and much of it is still insufficiently explored. From the carefully studied portions there emerges a world view which remarkably agrees in its main features with the concepts carved by the Aztecs into stones.[7]

The reference to "duality" may, indeed, represent with its abstract character some foreign element in the following passage from the *Annals of Cuauhtitlan*: "And it is told, it is said that Quetzalcóatl would invoke, deifying something in the innermost of heaven: she of the starry skirt, he whose radiance envelops things; Lady of our flesh, Lord of our flesh; she who is clothed in black, he who is clothed in red; she who endows the earth with solidity, he who covers the earth with cotton. And thus it was known, that toward the heavens was his plea directed, toward the place of duality, above the nine levels of Heaven."[8] The text, however, clearly conveys the genuinely Nahuatl conception of nature. She is a huge organism, embodying pairs of opposite forces which, as will be seen, keep her for ever in cyclic agitation. The expressions "starry skirt" and "radiance" refer to Ometéotl, the cosmic divinity of the Aztecs. It is both Lord and Lady, black (night) and red (day), and from it are generated all the natural forces worshipped in Nahuatl religion: "Mother of the gods, father of the gods, the old god spread out on the navel of the earth, within the circle of turquoise. He who dwells in the waters the color of the bluebird, he who dwells in the clouds. The old god, he who inhabits the shadows of the land of the dead, the Lord of fire and of time."[9]

The dynamic aspect of the universe is described as the result of the generation of four sons from Ometéol, who is both god and goddess.[10] Not surprisingly, the four sons cannot achieve their task, the stabilization of the cosmos. The four are in fact the personifications of contending forces which shall alternately play a dominating role. These four, strongly animistic powers are also identical with the basic notions of space (the four geographic directions) and with the basic units of time (day and night). Clearly, this represented a conceptual framework that could only stifle any advance toward a depersonalized representation of space and time. Such a trend probably never asserted itself in a cosmology which mirrored the endless battle of four gods, and held out no hope to either of them for final triumph, that is for a truly stable creation.[11] As M. León-Portilla, the leading interpreter of Nahuatl thought, put it, each of the four gods "endeavored to identify himself with the sun, so that he could rule the lives of men and direct the destiny of the world . . . Each god's period of ascendency constituted one of the ages of the world. But at the end of each age, war broke out and destruction followed."[12]

In a primitive phraseology the same story is told in the *Leyenda de los Soles*: "Here is the oral account of what is known of how the earth was founded long ago. One by one, here are its various foundations [ages]. How it began, how the first Sun had its beginning 2513 years ago – thus it is known today, the 22 of May, 1558. This Sun, 4-Tiger, lasted 676 years. Those who lived in this first Sun were eaten by ocelots. It was the time of the Sun 4-Tiger. And what they used to eat was our nourishment, and they lived 676 years. And they were eaten in the year 13. Thus they perished and all ended. At this time the Sun was destroyed. It was in the year 1-Reed. They began to be devoured on a day [called] 4-Tiger. And so with this everything ended and all of them perished."[13] The second Sun or world age is described in the same context as Wind-Sun, the third as Fire-Sun, the fourth as Water-Sun, causing in turn the destruction of the world by wind, fire, and flood.

What the legend tells about the fifth Sun or world age should be quoted in full: "This Sun, called 4-Movement, this is our Sun, the one in which we now live. And here is its sign, how the Sun fell into the fire, into the divine hearth, there at Teotihuacán. It was also the Sun of our Lord Quetzalcóatl in Tula. The fifth Sun, its sign 4-Movement, is called the Sun of Movement because it moves and follows its path. And as the elders continue to say, under this sun there will be earthquakes and hunger, and then our end shall come."[14] The belief that within the relatively short span of some 2500 years the world was destroyed and re-created four times could hardly generate a feeling of stability. Continuity could not be safely claimed in a world resembling a wheel, the four spikes of which, symbolizing the four main directions, served not so much as guideposts for orientation but rather as dim warnings about inevitably occurring cosmic destructions. For the Aztecs the flat earth was a quadripartite wheel or disc and was fittingly called *anahuatl* and *cem-anahuatl*, that is, ring and complete circle.

The Aztec outlook on the cosmos was, indeed, a full circle. Out of it, as was aptly noted by J. Soustelle in his perceptive study of Aztec cosmology,[15] no meaningful notion of space and time could emerge: "Mexican cosmological thought does not make a radical distinction between space and time; it refuses, in particular, to conceive of space as a neutral and homogeneous medium, independent of the process of duration. It fluctuates across heterogeneous and individual categories, whose particular characteristics succeed one another in a fixed rhythm and in a cyclic pattern. For it there are no space and time but only space-times, in which are submerged the phenomena of nature and human actions, all stamped with the qualities particular to each location and each moment." In the absence of a generalized notion of space and time, the basic conceptual foundation also was missing for a consideration of the succession of events as a rationally explorable chain in which each link causally acts and is also acted upon. The world picture of Aztec cosmology resembles, according to Soustelle, a screen on which a tireless mechanism flashes various colours in an inalterable succession. But the process in question "is not conceived as the result of a 'becoming' more or less rooted in duration, but rather as a sequence of sudden and total changes: today it is the East that dominates, tomorrow it will be the turn of the North; today we still live under propitious omens, but it will be without any transition that we shall come under the ominous cloud of evil days (*nemontemi*). The law of the universe is an alternation of distinct, neatly separated qualities, which dominate, disappear, and reappear without end."[16]

Man in that outlook was of necessity reduced to the status of flotsam and jetsam, and his interests appeared to be best served if he plunged headlong into the cyclic, rhythmic, and basically violent transformations of nature that surrounded him. This is why Aztec life took the form of carefully observed, rhythmically returning, gruesome rituals. The supreme aim of Aztec life was a participation in the rhythm of that violent action whereby their own gods lived. They in turn were representatives of the cataclysmic, cyclic forces of nature. The savagery of many of these rituals expressed their desire to experience in full the cruelty of nature. Such a desire was prompted by the paradoxical hope to find escape from cruelty by submerging into it. The willingness of the Aztecs to provide for acts of cruelty was so great that in the absence of wars they organized the so-called War of Flowers in which the best warriors of the various parts of the Aztec Empire decimated one another to provide fresh food, the hearts of fallen combatants, to their gods, representing the forces of nature.

Beneath the readiness of the Aztecs to cultivate cruelty lay an image of nature that could inspire no confidence either emotional or intellectual. This is well illustrated by the famous statue of Coatlicue,[17] known as "The Lady of the Serpent Skirt," which represented in the Aztec belief the mother of gods, or rather nature itself. The head of the figure consists of two serpents, its necklace is made of human hands and hearts, its hands and feet end in claws, and its skirt is a broad array of writhing snakes. Much the same impression is given by the Aztec altars adorned with skulls, and their sacrificial boxes for storing human hearts. Indeed, nature for the Aztecs was a source of fear which they ultimately came to honour and worship. Their view of nature generated in them an attitude that grew restless in the absence of political and economic frictions. Such was hardly a stance productive of culture and science, but the Aztecs welcomed conflicts as an "opportunity to vibrate to the deep rhythms of nature, rhythms which met in a celestial symphony in the Sacred War which the Sun fights each day as he, by his own death and sacrifice, ensures the life of man."[18]

Life and history became, therefore, trapped for the Aztecs in a vicious circle. The securing of life needed more human sacrifices, these in turn demanded the renewal of warfare. In such a framework proto-scientific achievements, such as paper-making,[19] wooden and clay stamps for the impressing of glyphs, and the storing of records in the royal archives,[20] could not produce an intellectually promising perspective. Tenochtitlán, the ancient Mexico City, with all its grandeur and magnificence,[21] described so impressively by Bernal Diáz del Castillo,[22] one of Cortés' comrades in arms, was not the cradle of a culture dedicated to the task of intellectual inquiry. With its gleaming pyramids, temples, plazas, palaces, causeways, bridges, walls, and material abundance, it was an elaborate labyrinth with no promising horizons or ideological encouragement for the cultural aspirations of the human mind.[23] The background of frightening cosmic cycles with its debilitating fatalism undermined not only the chances of the Aztecs for meaningful cultural advances, but also destroyed their political future. Permeated as they were with ideas deriving from the inexorability of cosmic cycles, they readily came under the sway of ominous predictions about the imminent end of the actual world period.[24]

The impact of such factors on the cultural and specifically scientific potentialities of the Aztecs is well illustrated by the events of the closing years of the Aztec Empire. The last Aztec ruler, Moctezuma II, who ascended the throne in 1503, successfully continued the Aztec policy of expansion. He collected tribute from 371 towns and administered justice with equity and promptness. Yet, during his last years as a ruler, Moctezuma's and his officials' attention became wholly dominated by anxiety as they looked forward to the coming of the year 1519. That year was known in the Aztec calendar as year 1-Reed (Ce-Acatl), as were the years 1467, 1415, and 1363, according to the 52-year cycle. The year 1-Reed, however, was also the year of birth of Quetzalcóatl, the ancient Toltec ruler of Tula, a temple-city about 60 miles north of the Aztec capital, Tenochtitlán. A sort of a cultural hero over all those lands, who after his death had grown into a mythical godlike figure, Quetzalcóatl opposed human sacrifices, an attitude that obviously made him a sworn enemy of the Aztecs. Defeated in a civil war, he withdrew toward the East and, according to a highly respected tradition, he disappeared over the sea vowing that he would return in a year corresponding to the year of his birth to re-establish his rule.

Though the Aztecs took over Tula long after Quetzalcóatl had become a legendary figure, by capturing Tula they fell heir not only to its riches but also to its omens. The first appearance in 1502 of bearded white beings with huge ships on the shores of the Gulf of Mexico only enhanced the suspicion that 1519 would indeed be the year of Quetzalcóatl's fateful return. Adding to all this were other ominous happenings – snowfall in Tenochtitlán, the birth of a child with two heads, – that brought further momentum to the growing sense of hopelessness among the Aztecs as the wheel of inexorable cycles was completing its fateful round. The years immediately preceding Cortés' march on Tenochtitlán in 1519 saw the Aztecs trapped in a frenzy of immolating human victims. Most of these were exacted from allied tribes living in the coastal areas, the scene of the ominous sightings of white men, Quetzalcóatl's reputed precursors. By adopting such a policy of despair the Aztecs merely drove more nails into their own coffins, as they pushed at the most critical juncture of their history some of their allies into alliance with the white man. Already demoralized, Moctezuma's powerful army proved no match, despite its great numerical superiority, for a small band of adventurers. It is, however, to be remembered that these, in addition to some strange weapons, were also armed with an outlook on nature, life, and history, which could not have been more different from that of Moctezuma and his people.[25] Cortés triumphed over an enemy already defeated, if not in body, certainly in mind and spirit. The political and cultural history of the Aztecs, especially in its closing phase, shows that they never came close to that attitude toward time and nature, the hallmark of which is confidence. Judging by their past it is most unlikely that the Aztecs would have ever succeeded in breaking the clutches of their dispiriting outlook on nature, had their political and cultural existence continued undisturbed for several more centuries.

*   *   *

A similar conclusion seems to apply to any analysis concerned with the scientific potentialities of Inca culture.[26] Considerations of this type should be offered with no small hesitation as the Inca scientific record is extremely spotty. What is well known about Inca culture is of social, not scientific, character. A closely knit political organization, excellent roads, bridges, and postal service to secure conquests and to facilitate ever new ones, extravagant amounts of gold and silver amassed by the rulers, these were the main impressions that Pizzarro and his fellow con-quistadors had gained on capturing the Inca cities. Had their conquest of the Incas been less brutal and had they had eyes for observing benevolently Inca cultural activities instead of suppressing them, it is unlikely that they would have noticed evidences of genuine scientific mentality.

It is, indeed, hardly justified to look for something seriously scientific in a culture that failed to develop writing of even the most rudimentary type. (It still remains to be seen that recent, ingenious efforts to decipher as possible script the geometrical patterns (*tocapus*) on priestly garments and wooden vessels can overcome serious ambiguities.) Among the Incas the fundamental vehicle of information, the spoken word, received very little support through tangible, symbolic means. Some details of Inca history were recorded in paintings kept in palaces open only to the ruling classes. Numerical details of specific information were registered on their famous quipus,[27] a sort of abacus made of strings, on which knots represented quantities. The quipus were not without some marvel of their own that should impress any historian of science. The knots were arranged in groupings that intimated a decimal notation. At the lower ends of the strings were the knots standing for unit numbers, higher up was the place for knots indicating tens, and above them the place for hundreds and even higher units. The quipus, however, were not instruments of calculation: they merely recorded numbers. Calculation was done by putting beans into compartments of flat boxes, one bean standing for each unit of the material to be counted. Once the counting was done, it was recorded on quipus. Needless to say, the quipus were useless without some explanation that the Incas could do only orally.

The duty of remembering the meaning of particular quipus rested with a special class of civil servants, the quipu-readers (*quipu-camayoc*). In the highly autocratic Inca Empire they remembered and repeated only what was in conformity with the interests of the Inca conquerors. What the Spaniards learned from them about Inca history and especially about the history of tribes vanquished by them was, therefore, as biased as most Spanish accounts of the comportment of the conquistadors. This circumstance ought to be kept in mind when one tries to reconstruct the cosmological and mythological lore of the Incas. Many of its details were engineered to tie the Inca rule to divine dispensation. There is something obviously artificial in the features of the supreme Inca god, Kon Ticsi Viracocha (popularly Kon-Tiki), as described in the official religion.[28] Viracocha seems to be a heavenly mirror-image of the absolute ruler, the Inca, which the latter certainly was. The main attributes of Víracocha also reflected the principal objective of the Inca: the stabilization of the empire by moulding it into a perennial immobility. In fact, the main guideline of good government as defined by the Inca was to keep the existing order as rigidly as possible. Innovation as a political device was strongly suspect.

Such was a rather wilful religio-political ideology which failed to take deep roots. It certainly did not supplant the older views in which the various parts of the earth and heavens formed a pan-organismic deity.[29] In that framework the earth was the universal life-giving mother, the mountains her breasts, the rivers her milk, and the gold her most recondite secretion. The sun was the universal father, and its eclipses were interpreted as its periods of sleep. The unofficial Inca myths of creation were crassly organismic in character, and this could only reinforce the extremely superstitious attitude of the Incas concerning strange objects in nature which they called *huacas*.[30] For the Incas those objects did not serve as pointers toward some basic lawfulness and rationality in nature. They failed to approach that initial level of intellectual maturity for which strange things are not stumbling blocks but rather aids for deciphering some of the orderliness of nature. The unusual appearance of a part of nature did not become for the Incas a source of wonderment, which is the first and foremost step in philosophy. Instead, they made idols of the *huacas* which then effectively acted as blinders on their interpretation of the world.

Equally incapable were the Incas to break out from the cycles of nature, though it should seem that the concept of a perennial, immutable Inca Empire would have encouraged an interpretation of the cosmos and history free of cycles. What is actually known is that the lunar months formed the basis of their ceremonial calendar while the solar year regulated their agricultural system. It is not known how the two were coordinated in the Inca calendar. Their knowledge of astronomy was distinctly primitive; their main astronomical instrument consisted of a row of four stone columns to which they related the positions of the sun. It is doubtful whether they were familiar with sundials.[31] Noteworthy is, however, Otfrid von Hanstein's suggestion[32] that the Incas gave special consideration to cycles of one thousand years (Great Sun-Year) and of ten thousand years (Sun-Cycle). Such interpretation of the record, if correct, clearly shows the hold of old cyclic notions of time and nature on their thinking. That this was actually the case receives a telling corroboration from a poignant circumstance of the sudden demise of Inca history.

As soon as the first rumours spread down the Pacific coast about the appearance of bearded beings (nobody was sure whether they were gods or humans), and about their "house in the sea," there was an upsurge of rumours and prophecies about the imminence of an all-engulfing catastrophe.[33] It looked, indeed, as if floodgates of pent-up visions about cosmic cycles had opened up overnight. All of a sudden a vital urgency seemed to be present in that old legend that told about Viracocha's descent on earth, about his walking among people teaching them how to live, and about his disappearing on the sea off the coast of Ecuador. Could it be that the Viracochas, the people whom the Incas once routed and whose god they expropriated, would be on their way back to revenge themselves?

The *huacas* were consulted with feverish haste, and people began to mutter that the twelfth Inca would be the last one. These ominous forebodings were given a gloomy hue by the sudden strike of devastating diseases that preceded almost everywhere the arrival of the bearded ones. In Quito the epidemics felled one high official after another. The Inca himself drew up his will stating that the bearded ones were Viracocha's people transformed into godlike beings, and voiced his willingness to meet more than half-way all their demands. The belief in an imminent visitation by the gods, accompanied by never before experienced epidemics, proved to be a menace more effective than a band of horsemen and their guns. At any rate, it could no longer be claimed that the Inca Empire was exempt from the remorseless turns of the wheels of fate.

★   ★   ★

Among the various civilizations of pre-Columbian America, it is the Maya that offers the best insight into the way in which a cyclic concept of cosmology coupled with an organismic concept of nature influences the chances of man's slow advance toward a scientific grasp of the external world.[34] Certainly, in the case of the Maya, the lack of enough time cannot be invoked as a main reason for their failure to advance to a distinctly scientific level of the understanding of nature. Unlike the short-lived Aztec and Inca civilizations, Maya culture represents a process stretching well over two thousand years. Its highest state of flourishing constitutes the so-called Great Period (A.D. 731–A.D. 987) of the Old Empire, which in turn comprises about six centuries (A.D. 317–A.D. 987) of Maya history. The beginnings of the first phase of Maya civilization, or the pre-Maya era, cannot be dated exactly, but undoubtedly go back as far as the third millenium B.C. It is, however, known from internal archaeological evidence that the Maya probably had recorded dates as early as 353 B.C., and certainly not later than 235 B.C. The six or five centuries following these dates are known as the third or last phase of the pre-Maya era, the end of which is assigned to the early fourth century A.D. The reason for this is that the years A.D. 320, and A.D. 328, are of major importance in Maya archaeology. The former is marked on the Leyden Plate, a small piece of jade, both sides of which are covered with carvings. The latter date is carved (in its Maya equivalent of course) on Stela 9, the oldest of those shaft-like stone monuments that throughout the Old Empire had been erected at regular, usually twenty-year, intervals.

The Maya represent in pre-Columbian America not only the longest continuous historical phenomenon. Theirs was also a political and social organization markedly different from the strongly centralized, authoritarian rule that characterized the Aztec and Inca Empires. The loosely knit system of Maya cities spreading across the Yucatan peninsula easily evokes the association of city-states in classical Greece. As was also the case with the Greeks, the climate left the Maya with ample leisure time. The main staple of Maya diet, the maize, reached the harvesting stage in about two months' time, and the remaining part of the year provided ample opportunity for the satisfaction of their cultural aspirations. This the Maya possessed in a remarkable degree.

The most palpable evidences of the cultural attainments of the Maya are their temples, set on the top of graceful pyramids, their palaces, plazas, ball-courts, and dance platforms, of which the easily accessible ones are today major tourist attractions. The skill and determination to produce these monuments had to be considerable especially in view of the limitations of Maya engineering. Some of those limitations should appear baffling indeed. The corbelled vault as used by the Maya indicates not only ingenuity but some mental inertia as well. As was put forward in a detailed analysis of Maya architectural techniques, the Maya seem to have been thinking "in terms of monolithic masses wherever possible, and took comparatively little interest in the manner in which individual stones were laid together."[35] This not only diminished the stability of their buildings, but also prevented them from following up the great possibilities embodied in some of their architectural creations that were identical for all practical purposes with true arches.[36] Their failure to realize this blocked their advance toward larger and more convenient buildings.[37]

Maya buildings are nevertheless an illustration of the ingenuity for which the absence of some basic tools is not a crushing handicap. The Maya, as was the case with all pre-Columbian Americans, had no wheeled vehicles. They also lacked familiarity with any practical use of the wheel whether as a pulley, a potter's wheel, or a water wheel. They had no domesticated animals capable of carrying heavy loads. In the absence of metal tools suitable for stone work, they had to rely on granite or basalt celts to carve out their building material from rock outcroppings. Apart from the use of wooden levers, ropes, and rolling logs, their main source of power was sheer human strength and determination. This they needed in more than ordinary measure to achieve the transportation of stones over long distances and marshy areas.

The Maya did not equal the Incas in the art of roadbuilding, nor are their pyramids as massive as those of the Aztecs. But Maya architecture shows superior artistic qualities as do their sculpture and painting. A fine evidence of Maya painting is the Codex Dresdensis, one of the three written Maya documents that survived both the humid climate of the Yucatan and the iconoclastic zeal of some narrow-minded missionaries.[38] The chief importance of these codices does not concern the arts, as such. They are a chief source of information on what represents the two highest intellectual achievements in pre-Columbian America. The first of these is the hieroglyphic writing of the Maya[39] that sets them intellectually high above the other erstwhile occupants of the American continent. At the same time it should be noted that Maya writing represents a less developed stage than the Egyptian hieroglyphic or the Sumerian cuneiform writing, and certainly cannot be compared with the most developed form of ideographs, the Chinese.

The other great intellectual achievement of the Maya is strictly scientific in character and naturally prompts one to indulge in conjectures why the Maya made but one outstanding scientific step. The step was the enormous feat of devising a positional type of counting which, in an indirect sense, included even the use of zero.[40] This the Maya did contemporaneously with the development of positional arithmetic in India, that is, during the first centuries of our era. For science in general and for astronomy in particular the significance of a positional notation of numbers cannot be emphasized enough. Adoption of the Hindu-Arabic decimal system with its positional notation in late medieval Europe was an indispensable condition for reaching a higher level of measurements and calculations which greatly helped the rise of physical and astronomical science during the sixteenth and seventeenth centuries. One would, however, look in vain for even a halting step along this direction in the cultural record of the Maya. Though Maya culture was in steady ascendance for six to eight centuries, the fortuitously devised place-notation of numbers was not accompanied by the development of arithmetical operations beyond those of addition and subtraction. There are no traces of multiplication, division, and of even the simplest fractions in the computational tables of the Codex Dresdensis.[41]

The same baffling disparity characterizes the accomplishments of the Maya in the field of astronomy. Experts on Maya history and culture who were not themselves astronomers readily hypothesized about a Maya Hipparchus.[42] Most general accounts of Maya culture abound in superlatives about the astronomical attainments of the Maya,[43] but a more complex picture emerges from the special studies such as

John E. Teeple's still classic monograph.[44] A chemical engineer by training, Teeple took up the study of Maya astronomical records in the 1920's. He proved beyond doubt the great care by which the Maya recorded the phases of the moon. Of the greatest importance was his "determinant theory" which he proposed as an explanation of how the Maya correlated the solar time and the 365-day year. Teeple, like others, was duly impressed by such facts that the length of the tropical year implied in the Maya calendars equalled in precision that of the Gregorian calendar.[45] The Maya recordings of lunar eclipses also seem to indicate familiarity with their regularity. Again, Maya data about Venus make almost inevitable the conclusion that the Maya achieved a remarkably exact determination of its synodic revolution (583.920 days as compared with the modern value of 583.935 days).

Feats like these raise further questions about Maya astronomy, in particular about their astronomical instruments and methods of observation. What is known in this respect is very meagre. To speak, as S. G. Morley did, of some round towers (caracol) as observatories, is indulging in a hyperbolic language.[46] Still, everything in Maya stargazing that stands critical scrutiny is of such value as to encourage the application of strict astronomical concepts and formulas as possible keys for disclosing the real meaning and organization of their observations. This was, indeed, done by several professional astronomers, Hans Ludendorff, Arnost Dittrich, and Maud W. Makenson, but their efforts failed to pay the expected dividend. After all, the investigator of Maya astronomy is beset with difficulties, some of which Teeple spelled out in the concluding part of his monograph. Thus, while we can be reasonably sure about the presence of eclipse records, the Maya glyph for eclipse is still not identified. Again, the astronomical recordings of various Maya cities prior to the introduction of a uniform moon numbering are still not satisfactorily reduced to one single system. In general, the interpretation of important points of Maya astronomy leaves wide room for disagreement. While some claimed the presence of Mars, Jupiter, and Saturn tables in the Codex Dresdensis, others, like Teeple, remained unconvinced. What Teeple remarked by way of introduction to his interpreting Stela 9 at Copan as a clue for the computation of the tropical year by the Maya, is particularly revealing of the situation:

There are probably over 2000 separate dates recorded on Maya monuments which would be an average of 5 or 6 or 7 for each day in the year. So if one starts with any assumed correlation, it is fairly easy to find Maya dates which coincide with his ideas of solstices and equinoxes. It would be just as easy, allowing a day or two leeway, to find 20 or 25 dates proving that the Maya celebrated Washington's Birthday or Yom Kippur. That does not look like a feasible method for producing convincing evidence. We have no real knowledge that the Maya were interested in equinoxes and solstices any more than they were in the Fourth of July. Possibly their check on the passing of a year was the day when the sun exactly overhead cast no shadow at noon; possibly it was the heliacal rising of some star. Inquiries made of people who have been long in the tropics, however, do not indicate that modern Indians are especially interested in equinoxes, solstices, vertical suns or heliacal rising of stars.[47]

The principal objection against reading scientific astronomy into Maya time-keeping comes from the single-minded devotion of the Maya to their sacred calendar, the tzolkin, consisting of 260 days. As Teeple aptly noted, it was based on "an arbitrary and orderly succession of days and months in regular order, going on forever without regard to any natural phenomenon."[48] And he added: "There is no astronomy in the Maya time count, except the passage of the day and a vague idea that a year is about 365 days long, and there is no accuracy about it except the accuracy of a machine that does not slip a cog and miss the count. To me it is simply a huge meter that records the passage of time on four different dials."[49] One is obviously tempted to probe into the deeper layers of this intellectual impasse and in this connection Makemson hit the mark in pointing out that Maya astronomers were rather astrologers, with no primary interest in systematic observations and in quantitative, or scientific correlation of the periods of the moon, sun, and planets. Beyond the practical objective to provide the correct date at springtime for the planting of maize, their overriding concern was the record-ing of celestial phenomena within the framework of their sacred 260-day year, the tzolkin. "Briefly stated," wrote Makemson, "it was the aim of the Maya astron-omer to find a figure which contained an even number of multiples of an astro-nomical cycle and a whole number of multiples of the 260-day period, in order that after a suitable interval the same phenomenon would be repeated on the same day of the tzolkin."[50]

As the phases of the moon, the solar year, the periods of the planets, and the various periods of eclipses are never the exact multiples of 24-hour periods, let alone of 260 days, Maya astrologers had to satisfy themselves with approximate solutions. When the discrepancy between the real time and the time measured in tzolkins became too large in a given cycle, they devised another, more inclusive one. A basic cycle was the so-called Calendar Round, consisting of 52 years (or 18,980 days which is the least common multiple of 260 and 365). They also had larger cycles of which special mention should be made of the so-called Initial Series, or Long Count, with a cycle length of 374,440 years. Within this cycle, which allowed for repeated corrections, the Maya were able to set up a leap-year system slightly more correct than the one provided by the Gregorian calendar. But this was an achievement that did not prompt the Maya in the direction of truly scientific thinking. Their mind was rather attracted to the magic of numbers and their arrangement in cycles. For the Maya the number 20 was particularly sacred. They invested it and the first nineteen numbers with divine halo (they are still worshipped as gods by some present-day Maya), and singled out the number 13 as harbinger of unusual good fortune. The combination or rather the product of the numbers 13 and 20 yielded, in turn, the sacred year of 260 days, and the Maya considered time as the perennial repetition of cycles each 260 days long. As J. Eric S. Thompson put it: "The cycle of 260 days is the core of the Maya calendar. It is a divinatory and sacred almanac which bears no known relation to any celes-tial phenomenon . . . It repeats through all eternity regardless of the positions of sun, moon, and stars."[51]

In the tzolkin every day had its peculiar transcendental significance as each day was one of the 260 possible combinations of two cycles consisting of 13 and 20

individual units respectively. The meaning attached to the day on which a particular individual was born, was believed to have set the pattern of his whole life.
Marriages, social activities, political decisions, such as declarations of war, were
carefully set for a propitious day. Their exactness in recording events and cataloguing them on an exactly recurring calendar wheel was motivated by their
concern that the god connected with the day of the particular event should safely
get the sacrificial offering on the right date. Individual happenings were of interest
for the Maya in so far as they were part of a repetitive pattern. The unique, the
irreversible, and the unrepeatable seemed to have no appeal for the Maya. This is
why they did not record in their own history the names of outstanding leaders,
sculptors, architects, and calendar makers.

With the advent of Maya book making there are traces of incipient interest
among the Maya in taking stock of the specifics of their history as shown by the
chronicle of Chilam Balam.[52] But in it one would look in vain for a distinct
trace of the idea of a progressive, unidirectional flow of time. Although Maya
calendars counted the time from about 3100 B.C., this date was not meant to be
an absolute, cosmic beginning. Maya recordings clearly evoke uncounted past
cycles of time when they refer to millions of years. Such large numbers hardly
indicate a precociously large-scale vision of the geological or cosmic past. They
rather suggest a resignation of being for ever trapped in the cyclic return of events.
For the Maya, the life and future of society were subject to an inevitable recurrence
of patterns. It is well known, for instance, that the Aztecs, who largely depended
on Maya calendars, expected the end of the world to come at the end of each
52 years, a time-span that contained 73 sacred years, or tzolkins. They called these
periods *ziuhmolpilli* (the Maya name did not survive), and on the last night of it
all inhabitants of Tenochtitlán spent the night on the surrounding hills in fear
of an all-destroying earthquake. When the sun nevertheless rose, they took it
amidst great rejoicing, as a sign that life was to go on along normal patterns for
another 52 years.

While similar details are not known about the Maya, their interpretation of
history was clearly dominated by the fatalism of cycles. During the New Empire
(987–1697) Maya priests began emphasizing periods of $256\frac{1}{4}$ years. It was the
equivalent of the product of 13 and 20, their two principal sacred numbers, the
20 standing here for 20 katuns or 20×7200 days. Obviously, their pigeonholing
of Maya history into such exact periods was not the result of an objective reading
of the course of events. It rather reflected their deep-seated urge to see even in
human history the inexorable rule of cycles in the form of alternating triumphs and
inevitable defeats. Here one is certainly faced with the inhibitory influence of a
cyclic conception of time and nature. Another evidence for the same is the two
ways in which the Maya applied their vigesimal arithmetic. One, which they
generally used in counting, was strictly vigesimal, the other, a modification of it,
found use in their calendars. In the former, the orders of magnitude followed as
$1, 1 \times 20 = 20, 20 \times 20 = 400, 400 \times 20 = 8000, 8000 \times 20 = 160,000$ and so on.
In the latter, the third order of magnitude was changed from 20×20 to $20 \times 18 = 360$, to conform approximately to the number of days in a year. Accordingly, the
fourth and higher orders of magnitude in the modified vigesimal system were

multiples of 360, that is $360 \times 20 = 7200$, $7200 \times 20 = 144,000$ and so forth. Behind this rather arbitrary variation of their system of counting lies the Maya preoccupation with odd time cycles which in turn diminished their incipient appreciation for scientific uniformity.

One may, of course, choose to rest satisfied with the glitter of appearance, and praise, as some admirers of Maya culture do, the enchantment of the Maya with the rhythm of time. It should, however, be obvious that grave question marks lurk behind picturesque descriptions like the following one: "Like a miser counting his hoard, the Maya priest summed the days that had gone and the days that were to come, stacking them in piles, juggling combinations to learn when the re-entering cycles of time would again pass abreast the turnstiles of the present."[53] In this beautiful evocation of the past one should rather catch a glimpse of the perplexing failure of the Maya to break out from the blinding confines of cycles and make a convincing step toward a truly scientific interpretation of nature. That they had not been successful in this respect illustrates that the emergence of science is a more extraordinary event than many children of the atomic age would imagine.

The failure of the Maya also brings into focus the extent to which engrossment with cyclic concepts undermines the chances of man's incipient gropings toward science. These cyclic concepts held complete sway over the Maya outlook on time, nature, as well as human history. Their wheel-like calendars, known as katun-wheels, are a telling illustration of this, and a similar story is told by Maya cosmology. According to Maya belief the world had already been destroyed three times, and the present world would be terminated by a flood coming through the mouth of a gigantic serpent girdling the whole world. This fusion of cyclic and organismic concepts in Maya thought is worth noting. Among its various manifestations is the invariable representation of the moon by the Maya as a woman. In the same organismic vein they attributed the eclipses, a major cyclic phenomenon of the heavens, to the periodically aroused appetite of ants to eat away part of some celestial bodies. The succession of four world catastrophes is in Maya thought like the four principal spikes of a wheel and their unceasing rotation reflects the life-rhythm of that huge animal, the world. Its symbol was in a crass animistic fashion either a serpent, an ubiquitous figure in Maya art, or an alligator on which, according to Maya belief, the world was resting.

Maya history provides startling illustrations of the defeatist psychological climate created by the combined impact of cyclic and organismic ideas. Students of the Maya have for a long time been baffled by the fact that time and again some of their best built cities were abandoned in a rush. Wars, famine, epidemics, natural disasters have usually played a part, but it seems that there was also a rather intangible factor at work. One cannot help wondering, for instance, on finding that in such widely separated places as Copan and Tikal the Maya stopped erecting stelae at moments that marked the end of some larger period of time as determined by their calendar. The last centuries of Maya history bring out clearly the nature of that recondite factor. In Morley's felicitous phrase it consisted in a "chronological coercion."[54] As time went on, the Maya came to consider as harbinger of major débâcles the Katun 8 Ahau, that is the twenty-year period (Katun) ending on the day 8 Ahau which in turn was one of the 260 possible combinations of the numbers 1 to 13 and of the twenty Maya day-glyphs of which the next to last was called Ahau. What had been at first a hindsight of their chroniclers and priestly scribes turned in the long run into an irresistible psychological conditioning for the Maya. The upheavals of Maya history that had taken place during the interval 1185–1204, and 256 years later between 1441 and 1461, seem to have been brought about not so much by the usually unpredictable turn of events, but rather by the wilful implementing of the law of cycles on the part of the Maya. The first interval saw the League of Mayapan fall apart with the subsequent rapid decline of the last Maya cultural renaissance. The second interval witnessed another civil war raging around Mayapan which ended in the destruction and abandonment of one of the chief centres of late Maya culture and national identity.

The wheels of time had ultimately become for the Maya the wheels of national fate, and they gave an increasingly submissive co-operation to its inexorable turns. Nothing illustrates this better than the last act of the drama of Maya history and culture. Following their victory over the Aztecs, repeated attempts were made by the Spanish to subdue the Maya city-states and christianize their inhabitants. In general, the Maya resisted valiantly though with clear anticipation of a final defeat. By 1546 much of Yucatan was in Spanish control. The Itza, the only notable Maya tribe to retain independence, owed its good luck largely to its rather inaccessible geographic position around the lakes of El Peten, about fifty miles inland from the once leading Maya city of Tikal. Their relative safety had not, however, diminished their awareness of the inevitable which they expected from the next return of the fateful period Katun 8 Ahau (1697–1716). In fact, when Fathers Fuensalida and Orbita visited in 1618 Tayasal, the last major Maya city located on the western end of Lake Peten Itza, to convert its inhabitants, the ruler, Canek, pointed out that the time predicted for the tribe's spiritual demise had not arrived yet. Or as the Spanish chronicler recounted the crucial part of the encounter:

> To this Canek replied: that although there had not yet arrived the time in which their ancient priests had prophesied that they were to leave off the adoration of their gods; because the age in which at this time they were was that which they call Oxahau, which is to say Third Age [Katun 3 Ahau]; these barbarians seem to have counted their ages backward, or to a determined number; which reached they forgot it, and return to begin a count anew [the *uudz* katun]: Because when they retired from Yucatan, which is already going on for three hundred years, they say that it was the Eighth Age [Katun 8 Ahau] and that there had not arrived so soon what was appointed for them [another Katun 8 Ahau]; and now [1618] they say it was the Third Age and that the time had not yet arrived.[55]

Armed with this information the missionaries waited patiently for almost eighty years when, in January 1696, Father Avendano set out to exploit the propitious time. He arrived a year and a half too soon as the Itzas explained to him on his arrival at Tayasal. On the other hand, the validity of Maya prophecies concerning the impending abandonment of their ancient faith was not questioned. But the expected conversion never became a reality. Fifteen months after the friar's visit, the wheel of fate turned full circle. In March 1698, Captain Martin de Ursua and his one hundred and eight men pushed to the shores of Lake Peten Itza, to force the Itzas into peaceful surrender. The Spanish withheld their fire during their crossing of the lake in a large galley even though exposed to the arrows of well over 2000 Itza canoemen. But near landing, one Spanish soldier, infuriated by an arrow wound, fired his arquebus. This also brought his comrades' restraint to an immediate end. The sound of a hundred or so firearms signalled not only the routing of the Itza canoe flotilla. It also triggered the pent-up defeatism of some 5000 Itza soldiers lined up on the shore. Without joining the battle, all took to flight leaving behind their families, homes, and everything they had built for centuries. It was the last and tragically mocking obedience of the Maya civilization to the tyranny of cycles.

NOTES

1. For a general orientation on Aztec history and culture there is a wide selection of works available. Of these the work by the noted American archaeologist, Herbert J. Spinden, *Ancient Civilizations of Mexico and Central America* (New York: American Museum of Natural History, 1917), is dated, but still contains much useful material. Among more recent publications special mention should be made of the very readable work by the former director of the University Museum of the University of Pennsylvania, George C. Vaillant, *Aztecs of Mexico: Origin, Rise and Fall of the Aztec Nation* (Garden City, N.Y.: Doubleday, 1953), and of the equally fascinating work by an active explorer of Aztec monuments, Laurette Séjourné, *Burning Water: Thought and Religion in Ancient Mexico* (New York: The Vanguard Press, 1961). The latter gives valuable interpretation of Aztec cosmological myths. When it comes to studies specifically devoted to the scientific attainments of the Aztecs, the picking is very meagre. Research on science in pre-Columbian America has been a rather neglected field, and what has been published relates largely to the achievements of the Maya. In the yearly bibliographies of *Isis*, the leading periodical of the history of sciences, a special listing of such studies was first given only in 1966. The highly unsatisfactory situation is also evidenced in the short summary of pre-Columbian science in R. Taton (ed.), *History of Science*, Vol. I, *Ancient and Medieval Science*, translated by A. J. Pomerans, Part III, chap. i, "Science in Pre-Columbian America" (New York: Basic Books, 1963), pp. 374–84. Pre-Columbian science received no systematic treatment in Charles Singer, E. J. Holonyard, and A. R. Hall (eds.), *A History of Technology*, Vol. I, *From Early Times to Fall of Ancient Empires* (Oxford: Clarendon Press, 1954), or in Vol. II, *The Mediterranean Civilizations and the Middle Ages* (Oxford: Clarendon Press, 1956). *La ciencia en la historia de México*, by Eli de Gortari (Mexico: Fondo de la Cultura Económica, 1963) has a lengthy and informative chapter on various aspects of pre-Columbian science in Mexico with a detailed bibliography.
2. On this, see Vaillant, *Aztecs of Mexico*, p. 199.
3. One of Cortés' companions, Friar Bernardino de Sahagún, gives in his *Historia general de las cosas de Nueva España* (Editorial Nueva España S.A.: Mexico, 1946), vol. III, p. 47, a detailed account of the various types of human sacrifices offered at the beginning of each of the 18 Aztec months of the year. For a long excerpt in English, see Séjourné, *Burning Water*, pp. 12–13.
4. Such is Spinden's opinion (*Ancient Civilizations of Mexico and Central America*, p. 189). The rather recently discovered National Stone shows a similar calendar stone in a vertical

position, as part of a huge monolithic block, a circumstance that may cast some doubt on Spinden's interpretation. See Vaillant, *Aztecs of Mexico*, Plate 52.

5. For a description, see Spinden, *Ancient Civilizations of Mexico and Central America*, pp. 191–192, and Vaillant, *Aztecs of Mexico*, pp. 163–64.

6. The list of Aztec hieroglyphs for movement conveys forcefully the essentially cyclic character of motion as conceived by the Aztecs. The figures closely resemble a water wheel with four arms, and one of the glyphs is an almost exact replica of the swastika, the classic symbol of cyclic returns. For the diagrams, see Séjourné, *Burning Water*, p. 97.

7. A very informative account of such studies is Miguel León-Portilla, *Aztec Thought and Culture: A Study of the Ancient Nahuatl Mind*, translated from the Spanish by Jack Emory Davis (Norman: University of Oklahoma Press, 1963). Subsequent quotations from old Nahuatl texts will be from this work.

8. *Ibid.*, p. 29.

9. *Ibid.*, p. 32.

10. *Ibid.*, p. 33.

11. "The gods battle and their struggle for supremacy is the history of the universe, their alternative triumphs become so many other creations," as was aptly noted by A. Caso in his *La religión de los Aztecas* (Mexico City: Enciclopedia Illustrada Mexicana, 1936), p. 11.

12. León-Portilla, *Aztec Thought and Culture*, p. 36.

13. *Ibid.*, p. 38. The significance of various numbers and calendar terms occurring in this passage will be discussed later in this chapter.

14. *Ibid.*, p. 39.

15. *La pensée cosmologique des Anciens Mexicains: Représentation du monde et de l'espace* (Paris: Hermann, 1940), p. 85.

16. *Ibid.*

17. For a description and illustration, see Vaillant, *Aztecs of Mexico*, p. 163.

18. *Ibid.*, p. 222.

19. On the technique and impressive extent of Aztec paper-making, see Victor W. von Hagen, *The Aztec and Maya Papermakers* (New York: Augustin, 1943).

20. For details on this point, see Victor W. von Hagen, *The Aztec: Man and Tribe* (rev. ed.; New York: The New American Library, 1961), p. 192.

21. For much of this architectural accomplishment the credit should go to the Toltecs whose name in Nahuatl means "master builders." Interested more in cultural pursuit than in warfare they could not stem the advance of the belligerent Aztecs.

22. As he points it out, some of his fellow soldiers familiar with the splendour of Renaissance Rome and Venice ranked the Great Temple patio above everything they had seen before. (*Historia verdadera de la conquista de Nueva España* [Ediciones Mexicanas S.A.: Mexico, 1950], p. 164.)

23. Even Séjourné (*Burning Water*, pp. 45–46) viewed as idle speculation the possibility that the Aztec culture might have evolved toward the level reached by ancient Greece. For this to happen, according to her, the Aztecs should have been left spared of contact with Europe. In addition, there ought to have been a parting with Aztec barbarism and a return to former, more humanistic religious ideals as probably cultivated by the Toltecs.

24. On the cosmic part of this, see Séjourné, *Burning Water*, p. 39. On the historical part, see von Hagen, *The Aztec: Man and Tribe*, pp. 122–23, 200–01. Most of these details were recorded by early Spanish chroniclers the chief of whom was Sahagún. Many of his fascinating stories, to be taken of course with caution, are paraphrased in that still informative, nineteenth-century account of the conquest by William A. Prescott, *History of the Conquest of Mexico* (8th ed.; New York: Harper & Brothers, 1851), vol. I, pp. 122–24, 312–15.

25. This point is often noted by authors of books on Aztec history and culture, but never analysed in depth, or traced back to its ultimate source: the presence or absence in a culture of the belief in a transcendental, personal Creator of the universe. See, for instance, von Hagen, *The Aztec: Man and Tribe*, pp. 45, 196; or Vaillant, *Aztecs of Mexico*, pp. 223, 239. Séjourné, who likes to extol the "prodigious, creative power" of the Aztecs, also admits

F

that such a power could have been triggered for real cultural advancement only by a more enlightened spiritual conception of the cosmos, the germs of which were present, according to her, in the Aztec tradition. (*Burning Water*, p. 44.)

26. Needless to say, there are no monographs devoted to Inca science as such. In almost all modern accounts of Inca culture and history there is a discussion of Inca engineering (roadbuilding, architecture), and of Inca calendars. Among the works to be consulted the foremost is John Howland Rowe's long essay, "Inca Culture at the Time of the Spanish Conquest," in Julian H. Steward (ed.), *Handbook of South American Indians*. Vol. 2, *The Andean Civilizations*. Smithsonian Institution. Bureau of American Ethnology. Bulletin 143 (Washington, D.C.: United States Government Printing Office, 1946), pp. 183–330. Rowe's essay gives a balanced synthesis and summary of all previous research on Inca culture with full bibliographical documentation. One shall read with profit among the more recent publications Burr C. Brundage, *Empire of the Inca* (Norman: University of Oklahoma Press, 1963); Louis Baudin, *A Socialist Empire: The Incas of Peru*, translated from the French by Katherine Woods, edited by Arthur Goddard (Princeton, N.J.: Van Nostrand, 1961); and Victor W. von Hagen, *Realm of the Incas* (rev. ed.; New York: The New American Library, 1961). The latter also edited the English translation by Harriet de Onis of *The Incas of Pedro de Cieza de León* (Norman: University of Oklahoma Press, 1959) which contains the two *Crónicas* of Cieza based on his seventeen-year-long stay among the Incas from 1541 to 1558.

27. On the quipus the basic studies are L. Leland Locke, *The Ancient Quipu or Peruvian Knot Record* (New York: The American Museum of Natural History, 1923), and Erland Nordenskiöld, *The Secret of the Peruvian Quipus*. Comparative Ethnographical Studies. Göteborg Museum. Vol. VI, Part I (Göteborg: 1925).

28. See Rowe, "Inca Culture at the Time of the Spanish Conquest," p. 316.

29. See Brundage, *Empire of the Inca*, pp. 21–22, 44.

30. Within a radius of about 20 miles around Cuzco the number of *huacas* was about 350 at the time of the Spanish conquest. The most important of these was a spindle-shaped unwrought stone on Huanacauri hill near Cuzco. See Rowe, "Inca Culture at the Time of the Spanish Conquest," p. 296.

31. Contrary to the interpretation of the *Inti-huatana*'s by Sir Clement R. Markham, a leading Peruvian explorer around the turn of the century (*The Incas of Peru* [London: Smith, Elder and Co., 1910], p. 116), these round-shaped platforms were not astronomical "observatories" even in the sense of elaborate sundials. The account of Inca "astronomy" by Jean du Gourcq, "L'astronomie chez les Incas," *Revue scientifique*, 52 (1893): 265–72, is a typical example of a romantic amplification on some very meagre data. More sober approach to the question is found in the essay of Erland Nordenskiöld, "Le calcul des années et des mois dans les quipus péruviens," *Journal de la Société des Américanistes de Paris*, 4 (1926): 16–23.

32. Otfrid von Hanstein, *The World of the Incas: A Socialistic State of the Past*, translated by Anna Barwell (London: George Allen and Unwin Ltd., 1924), p. 238.

33. See Rowe, "Inca Culture at the Time of the Spanish Conquest," p. 294, and Brundage, *Empire of the Inca*, pp. 262–63 and 306.

34. There is no monograph on Maya science as such. In almost all general works on Maya history, culture, and archaeology, there are chapters that deal with the scientific accomplishments of the Maya. In addition to Spinden's work quoted above (note 1 in this chapter) one should also consult his *Maya Art and Civilization* (rev. ed.; Indian Hills, Col.: Falcon's Wing Press, 1957). Among the general accounts of Maya culture and history are *The Ancient Maya*, by Sylvanus G. Morley (Stanford, Calif.: Stanford University Press, 1946) and *The Rise and Fall of Maya Civilization* by J. Eric S. Thompson (Norman: University of Oklahoma Press, 1954).

35. Ralph L. Roys, "The Engineering Knowledge of the Maya," in *Contributions to American Archeology*, Vol. II, No. 6 (Washington, D.C.: Carnegie Institution of Washington, 1934), pp. 27–105; quotation on p. 88. The study is somewhat pedantic in places, and its general conclusion is that the architectural achievements of the Maya are "roughly commensurate with their general cultural limitations" (p. 35).

36. *Ibid.*, pp. 78–80.
37. Roys suggests that the wide availability of lime cement inhibited their thinking against exploring the advantages of cutting exact pieces of stones and against perceiving forms of stone as elements of a true arch.
38. The three surviving codices are the Dresdensis, the Tro-Cortesianus, and the Peresianus. All three are made of the bark of the copo tree which the Maya pounded into pulp, and mixed with some natural gum as a binding substance. The codices form one long sheet folded like a screen into "pages" of about 9 by 5 inches. Here, a brief mention should be made of the "Grolier Codex," displayed during June 1971 at the Grolier Club in New York. Consisting of only 11 pages out of an original 20-page booklet, the chief value of the codex, provided its authenticity is firmly established, is a table of glyphs, which can be interpreted as a perpetual calendar based on the synodical revolutions of Venus.
39. For the best comprehensive modern study, see J. Eric S. Thompson, *Maya Hieroglyphic Writing: An Introduction* (new ed.; Norman: University of Oklahoma Press, 1960). As the chief objective of Maya writing was the fixing of events in various time-series, or cycles, this book is invaluable for anyone trying to see the extent to which the Maya world view was caught up in speculations about small and large cycles of time.
40. Concerning the controversy on the meaning of zero, see the balanced survey of the evidences and opinions in the foregoing work by Thompson, pp. 137–38.
41. J. Eric S. Thompson, "Maya Arithmetic," in *Contributions to American Anthropology and History*, Vol. VII, No. 36 (Washington, D.C.: Carnegie Institution of Washington, 1942), pp. 37–62. The utter primitiveness of Maya arithmetic can be seen in all extensive accounts of the subject, such as Charles P. Bowditch, *The Numeration, Calendar Systems and Astronomical Knowledge of the Mayas* (Cambridge: University Press, 1910). The actual counting was done in all probability by some form of abacus, consisting perhaps of compartments in flat boxes with grains of maize or beans representing the various units of the vigesimal system.
42. See, for instance, Morley, *The Ancient Maya*, pp. 46–47.
43. See, for instance, H. J. Spinden, "Ancient Mayan Astronomy," *Scientific American*, 138 (Jan. 1928):9–12.
44. John E. Teeple, "Maya Astronomy," in *Contributions to American Archeology*, Volume I, No. 2 (Washington D.C.: Carnegie Institution of Washington, 1931), pp. 29–116.
45. Recently this was called into question by David W. Chambers in a paper in which he also concluded that "until additional information is uncovered, it seems improper to ascribe knowledge of the Metonic cycle [19 years = 235 lunations] to the Maya." See his article, "Did the Maya Know the Metonic Cycle?", *Isis*, 56 (1965):348–51. Chambers was, however, reluctant to consider the excellent precision of the Maya in several astronomical matters as a result of chance.
46. See his *The Ancient Maya*, pp. 327–28 and 333. What is known with reasonable certainty about these structures is that the direction of their corridors lies toward the point of sunset at equinoxes and solstices.
47. "Maya Astronomy," p. 70.
48. *Ibid.*, p. 36.
49. *Ibid.*, p. 41.
50. Maud W. Makemson, "The Enigma of Maya Astronomy," *Dyn*, 1 (1944, Nos. 4–5): 52.
51. See his *Maya Hieroglyphic Writing*, pp. 102–03.
52. *The Book of Chilam Balam of Chumayel*, translated by Ralph L. Roys (new ed.; Norman: University of Oklahoma Press, 1967).
53. Thompson, *Maya Hieroglyphic Writing*, p. 1.
54. Morley, *The Ancient Maya*, p. 444.
55. *Ibid.*, p. 445.

CHAPTER FOUR

The Shadow of Pyramids

An aerial view of the great pyramidal constructions of pre-Columbian Central America inevitably evokes in one's mind the classic land of pyramids, Egypt. As architectural styles are powerful expressions of cultural orientations, important traits of ancient Egyptian life and mentality can be expected to resemble some of the principal aspects of Maya, Aztec, and Inca civilizations. Most noticeable of those traits is the theocratic and more or less benevolent despotism that set the tone of social life under most dynasties in ancient Egypt. Its hold on the people was remarkably durable, tying into a closely knit unit all aspects of life through many generations. The typical Egyptian was hardly more than a cogwheel in a huge and steady social machinery in which every part lived to his duty of performing the work-quota assigned to him. As was the case with the Maya, the co-ordination of so many labourers and artisans largely depended on diligently applied simple methods some of which constitute what may be called ancient Egyptian science.[1]

Like the early students of Maya science, Egyptologists of the nineteenth century gave exaggerated accounts of the scientific attainments of ancient Egypt. It took the work of the last two generations of Egyptologists to bring evaluation more in line with the actually known record. This shows, in close resemblance to the achievements of the Maya, several impressive but disconnected insights against a vast background of practical skill and grossly erroneous beliefs. To secure the proper distribution of grain and other basic commodities of life, ancient Egypt had to rely on a laborious system of stock-taking and bookkeeping. The practical task facing them was gigantic in view of their rudimentary methods of calculating. This disparity between tasks and means persisted throughout Egyptian history. While practical needs often spark the genius of inventiveness, in the case of Egyptians the toil of cumbersome bookkeeping seems to have exercised an oppressive influence that must have weighed heavily on Egyptian culture. As J. Vercoutter aptly put it, the burdens of bookkeeping kept Egyptian arithmetic in a bondage from which it had never freed itself.[2]

The redeeming features of Egyptian arithmetic are few and far between. One of them is the decimal system of counting with special glyphs for all powers of 10 up to 1 million. The use of zero was, however, only implicit, and the idea of place notation was entirely absent. To write 50, for instance, Egyptian scribes repeated five times the inverted U, the sign of 10. The unwieldy features of such a notation should be obvious from the twenty-seven and thirty-nine hieroglyphic signs needed to write 999 and 387,975 respectively. But whatever number was recorded by ancient Egyptians, it was never a number in itself. Numbers in Egyptian records almost invariably refer to quantities of given items, such as 5 stones, 8 loaves, and the like. The Pythagorean type of statements about numbers as such was entirely foreign to them. They even failed to reach the level of using

explicitly the simplest mathematical operators, the signs of addition and subtraction. In all appearance they felt that once the practical problem had been solved
there was no need for theoretical formulation and refinement. A good example is
their use of fractions. With the exception of $\frac{2}{3}$, all the fractions recorded in the
various papyri and stone monuments are unit fractions. Odd as it may be, ancient
Egyptians never used the fraction $\frac{1}{3}$ which they obtained by explicitly halving $\frac{2}{3}$.
Quantities like $\frac{3}{4}$, $\frac{4}{5}$, $\frac{5}{7}$ and so forth had no appeal to them possibly because answers
to their practical problems could be obtained by relying on unit fractions alone,
although the use of other types of fractions would have greatly facilitated their
work. Their stubborn clinging to unit fractions suggests their failure to recognize
that when a unit is divided into eight equal parts, then $\frac{1}{8}$ is the complement of $\frac{7}{8}$.
To perform the most elementary types of division the Egyptian scribe had, there-
fore, to fall back on the equivalent of the lengthy method of proportional division,
or preferably on practical tables dealing with concrete situations.

The preference given to concrete terms over abstract ones is a main characteristic
even of the Rhind Papyrus, the most advanced form of all extant Egyptian mathematical documents. It was its scholarly translator, T. Eric Peet, that singled out as
"the outstanding feature of Egyptian mathematics . . . its intensely practical
character."[3] Following his further investigations of the matter he did not feel the
need of modifying a conclusion that seemed rather severe to some students of
Egyptian mathematics.[4] Much speculation was stirred in particular by a problem
recorded in a somewhat damaged form in Papyrus Berlin 6619 which reads as
follows: "Divide 100 into two parts, so that the square root of one part be three
fourths of the square root of the other."[5] The answer is 64 and 36, or the squares
of 8 and 6, and it can be readily obtained from solving two equations, $x^2 + y^2 = 100$,
and $y = \frac{3}{4}x$. Expressed in algebraic formalism the problem may suggest familiarity
on the part of some ancient Egyptians with the solution of second-degree equations.

Such an interpretation, however, presents one with the question of how a
relatively advanced accomplishment can find a natural place in the rudimentary
realms of Egyptian mathematics wholly dominated by the practical aspects of
counting. Mention has already been made of the absence of the abstract use
of numbers in Egyptian arithmetic. Algebra is in turn difficult to imagine without
abstract quantities, or within an arithmetic restricted to concreteness and practicality, of which ancient Egyptian arithmetic forms a classic case. A principal feature
of it is the total absence of generalized statements, a point which has been repeatedly
emphasized by O. Neugebauer, the foremost living authority on Egyptian mathematics.[6] More importantly, one would look in vain for traces of generalization
in the steps that yielded in the foregoing problem the correct answer for the ancient
Egyptian scribe. As Peet aptly noted, the actual steps were used by the scribe simply
because for him there was no other way of obtaining the answer.[7]

There are, indeed, grave reasons against seeing the practical solution of a second-
degree equation as a piece of algebra in a computational framework characterized
by the persistent mixing of qualities with quantities. As the task of ancient Egyptian
mathematicians consisted in treating concrete problems of calculation, the various
classes of goods to be added or divided intruded everywhere into their procedures
and this effectively barred the generalized view from their mental horizon. To

leave ample room in their arithmetic for the qualitative classification of items was a necessity for them in the absence of currency in ancient Egypt. At the same time their failure to perceive the usefulness of currency indicates the very same aversion on their part for generalized notions. Ancient Egyptian arithmetic brings out forcefully the truth of Plato's remark that they were a nation of shopkeepers.[8] The arithmetic used by them was practice, not science.

The Rhind Papyrus, which in its title promises "rules for enquiring into nature and for knowing all that exists, [every] mystery . . . every secret,"[9] leaves the modern reader looking for the origins of science with a sense of keen disappointment. A similar impression is in store for him on studying Egyptian geometry. He is to find that there is no foundation for statements that are variations on Herodotus' often quoted claim that the Greeks learned their geometry from the Egyptians.[10] Contrary to the impression given by Herodotus, his compatriots could study only the art of practical geometry in Egypt. Moreover, the Greek cultivation of geometry rapidly transcended the "earthy" features of Egyptian geometry, or rather land surveying, which always stayed on the level of procedures by trial and error. It remains a matter of opinion that in calculating areas of squares, triangles, and trapeziums ancient Egyptians had relied on some generalized reasoning. Actually, the texts contain not even the slightest trace of general formulas or reasonings. A classic illustration of this is Problem 14 of the Moscow Papyrus that gives correctly the volume of the frustum of a pyramid of given square basis and height.[11] The steps of calculation do entitle the modern scholar to arrive at the correct generalized formula, $V = (h/3)(a^2 + ab + b^2)$, but this does not prove that the author of the Papyrus had such a generalized knowledge in the back of his mind. One should be on guard at the same time against slighting the value of practical methods. The Egyptian way of calculating the area of a circle by squaring eight-ninths of the diameter is worthy of admiration as it implies a value 3.1605 for π, a rather good approximation. Yet, the accomplishment cannot be considered science, but only a piece of protoscience in which statements and procedures are not generalized and supported by proofs.

This lack of generalization is palpably evident in the wholly unsystematic character of the Egyptian units of measurements. A particularly telling case is their division of the full day into two parts, day and night, each with twelve hours. While the division of the day into twenty-four parts meant accepting a steady system, for ancient Egyptians the length of day and night hours varied seasonally. Such an attitude toward measuring time hardly represented a fertile ground for an ordered investigation of celestial phenomena. The diagonal star clocks found on various coffin lids and on the cenotaphs of some Pharaohs are a case in point. They show the effort to correlate the position of stars to the cyclically changing length of night hours.[12] This not only indicates an inversion of the proper order of the terms of reference, but also gives a glimpse of the "otherworldly" character of Egyptian astronomy. It stood not so much in the service of the living as of the dead. The various documents and texts of ancient Egyptian astronomy have "a basically non-scientific character," and their investigation by modern astronomical concepts yields at best purely qualitative results.[13]

A cursory look at the calendars in use by ancient Egyptians brings this out forcefully. Their three calendars, two lunar and one civil, remained without co-ordination during much of Egyptian history.[14] They looked with suspicion at the reduction of the various calendars to a common denominator. Actually, the crown-ing ritual of the Pharaohs contained an oath barring any attempt at calendar reform.[15] The oath was, in fact, an official vote cast on behalf of a primitive outlook in which the most different considerations could live peacefully side by side.[16] The civil calendar consisted of 365 days, a length of time which, in view of the "exceedingly crude level"[17] of Egyptian astronomy, was most likely derived by averaging out the time elapsed between subsequent recurrences of the inundation of the Nile. The civil calendar was in turn divided into twelve months of thirty days to which they added five so-called epagomenic days. As they cared not to bring the civil year of 365 days into line with the astronomical year, the beginning of the former gained on the latter a full month every 120 years. Consequently, the three main Egyptian seasons, Winter, Summer, and Inundation, had no fixed places with respect to the twelve months.

As the yearly rising of the Nile closely coincided with the time when Sirius (its Egyptian name Sepedet was rendered by the Greeks as Sothis) had become visible above the horizon shortly before sunrise, this star received their special attention. Perhaps they had even recognized at very early times that the Sothic cycle has a period of about 1460 years, or the amount of time needed for the repeated coinci-dence of their civil New Year's day with the heliacal rising of Sirius. In support of this one may quote several datings in Egyptian sources of the star's heliacal rising in terms of the civil year. The allusion of Censorinus (fl. A.D. 240) to the Sothic cycle as the "Great Year" of the Egyptians[18] was obviously a reading into ancient Egyptian thought of something with which the Romans of Censorinus' times were preoccupied. The same is also true of the reference of Tacitus[19] and of Pliny[20] to a "Phoenix cycle." It was again the late Hellenistic times that witnessed the composi-tion of "The Book of Sothis or the Sothic Cycle,"[21] which its author tendentiously attributed to the noted Egyptian historian, Manetho, who lived in the third century B.C. It is even rather doubtful that the Egyptian calendar was systematically based on the Sothic cycle.[22] This would have presupposed a regular recording of the observations of Sirius over several centuries. Evidences of this are lacking.[23]

Whatever the extent and scientific character of the ancient Egyptians' acquaint-ance with the Sothic cycle, their addiction to cyclic measures of time is unques-tionable. Of great importance to them was the equivalence of 309 lunations to 25 years. This cycle provided them with a convenient framework for fixing their lunar festivals. The way this was done is illustrated in the Carlsberg Papyrus, No. 5, where one finds the 309 months fitted into 16 ordinary years of 12 months and 9 long years of 13 months. Such a procedure throws a sharp light on the over-whelmingly ritualistic conception of time as conceived by ancient Egyptians. Their astronomical texts have one overriding concern: the fixing of the hour and season of religious festivals. These in turn constituted endlessly recurring cycles. Ancient Egyptian man considered it the chief duty in life to identify himself as closely as possible with this cyclic pattern. After all, it was his basic religious conviction that the supreme good in life was to experience the rhythmic harmony of nature.[24] For him eternity consisted in assimilating himself to major cyclic motions in nature through symbolic means. Such motions could be the daily round of the sun, or the rotation of circumpolar stars. Souls that supposedly had reached the realm of those stars were called *akhu* or transfigured spirits. The wish to be absorbed in the great rhythm of the universe is the great theme that unites the countless details of the journey of the souls of the dead as fancied by ancient Egyptian religion.

The cyclic recurrence of the same circularity both in the sky and in nature was a supreme proof for the Egyptians that the cosmos was changeless in its ultimate nature. In such an outlook single events and single processes were of little if any significance. They simply could not serve as the carriers of special intellectual content. For the ancient Egyptian the understanding of existence, individual and cosmic, meant an immediate, general overview, which in the end prevented him from achieving an objective analysis of the parts of nature. A case in point is the failure of ancient Egyptians to achieve at least a rudimentary form of a systematic study of the animal world with which they were wholly obsessed.[25] Theirs was an almost endless line of mawkish gods and goddesses having the head of a bull, an ass, a hawk, a cat, a ram, a cobra, a goose, a scorpion, or a vulture. In ancient Egypt there were not only sacred trees but also sacred fishes and sacred insects. For the ibis, the sacred bird of Thoth, a huge sanctuary was kept at about 25 miles north of Hermopolis, but for purposes other than the satisfaction of natural human curiosity about their ways of life. Obsession with animals resulted in representations that boldly fused the bodies of different animals. The mythical animal, called *sefert*, had the head of an eagle and the body of a lion with huge wings. The *sedja* had a leopard's head, the fore-part of a lion, the hind-part of a horse, a tail ending in a lotus flower, and eight teats to complete this repulsive exercise in fantasy. The *ammita*, or the "eater of the dead," carried the head and the hind-part of a crocodile and the body of a lion. The *seba* was a serpent with twelve human heads growing out of its back. The Egyptians fancied lion-headed giraffes with the same ease as they did lions with human heads both at the neck and the tail.

Evidently, none of this had anything to do with observations even in a proto-scientific sense. Different animals, or men and animals, were fused with no second thought within a mentality for which animal life was a chief expression of the unchanging aspect of the universe. This was obviously the result not of painstaking observation but rather the objectivization of innermost preferences in understanding. It is most important to keep this in mind in trying to understand the ancient Egyptian's enormous admiration for and awe of animals. They appeared to him as being free from the unpredictable, individual features and actions that characterize man. In these features the ancient Egyptian could see no redeeming value. On the other hand, animals of the same species looked closely alike, and generations of all species followed one another in a most steady pattern. Animals, therefore, came to stand as the perfect embodiments of the rhythmic repetitions in the matrix of changelessness.

No wonder then that most parts of the world, great and small, were pictured by ancient Egyptians as animal gods. In fact, the whole world was for them one huge animal, often depicted as the primeval serpent.[26] Its bending into a circle symbolized for them the limits of the cosmos. The rhythmic undulations of its wave-like advance were imagined as the source from which time and space had emerged.

The high points of its convoluted back were symbols for the cardinal points. Some of the Egyptian creation-hymns could hardly be more explicit in their account of that animistic, organismic, and rhythmic world view. In a hymn from the Coffin Texts the Serpent exclaims:

I extended everywhere, in accordance with what was to come into existence,
I knew, as the One, alone, majestic, the Indwelling Soul, the most potent of the
gods.
He [the Indwelling Soul] it was who made the universe in that he copulated with
his fist and took the pleasure of emission.
I bent right around myself, I was encircled in my coils, one who made a place for
himself in the midst of his coils.
His utterance was what came forth from his own mouth.[27]

The passage, composed during the Middle Kingdom (Dynasties XI–XIII, c. 2080 B.C.–1640 B.C.), presents a specific reflection on generally accepted accounts of the origin and development of the world. The following example[28] of such accounts from the same period of Egyptian history is in the form of a monologue of the creator of the world: "I am he who came into being as Khepri (i.e., the Becoming One). When I came into being, the beings came into being, all the beings came into being after I became." At this point heaven, earth, and animals did not yet exist. Whatever existed was still inside the primeval ocean: "I, being in weariness, was bound to them in the Watery Abyss," states the creator of all. The distinct coming into existence of every being is attributed to two causes. The immediate cause is the creator of all who declares: "I planned in my own heart, and many forms of beings came into being as forms of children, as forms of their children. I conceived by my hand, I united myself with my hand, I poured out of my own mouth, I ejected Shu, I spat out Tefnut." These two deities whose role would be the actual generating of heaven and earth, are described as products, together with the creator of all, of the Watery Abyss, the ultimate entity in Egyptian cosmogony. As the creator of all goes on with his monologue: "It was my father the Watery Abyss who brought them up, and my eye followed them (?), while they became far from me. After having become one god, there were (now) three gods in me. When I came into being in this land, Shu and Tefnut jubilated in the Watery Abyss in which they were." Clearly, the gods, including the creator of all, were not assigned a commanding position over nature. They were the manifestations of nature and of her most fundamental aspect, the primeval waters. No wonder, that in the same context one finds attributed the creation of men, and of the sun and the moon, to the tears of the creator of all. What brings the whole account to a conclusion is a cosmogony completely lost in mythology: "Shu and Tefnut produced Geb and Nut; Geb and Nut produced out of a single body Osiris, Horus the Eyeless One, Seth, Isis, and Nephthys, one after the other among them. Their children are numerous in this land." That ancient Egypt was overcrowded with deities was the only truth in this cosmogony.

The world view expressed in the passages quoted above is a faithful replica of cosmogonical details contained in the Pyramid Texts most of which date from the latter half of the Old Kingdom (Dynasties III–VI, c. 2660 B.C.–2180 B.C.), the finest phase of Egyptian culture. The three main Egyptian cosmogonies, those of Heliopolis, Hermopolis, and Memphis, which antedate even the Old Kingdom, are also deeply steeped in animistic pantheism. The universal chaos exemplified by the murky floods of the Nile stood invariably for the ultimate living entity. In Heliopolis it was Atum that was believed to have emerged out of that chaos as the first distinct entity in the form of a mound. In Memphis, Ptah, the first of the gods, was also identified with the primeval dryland. The Hermopolitan cosmogony centred on the activities of a group of eight gods of which Nun, the symbol of the huge watery waste, represented the creative principle. Of the eight, four were male deities, four females. The males were depicted with frogs' heads while the females had serpents' heads.

The immediate background of this imagery was the amphibious life with which the mud was teeming as the waters of the Nile began to recede. The Nile itself was believed to be the concrete manifestation of the primeval ocean. The Nile's yearly inundations served as supreme evidence of the inevitable restoration to life through mystical immersion into the Watery Abyss. Ritual texts abound in such references to the Nile. Its two principal sources were pictured as the two main holes through which the Watery Abyss breaks through the mountainous crust of the earth. Immersion into this primeval water was the condition both for the mighty sun as well as for the puny individual to regain the strength of life. The final destination on the long and elaborate journey, which the dead had to undertake, was the primeval ocean, the unfailing source of life's rebirth. This consideration is particularly well expressed in a prayer recited during embalming. The ultimate salvation of the deceased is hoped for as the prayer evokes the rejuvenation of Osiris through exposure to the ever living waters of the Nile: "Thou drinkest from them, thou art satiated from them. Thy body fills itself with fresh water, thy coffin is filled with the flood, thy throat overflows. Thou art the Watery Abyss, the oldest, the Father of the Gods."[29]

The prayer fittingly enough starts with a reference to Osiris, the classic Egyptian symbol of the cyclic death and renewal of nature and of everything, divine and human, generated by nature. While the fancy of ancient Egyptians was not exercised about world catastrophes and rejuvenations, the cyclic throbbing of cosmic life was implied in their pantheistic conception of nature. Their gods were rooted in the strength of the primeval mound which in turn was imagined to be the result of the ever active, innate life of the primeval chaos. This life manifested itself in an upward stirring, which was the symbol of light, life, land, and consciousness. This elemental concretization of the vague, ultimate being is the deity, Atum, or the 'Complete One'. He is at the same time also the source of light, or sun-god, and the origin of the world mound rising out from the primeval waters, and the producer of all forms of life appearing on the slopes of what later became formalized as the world pyramid.

This pantheistic nature of Atum is clearly expressed in Utterance 598 of the Pyramid Texts: "Hail to you, O Atum! Hail to you, O becoming One who came into being of himself! You rose up in this your name of High Hill, You came into being in this your name of Becoming One."[30] Whatever happens subsequently in the world is the continuation of this primeval self-generation inside the eternal and infinite cosmic waters. The animistic trait of this world view is strikingly conveyed in the description of Atum as bisexual, and even more grossly in the various descriptions of how the great "He-She" produced the first creatures, Shu and Tefnut, either by masturbation, or by spitting, or by expectoration.[31] Masturbation had remained a most popular creation motif throughout ancient Egyptian history, and should bring home to the modern mind the extent to which ancient Egyptian world view was trapped in a crudely biological and animistic framework of thought.

Some further details are worth recalling in this connection. Shu and Tefnut, the first male and female, were the symbols, or gods, of the air and of matter (or stability and order) respectively. Following their generation they remained interlocked in mutual embrace with Atum. Later Shu and Tefnut took on the work of cosmic procreation and produced Geb, a male, and Nut, a female, whose body lay over the reclining body of Geb. Their separation, which followed their sexual union, resulted in the distinct earth, Geb, and the distinct sky, Nut. Also, as the body of Nut began to rise as a huge arch over Geb, she gave birth to the stars that went on to perform their daily cyclic sailing across Nut's belly, the specific symbol of the starry heavens.

The papyri buried with the priests of Amen-ra at Thebes contain interesting diagrammatic illustrations of this world picture. This is all the more significant as Theban cosmology was formulated during the New Kingdom (Dynasties XVIII-XX, c. 1579 B.C.–1075 B.C.) and shows, therefore, the lengthy hold on the ancient Egyptian mind of the basic ideas of the world view codified in the much earlier cosmogonies. One of the diagrams depicts Nu forming an arch with her body as only her toes and fingers touch the ground. Under her is lying the body of the earth-god in the form of a snake-headed man. The second diagram is more elaborate. There Nut's body spangled with stars is helped to remain in its arched position by the standing figure of Shu whose hands extend to her breasts and loins, intimating the idea of animistic vitality dominating the universe. An equally significant part of the diagram is the earth-god whose horizontally extended feet and arms touch the fingers and toes of Nut bridging thereby the flow of vitality into a full circle. The third diagram is the most complex as it contains additional gods together with the drawing of a beetle and of the head of a ram, two of the practically countless animal images on ancient Egyptian representations.

Only picturesque, grotesque, and often morbid details were added throughout the long centuries of Egyptian history to the classic original narratives of the world's "creation". The basic ideas were carefully retained as can be seen from the "Legend of the Creation of the World by the Lord of All."[32] It was found in a papyrus text written during the beginning of the Hellenistic phase of Egyptian history, in the late fourth century B.C. The text clearly reflects the turgid cosmological ideas of an earlier phase of Egyptian history, the so-called Late Period (Dynasties XXI-XXXI, 1075 B.C.–332 B.C.). In fact, it faithfully repeats some expressions that seem to have become stereotyped in cosmogonies recorded during the New and Middle Kingdoms. In the Legend the god-of-all describes himself as the one who "came into being as KHEPRI," that is, the being whose nature is becoming. He is not, however, prior to the primeval matter but only to concrete forms of it:

"I came into being out of the primeval matter, taking manifold forms from the beginning." The process of the self-unfolding of the "god-of-all" is recounted in heavily biological terms:

Very many beings came into being (or assumed forms),
Beings from the beings of the children of the beings of their children.
I, even I, cohabited with my fist.
I obtained the joys of love with my shadow,
I ejected seed from my own mouth.
I evacuated in the form of Shu
I micturated in the form of Tefnut.
My father Nu [the watery abyss] made them to rest,
My eye [the sun] followed them through long periods of times,
They went away from me
After I came into being as the One (or Only) God,
Three gods proceeded from me when I took form on this earth,
Shu and Tefnut rejoiced in the inert waters in which they were,
They brought back my eye which followed after them.

Needless to say, the legend's account of the "creation" of humans goes along the same physiological lines, and so does the explanation of the coming into being of the second main celestial body, the moon:

Now after this I gathered together my members,
I shed tears upon them.
Men [and women] arose from the tears which came forth from my eye.
After my eye [the sun] had come back, it raged like a panther at me when it
    found that I had made another eye [the moon] in its place, and that I had
    endowed it with splendour.

In such an animistic concept of the world, cyclic notions played a basic role. The pointed reference to "long periods of time" in the foregoing passage is a faithful echo of assertions made in much earlier documents such as the Coffin Texts. In one of these, Atum, the High God, describes himself as the Primeval Waters who was alone when his name came into existence and whose first distinct form was a circle.[33] It is only natural that a circle should rotate, and this is the frame of reference in which the High God's first activities are accounted for:

His cycle with Shu was the circling of Command and
    Intelligence,
    asking his (Intelligence's) advice;
    and Command and Intelligence said to him –
    "Come, then, let us go and create the names of this coil
    according to what has come forth from his heart."
    And that was the cycle with Shu, the son whom he
    himself had borne.[34]

It should not be surprising that such a world view was not much of an intellectual spark. Its light was merely a will-o'-the-wisp creating hopeless illusions about the true likeness of the outer world. These illusions could not be effectively dispelled when Amenophis IV (Akhenaton), husband of the renowned Queen Nefertiti, decided, around 1370 B.C., to enact the most sweeping changes in Egyptian cultural and religious history. His efforts to establish the sun-god (Aten) as the supreme and sole deity included the feverishly rapid construction of a new capital on the site of the modern city of Tell el-Amarna. Ideologically, the seeds of religious innovation were already sown during the reign of Amenophis III who commissioned the twin brothers, Suti and Hor, to carve into stone new hymns to the sun.[35]

In these hymns the sun was attributed supremacy over everything. The new cult allowed worship only for the sun's physical orb, and insisted on an exclusive representation of the sun which consisted of a round disc branching out in rays. The rays in turn ended in figures of human hands symbolizing the utter dependence of anything living on the sun's warmth and light. Yet, for all the encomiums heaped on it, the new supreme deity could at best perennially produce himself. The future object of the glittering worship in Akhenaton's new capital could determine its own birth but had to be born again and again. It was "the primordial being," but only in the sense that he "himself made himself."[36] In a sense all phenomena depended on its heat and light, but for all its exclusive divine glory, the sun, "the Creator of all," remained a "Great Falcon, brilliantly plumaged, brought forth to raise himself on high of himself."[37]

The unfailing intrusion of animism into the elevated tone of Akhenaton's "monotheism" should at once dissipate any idea about a real departure from the traditional features of Egyptian world view. It remained, in spite of Akhenaton's seemingly resolute break with the past, trapped in the morass of a pantheistic, animistic matrix. Such a "monotheism" had no inner strength to provide a radically new look at nature. The hold of the traditional world view proved to be insuperable. Within its framework only a very limited advance could be made toward a rational interpretation of the physical world. The meagre possibilities along these lines seem to have all been exhausted during the relatively short cultural renaissance of the Old Kingdom. The intellectual curiosity of ancient Egyptians came to a baffling standstill before the second millenium was well under way. The Rhind Papyrus can hardly be assigned a later date than the twentieth century B.C. Egyptian medicine, the most admired branch of Egyptian know-how in antiquity, reached its high-water mark even earlier.

Egyptian medical art is a classic case of the bondage in which a fledgling branch of knowledge could be held by a heavily animistic world view. The physicians of ancient Egypt have to their credit the first rudiments of a medical terminology, the first use of bandages, the pioneering of anatomical investigations, and the first listing of some truly useful, naturally available drugs. These feats exercised a crucial influence, both on Greco-Roman and Arabic medicine. But whatever gems appeared within Egyptian medical practice remained in each case intermingled with an enormous mass of senseless, magical, and animistic notions. The marvellous recognition of the parallelism between the various lesions of the brain and the functional losses in different parts of the body did not give second thoughts to ancient Egyptian physicians. They persisted in their attachment to the belief that all sickness, even external injuries, had as their cause the presence of evil spirits and forces.

Hence the ubiquitous presence of incantations and formulas of magic in their medical papyri, the oldest of which, dating from the early part of the second millenium, are extracts from much earlier texts. It was only in the case of wounds inflicted externally that in addition to magical rites rational procedures too were applied. Other external symptoms of internal diseases, such as boils and swellings, were treated more by charms than by strictly medical means. Magic was the exclusive remedy for such internal pains as headache and fever. Very typically, the most enlightening sections of the Edwin Smith Papyrus, which was claimed to be the first scientific medical treatise,[38] are restricted to the treatment of fractures and wounds. The animistic proclivities of Egyptian medicine were also evident in the pharmaceutical use of entire small animals which were chopped up or pulverized, and administered in meticulously measured doses. Again, the few passages in which there seems to be a sequence of examination, diagnosis, and treatment are hidden in a jumble of magic formulas. In fact, not even a bandage was supposed to be removed without the recitation of some ritual phrase. Obviously, no cure could be assigned for scorpion bite as seven scorpions served as bodyguards of Isis. On the other hand the most repulsive and indigestible materials were fed all too often to the sick. The rationale behind this lay with the belief that the cause of sickness was an animistic force or spirit, which would rather leave than remain exposed to bad smells or nauseating food.

No wonder that within such an approach Egyptian physicians failed to recognize the true bearing of some precious items of information, such as pulsebeat. Not that valuable leads to a major breakthrough in this particular direction had been lacking. In the longest and most famous Egyptian medical text, the Ebers Papyrus, it is noted that the action of the heart can be felt at the same time on the head and the limbs by fingers placed on arteries. There is something deeply tragic in the promisingly eloquent passage introducing in the same Papyrus the section on heart and vessels: "The beginning of the science of the physician; to know the heart's movement and to know the heart."[39] Such an auspicious beginning, the emphasis on the all-important role of the heart in the body, was to end in a miscarriage. The section immediately following refers not only to the physician but also to the exorcist, and in a typical animistic vein describes the pulsebeat as the heart's speech to the vessels.

The promising initiatives and subsequent stagnation that characterized Egyptian medicine form also the pattern of the history of technical skill in the Nile valley. R. Engelbach, the author of a very perceptive and competent study of the mechanical techniques in ancient Egypt, felt impelled to note at the very outset that "there was little radical progress from the Third Dynasty until nearly Ptolemaic times,"[40] that is, from about 2650 B.C. until 330 B.C. Such a situation is all the more puzzling as in some respects ancient Egyptian craftsmen displayed great ingenuity in putting to best use the relatively few types of tools at their disposal. Their simple but effective method of producing sheets of paper from the leaves of the papyrus plant represented a major advance over the far more cumbersome method of preparing animal skin for writing purposes. Egyptians were the first to produce plywood with as many as six layers of different woods. Their carpenters developed effective methods of joining pieces of wood even in such intricate constructions as the hulls of boats. The decorative crafts, such as inlaying, veneering, and overlaying, show themselves on an astonishingly high level in the countless artifacts found in the tombs of the Pharaohs of the XVIIIth Dynasty (sixteenth century B.C.). In this connection the riches of the burial chambers of Tut-ankh-amen have possibly received the widest publicity. The outlay of some of those chambers reveals a highly recondite form of architectural planning. Its secretive character withstood even the space-age efforts to locate the burial chamber in the Great Pyramid by very sensitive cosmic-ray methods.

But the real technological mystery of the pyramids lies in their enormous proportions. In this respect too, nothing essentially new is offered by pyramids built after the Old Kingdom. This highly significant detail is another evidence of the general pattern of ancient Egyptian cultural history. In general, the engineering technique employed in the construction of pyramids[41] represents a strange mixture of ingenuity and backwardness. The lack of familiarity of ancient Egyptian engineers with the pulley provides a striking illustration of this. Even with pulleys, the intricate placing of hundreds of thousands of huge blocks of stone, at considerable heights, is an enormous feat. Without pulleys, such a task should appear well-nigh insuperable. Again, the builders of pyramids could not rely on wheeled vehicles for the transportation of heavy stone blocks. This had to be done on wooden sleds. There was, in addition, the problem of quarrying, shaping, and polishing the same blocks, with no recourse to hard metal tools. Yet, for all the primitiveness of his instruments and methods, the Egyptian stone cutter and mason worked marvels in fitting huge blocks of stone together. The average separation of the casing blocks at the foot of the Great Pyramid is only $\frac{1}{50}$ inch. Unfortunately, no longer can be seen those marble plates of extraordinary smoothness that turned the sides of the pyramids into almost perfect mirrors, a spectacle of which ancient authors spoke with awe.

Much, though rather misplaced, admiration is still accorded by some obscurantist authors to the all-over masterplan that determined the construction of pyramids. One of the early proponents of the special significance of the ancient Egyptian cubit was none other than Sir Isaac Newton, whose great mind had become trapped in the latter half of his adult life in cabbalistic speculations.[42] Two and a half centuries later it was another astronomer, though of less stature, who tried to give scientific respectability to speculations about the pyramids as embodiments of exact geophysical and astronomical data.[43] Still, if the pyramids represented anything scientific, it was not science in the modern sense. Nor were the pyramids ultimately a reflection of the social and political system of ancient Egyptian life.[44] They stood for far more than a gigantic slope accommodating the rays of the Sun-god on their descent upon the earth.[45] The pyramids were more than superbly constructed facilities to secure the king's safe ascent to the divine realm. Their true meaning is only hinted at when Egyptologists routinely note the ubiquitous role of pyramidal forms in the cult of the Sun-god.[46] In their deepest meaning the pyramids were symbols of a conception about the world that nipped in the bud all scientific endeavour. Such was the shadow which they cast on Egyptian religiosity and world view. They are monumental reminders of the primeval mound which in turn is intimately tied to the animistic world view that has no room for a concept of time which implies the unique meaning of each moment.

Caught up in an animistic, cyclic outlook, the Egyptians had grown insensitive both to scientific reflections and to historical ones as well. With each dynasty they readily started a new chronology as if documenting the incompatibility of the cyclic and one-directional flow of time. Though they recorded data with extraordinary diligence, they had as little appreciation of history as did the Hindus who hardly recorded anything.[47] Their common failure to reach the level of both scientific and historical thinking is not a coincidence. Science and historiography are but different types of a causal and rationally confident probing into the space-time matrix in which external events, physical or human, run their irrevocable courses. To achieve science one has to recognize that these courses are not returning on themselves in a blind circularity in close analogy to the workings of perpetual motion machines.

Among those who made the decisive steps toward the recognition of this, there were no ancient Egyptians. Much of their intellectual history had been a long stagnation in the morasses of an animistic and cyclic world view, which in turn rested on their conception of the Watery Abyss as the ultimate entity. From its dark, pantheistic depths and from its utterly unpredictable stirrings there could not emerge an unambiguous and effective pointer suggesting the presence of clear, rational laws in the universe. The actual world was not seen to display what the source, the murky chaos, was itself lacking. Consequently, the long peaceful centuries of Egyptian history did not act as a cultural stimulant. To be sure, advance towards science was blocked by other causes as well among which historians of science usually mention the geographic isolation of Egyptians of old. Few of these causes seem, however, to be truly independent of the organismic and cyclic world view that dominated their thinking. Their proverbial conservatism,[48] their social, political, economic, and industrial patterns were to a noticeable degree the derivatives of their animistic and cyclic interpretation of human and cosmic existence. Such an outlook was hardly to shift to more rational tracts when the Persian conquerors exposed Egypt, in the sixth century B.C., to a direct contact with Babylonian culture. The new exposure was not to be a fertile one. The world view and culture symbolized by the pyramids could not get much corrective from a culture which had as its chief expression a pyramid-like structure, the ziggurat.

NOTES

1. On Egyptian science the best surveys are the chapters on mathematics and astronomy by J. Vercoutter and on medicine by G. Lefebvre in R. Taton (ed.), *History of Science*, Vol. I, *Ancient and Medieval Science*, translated by A. J. Pomerans (New York: Basic Books, 1957), pp. 17–64. *The Legacy of Egypt*, edited by S. R. K. Glanville (Oxford: Clarendon Press, 1942) and Pierre Montet's *Eternal Egypt*, translated from the French by D. Weightman (New York: The New American Library, 1968) offer a most informative view of ancient Egyptian history and culture with chapters on science and technology.

2. In his essay in Taton, *History of Science*, vol. I, p. 19.

3. *The Rhind Mathematical Papyrus*, Introduction, transcription, translation, and commentary by T. Eric Peet (Liverpool: University of Liverpool Press, 1923), p. 10.

4. See T. Eric Peet, "Mathematics in Ancient Egypt," *Bulletin of the John Rylands Library*, 15 (1931): 409–41; especially pp. 437–39. The scientific character of Egyptian mathematics is defended in O. Gillain, *La science égyptienne: l'arithmétique au Moyen Empire* (Bruxelles: Édition de la Fondation Égyptologique Reine Élisabeth, 1927), pp. 308–11; in K. Vogel, *Die Grundlagen der ägyptischen Arithmetik in ihrem Zusammenhang mit der 2:n-Tabelle des Papyrus Rhind* (Munich: F. Straub, 1929), pp. 183–84; in H. Wieleitner, "War die Wissenschaft der alten Aegypter wirklich nur praktisch?" *Isis*, 9 (1927):11–28.

5. For its detailed discussion, see O. Neugebauer, "Arithmetik und Rechentechnik der Ägypter," *Quellen und Studien zur Geschichte der Mathematik*, Abt. B: *Studien*, 1 (1929–31): 306–11.

6. See his *Die Grundlagen der ägyptischen Bruchrechnung* (Berlin: Julius Springer, 1926), p. 17.

7. "Mathematics in Ancient Egypt," p. 423.

8. See *Laws* 747c. Of course, by Plato's time money was widely used in Egypt, and he refers both to the Egyptians' love of money (*Republic* 436c) and to their expertise in geometry (*Laws* 819c–d).

9. See the translation by Peet, p. 33. In another, avowedly free translation the same part of the title reads: "The entrance into the knowledge of all existing things and all obscure secrets." *The Rhind Mathematical Papyrus*, Vol. I, Free Translation and Commentary by Arnold Buffum Chace (Oberlin, Ohio: Mathematical Association of America, 1927), p. 49.

10. *Herodotus* with an English translation by A. D. Godley (London: Heinemann, 1926), vol. I, p. 399; in the context (Book II, sec. 109) Herodotus speaks of the complaints of those Egyptians whose taxes remained unchanged though the size of their plots diminished through the changes of the Nile's river-bed following each inundation. Because of their complaints, "the king would send men to look into it and measure the space by which the land was diminished, so that thereafter it should pay the appointed tax in proportion to the loss. From this, to my thinking, the Greeks learnt the art of measuring land [geometry]; the sunclock and the sundial, and the twelve divisions of the day came to Hellas not from Egypt but from Babylonia."

11 "Mathematischer Papyrus des staatlichen Museums der schönen Künste in Moskau," edited with commentary by W. W. Struve, in *Quellen und Studien zur Geschichte der Mathematik*, Abt. A: *Quellen*, 1 (1930):134–45.

12. See O. Neugebauer and R. A. Parker, *Egyptian Astronomical Texts*, Vol. I. *The Decans* (Providence, R.I.: Brown University Press, 1960), p. 97, and Vol. II, *The Ramesside Star Clocks* (1964), pp. 9–18.

13. *Ibid.*, vol. I, p. 95.

14. See R. A. Parker, *The Calendars of Ancient Egypt* (The Oriental Institute of the University of Chicago. Studies in Ancient Oriental Civilization, No. 26; Chicago: University of Chicago Press, 1950), p. 56.

15. *Ibid.*, p. 54.

16. See O. Neugebauer, "The Origin of the Egyptian Calendar," *Journal of Near Eastern Studies*, 1 (1942):402–03.

17. The expression is of O. Neugebauer, *The Exact Sciences in Antiquity* (2d ed.; Providence, R.I.: Brown University Press, 1957), p. 80.

18. In his *De die natali*, chap. 18; see edition by F. Hultsch (Leipzig: B. G. Teubner, 1867), p. 38.

19. In his *Annals* Tacitus scorns (Book VI, chap. 28) efforts to partition Egyptian history into the alleged 1461-year periods of the sacred bird of the Sun.

20. In his *Natural History* (Book X, chap. 2) Pliny reports Manilius' opinion that the life-span of the phoenix coincides with the Great Year.

21. Appendix IV in *Manetho*, with an English translation by W. G. Waddell (Loeb Classical Library; Cambridge, Mass.: Harvard University Press, 1964), pp. 234–49.

22. The chief difficulty of a possible connection of the "era of Menophres" with the Sothic cycle lies in the fact that no Pharaoh with this name is known to have existed during the

G

fourteenth century B.C., and especially among the first representatives of the XIXth Dynasty. One of the titles of Paramses bears the only resemblance with "Menophres," and it is on that slender reed of evidence that the French Egyptologist, P. Montet, based his view that the astronomer Theon (fl. A.D. 130) did in fact report a long-standing tradition of Egyptian historiography. (*Le drame d'Avaris: Essai sur la pénétration des Sémites en Égypte* [Paris: Librairie Orientaliste Paul Geuthner, 1941], pp. 111–12. See also his *Eternal Egypt*, pp. 260–61).

23. The principal twentieth-century supporter of the Sothic cycle as the basis of Egyptian chronology was Eduard Mayer in his *Aegyptische Chronologie* (Berlin: G. Reimer, 1904). The contrary position is strongly put forward in O. Neugebauer, "Die Bedeutungslosigkeit der Sothisperiode für die ältere aegyptische Chronologie," *Acta orientalia*, 17 (1938): 169–95.

24. As emphasized by H. Frankfort in his *Ancient Egyptian Religion: An Interpretation* (New York: Columbia University Press, 1948), pp. 29 and 106–09.

25. For descriptive details and illustrations of Egyptian animal worship, see Sir E. A. Wallis Budge, *From Fetish to God in Ancient Egypt* (London: Oxford University Press, 1934), chap. ii. For some penetrating remarks on the philosophical background of such a cult, see Frankfort, *Ancient Egyptian Religion*, pp. 8 and 13.

26. For texts and drawings on this, see R. T. Rundle Clark, *Myth and Symbol in Ancient Egypt* (London: Thames & Hudson, 1959), pp. 51, 81, and 244–45.

27. *Ibid.*, p. 51.

28. The following quotations are from the English translation of the cosmogonical section of the "Book of Overthrowing Apopis," in Alexandre Piankoff, *The Shrines of Tut-Ankh-Amon* (New York: Pantheon Books, 1955), p. 24.

29. Quoted *ibid.*, p. 26.

30. Clark, *Myth and Symbol in Ancient Egypt*, p. 38.

31. *Ibid.*, p. 42.

32. The text in both of its versions is given in translation in Budge, *From Fetish to God in Ancient Egypt*, pp. 432–36. Subsequent quotations contain slight modifications of the text as given by Budge.

33. *Myth and Symbol in Ancient Egypt*, p. 74.

34. *Ibid.*, p. 75.

35. For lengthy passages from these hymns carved on Stela No. 826 of the British Museum, and for a discussion of their cultural and historical context, see James H. Breasted, *Development of Religion and Thought in Ancient Egypt* (New York: Charles Scribner's Sons, 1912), chaps. ix and x.

36. *Ibid.*, p. 317.

37. *Ibid.*, p. 316.

38. In his Foreword and General Introduction to *The Edwin Smith Surgical Papyrus* (Chicago: University of Chicago Press, 1930), James H. Breasted, the noted American Orientalist, kept insisting that the author of the papyrus was "the first scientific observer known to us," and that the papyrus itself was "the first scientific document." (See vol. I, pp. xiii, 12, 13, and 15).

39. As quoted by Warren R. Dawson in his contribution, "Medicine," to *The Legacy of Egypt*, p. 189.

40. "Mechanical and Technical Processes. Materials," in *The Legacy of Egypt*, p. 120.

41. The mechanical and technical skill needed for building pyramids with primitive tools is well portrayed in Engelbach's foregoing essay in *The Legacy of Egypt*; see especially pp. 148–54.

42. See "A dissertation upon the *sacred* cubit of the *Jews* (Hebrews rather, or Israelites) and the *Cubits* of the several Nations; in which, from the dimensions of the greatest Egyptian pyramid, as taken by Mr. *John Greaves*, the antient cubit of *Memphis* is determined. *Translated from the* Latin of *Sir* Isaac Newton, *not yet published*," in John Greaves, *Miscellaneous Works* (London, T. Birch: 1737), vol. II, pp. 405–33. Greaves, professor at Oxford, died in 1652.

43. The scientist in question was Charles Piazzi Smyth, Astronomer Royal of Scotland. Of his numerous publications on this topic mention should be made of *Our Inheritance in the Great Pyramid* (London: A. Strahan, 1864) and *Life and Work at the Great Pyramid* (Edinburgh: Edmonston & Douglas, 1867).

44. As contended, for instance, in E. Baldwin Smith, *Egyptian Architecture as Cultural Expression* (New York: D. Appleton Century Co., 1938), pp. 240–56.

45. Contrary to the claim of Breasted in his *Development of Religion and Thought in Ancient Egypt*, p. 72.

46. *Ibid.*, pp. 70–72; see also I. E. S. Edwards, *The Pyramids of Egypt* (London: Max Parrish, 1961), p. 236.

47. "Indian man forgot everything, but Egyptian man forgot nothing," in O. Spengler's striking characterization. *The Decline of the West*, translated by Charles Francis Atkinson (New York: Alfred A. Knopf, 1957), vol. I, p. 12.

48. O. Neugebauer spoke of ancient Egyptians as ones "who are considered to be the most conservative race known in human history" ("The Origin of the Egyptian Calendar," p. 397). Years later he took the view that "there was as little innate conservatism in Egypt as in any other human society" (*The Exact Sciences in Antiquity*, p. 71).

CHAPTER FIVE

The Omen of Ziggurats

Had the ancient inhabitants of the Tigris and Euphrates valley[1] had easy access to granite or other durable building materials, the Mesopotamian skyline would now show a striking similarity to the horizons of Egypt. The traveller would see the flat landscape interrupted by the contours of pyramid-like structures, the ziggurats, or etymologically speaking, structures that are "high and pointed." This feature of the ziggurats exists today in name only. Built for the most part of sun-dried bricks, they could not equal either in height or endurance the Egyptian pyramids. The ziggurats, with the exception of that of Ur, are a far cry even from their erstwhile dimensions. Often reduced to heaps of brick, not a few of them are still buried under dry mud or drifting sand, and can be distinguished only by the archaeologist's trained eyes from the countless natural mounds produced by wind and inundations.

Like the pyramids of Egypt, the ziggurats of Mesopotamia had been a prime architectural objective over many centuries. The prototypes of ziggurats were constructed early in the Sumerian period that spanned from the twenty-eighth to the twenty-fifth centuries B.C. Intensive repair work was still done on ziggurats early during the Persian occupation (538 B.C.–323 B.C.) that put an end to Mesopotamian independence. The ziggurats were a symbol of cultural unity in ancient Mesopotamia not only in time but also in space. Ziggurats were built and kept in repair throughout the Tigris and Euphrates valley regardless of which group of city-states dominated much of the land. This cultural unity is all the more remarkable as it came about through an amalgamation of Sumerian and Semitic elements. Historically speaking, the centre of political power showed a gradual shift in the upriver direction, from the Sumerian cities of Ur and Uruk to Babylon, and from there to Assur. Although power often changed hands between these last two, the culture, the world view, the religious outlook, the political organization, the industrial crafts, the intellectual and scientific heritage retained much the same character.

Cultural life in ancient Mesopotamia had seen only one truly creative phase which coincides with the reign of the Hammurabi dynasty in Babylon during the nineteenth–seventeenth centuries B.C. The same period also represented the high-water mark of the political ascendancy of Babylon, and for this reason ancient Mesopotamian culture is often spoken of as Babylonian, especially when its scientific component is considered. It should, however, be kept in mind that Babylon took over almost in entirety the old Sumerian culture following the capture of the Sumerian city-states by the Semitic prince, Sargon (2371 B.C.–2316 B.C.). After the Hammurabi period cultural activity consisted largely in compilation and recovery not only in Babylon but even more so in Assur, and finally during the Hellenistic or closing phase of Mesopotamian history. Of extensive recovery work ancient Mesopotamia often stood in great need. While Herodotus' often quoted saying, "Egypt is a gift of the river [Nile],"[2] well expresses the favourable and dependable climatic conditions of the Nile valley, weather conditions have always been very erratic in the Tigris and Euphrates valley with frequently disastrous consequences. Unexpected torrential rains, followed by savage floods, could destroy in a matter of hours all buildings and irrigation works over large areas and reduce life to a mere subsistence level.

The inclement climate was only one of the several adverse factors that prevented in ancient Mesopotamia a cultural development beyond a certain level and finally caused its ultimate decay. The ruinous condition of ziggurats is a fitting symbol of this sad cultural predicament. Most of those ziggurats that are entirely or in part excavated show only the remnants of the first or ground tier from their usually three-tiered construction. The same holds true of the temple complexes at the centre of which stood the ziggurat. Their original splendour can only be conjured up by the archaeologist's imagination. Again, it takes one's mental eyes to evoke the glitter of blue glazed tiles that covered the small temples forming the very top of ziggurats. In addition to these small sanctuaries the ziggurats also contained a larger temple on their ground floor. The simultaneous presence and relative position of two temples suggest that they formed a symbolic staircase for the periodic descent of deity from his heavenly abode.[3]

While such an interpretation seems to be well founded, it leaves untouched the deeper meaning of ziggurats. Although it is true that they were not burial places, it does not, therefore, follow that "the ziggurats were in no sense parallel in function to the pyramids of Egypt and any external similarities between examples of the two were accidental."[4] The deepest significance of the pyramids, as was shown in the preceding section, far transcended the scope and meaning of royal burials. The pyramids were symbolic replicas of the primeval cosmic mound, and were meant to convey a world view whose chief characteristic derived from organismic and cyclic notions. As to the ziggurats, there are strong indications that they also represented the primeval mound with connotations similar to those noted in connection with the pyramids. The upper and lower temples of the ziggurats can be interpreted as the places where the periodic death and rise of the local deity took place. Furthermore, the upper temple possibly served also as the place where the consummation of "sacred marriage" occurred.[5]

Later more will be said about the organismic and cyclic conceptual framework embodied in the ziggurats which also form a monumental synthesis of ancient Mesopotamian technology. The planning, building, and decorating of the ziggurats implied craftsmanship and practical know-how extending over wide areas that certainly contained protoscientific elements. The planning and layout of large architectural constructions presupposed systematic measurements which in turn implied at least some practical geometry and its application on a grand scale, especially if one considers the temple complex and city surrounding the ziggurat. Herodotus, a usually reliable source, speaks in superlative terms of the grandiose planning of the ancient city of Babylon. In his words, "it was planned like no other city whereof we know."[6] Herodotus singles out in particular the hundred gates interrupting the massive walls and on the top of the walls the thoroughfare formed by two rows of single-room houses and wide enough for four-horse chariots.[7]

The actual construction of the ziggurats presupposed the art of bricklaying, the handling of bitumen, and of wattled reeds. As hard stone and good timber were scarce in Mesopotamia, bricks were held together by bitumen that was also used for waterproofing baths, drains, and boats. Bitumen served as drug and fuel, and was utilized also in sculpture and inlay work. The decoration of the ziggurats implied expertise in a large number of crafts. Among them were ceramics, as shown by the glazed bricks, and metallurgy, as illustrated by various ceremonial objects unearthed during the hundred years of relentless archaeological explorations. From the archaeological finds it is possible to reconstruct a well-developed chemical technique.[8] Its tools were pestles, mortars, crucibles, drip bottles, mills, strainers, and apparatus for filtering, distillation, and extraction. Among the chemicals produced were various acids, in particular sulphuric acid and perhaps *aqua regia*, various salts such as washing soda, common salt, borax, saltpetre, and copper sulphate. In addition to the common metals and metallic acids, ancient Mesopotamians were familiar with the uses of mercury, sulphur, arsenic, and some of their compounds. Among the chemical processes the tanning of leather, the dyeing of wool in shades of yellow, blue, black, red, and purple, and the preparation of soap (though probably not of true soap which is produced when an oil is boiled with a caustic alkali) are worth mentioning.

Planning, building, and decorating on a large scale presuppose a fairly advanced form of recording and storing information. The cuneiform writing of ancient Mesopotamia should convey forcefully the intellectual acumen of their users. From the viewpoint of technical execution the clay tablets on which the cuneiform writing was recorded are often masterpieces of manual skill, and this holds even more so of the cylindrical seals by which standard series of symbols were imprinted into the wet clay. Unlike the hieroglyphic writing of ancient Egyptians, cuneiform writing far transcended its original pictographic stage. The post-Sumerian or Babylonian and Assyrian writing used not only a large number of arbitrary symbols for words and syllables but a purely alphabetic system as well. An unequivocal spelling of words was not, however, always possible because of the non-Semitic character of the Sumerian cultural heritage that had become an integral part of Babylonian and Assyrian learning.

Ambiguities touch even on the reading of tablets with purely or largely mathematical content. In addition, the number of tablets already deciphered represents only a fraction of the total of some half million tablets lying still undeciphered in museums. This must be kept in mind when one speaks, for instance, of Babylonian mathematics. The available material corresponds to small sections of a broad landscape much of which cannot be seen. This is well illustrated by the fact that while there exist tablets with computations of the volumes of frustra of cones and of pyramids, computations of the volumes of cones and pyramids are not referred to in any of the known texts. Consequently, efforts to establish the chief characteristics of Babylonian mathematics[9] should remain tentative for the time being.

One fact, however, entitles the student of Babylonian mathematics to offer with some confidence a few general conclusions. All the available texts come, surprisingly enough, from two relatively short periods of Babylonian history. One of them is known as the age of the Hammurabi dynasty, the other is the Seleucid era, or the last three centuries B.C. For all that separation, the more recent group of texts contains little new as compared with the old. As very little is known about Sumerian mathematics, one has to assume that Babylonian mathematics reached its highest stage of development within a relatively short period of time. It should seem even more puzzling that a period of short flourishing had been followed by about thirteen centuries of neglect, and that a second period of flourishing had achieved only the recovery of the attainments of the first.[10] As an explanation of this, Otto Neugebauer recalled that all great, creative periods of mathematics alternated with long periods of stagnation.[11] There is a great deal of truth in this, but Neugebauer's remark cannot apply in the same sense to the development of mathematics before and after 1400, or about the time of the full recovery of ancient Greek mathematical texts in late medieval Europe. Since 1400 the periods of stagnation in mathematics had been marked by a steady though not spectacular progress, while prior to 1400 the periods of stagnation meant standstill, and they were more often than not stages of sharp decline and at times stages of almost complete demise. Such was at least the case, with the exception of Greece, in the various ancient civilizations. In other words, the fate of mathematics in ancient Mesopotamia is more than a problem of progress and stagnation. It is rather one of the several historical instances illustrating the repeated inability of scientific thinking to reach a stage in which its progress becomes self-sustaining.

To see this problem in its acuteness one need only recall a few details about the principal attainments of Babylonian mathematics. It is no exaggeration to say that in more than one way the calculators of Hammurabi's age were three thousand years ahead of general mathematical development. Their skill is amply documented by the positional annotation of numbers, by the tables of squares and cubes, of square roots and cube roots, of sums of squares and cubes. These latter were used for the solution of certain types of cubic equations and of exponential functions encountered in computing compound interest. The attention paid by old Babylonian calculators to squares is best evidenced in their tables of Pythagorean numbers and their ratios. As a matter of fact the Babylonians should be credited with much of the early speculations about numbers which ancient tradition and modern histories of science assign to Pythagoras and his school. While the Babylonians apparently failed to recognize the notion of prime numbers, their skill in solving quadratic equations was very impressive.[12] It is evident from their handling of concrete problems that they were fully aware of the proposition that the product of the sum and difference of two quantities is equal to the difference of their squares.

This is not to say that they used algebraic notation or generalized formulas as such. They never stated propositions in a general form. As B. L. van der Waerden aptly noted, "the question which the Babylonians asked, was always to calculate something, never to construct or to prove something."[13] For all that, their advance toward abstract, generalized mathematical thinking was considerable. This should be clear from even a briefest summary of their chief attainments in algebra: solution of equations with one unknown; solution of systems of equations with two unknowns; solution of arithmetic progressions such as $1 + 2 + 4 + 8 + \ldots 2^n$, and $1^2 + 2^2 + 3^2 + \ldots n^2$; finding Pythagorean numbers from the practical equivalents of the formulas $z/y = \frac{1}{2}(\alpha + \alpha^{-1})$ and $x/y = \sqrt{((z/y)^2 - 1)}$. While in the case of the Egyptians, the obsession of their scribes with concrete items evidences a lack of ability or appreciation for abstract and generalized thinking, the Babylonians' emphasis on concrete calculations reflects an implicit striving for the world of pure numbers. In a deep sense, Babylonian mathematicians did not care much for the real world in spite of all the concreteness of their work. With a surprising abandon they multiplied areas with no apparent concern as to the relation of such a procedure to reality. Again, cuneiform tablets contain numerous examples relating to wages, workmen, and working hours or days. Time and again the solution yields a non-integer for the number of workmen to be determined.

The same disregard for the real world is also prevalent in Babylonian geometry with rather disastrous results. While arithmetic and algebra can flourish without attention being paid to the actual physical world, geometry cannot conceivably make a good headstart by adopting a similar stance. Thus, one would look in vain in the Babylonian geometrical texts for something similar to the lucidity and generality of the basic propositions of Euclid's geometry. Babylonian geometry remained rudimentary in its dealing with the practical determination of areas and volumes of elementary shape. The tablets only rarely mention such geometrical terms as width, length, and height, and are usually vague about their respective orientation such as perpendicularity. Babylonian geometry is not so much geometry as it is rather a branch of numerical solutions of concrete problems connected with the construction of various earthworks such as canals, dams, and ditches. While Babylonians of old used the concept of similarity of forms, they never discovered generalized geometric relations embodied in the three-dimensional surroundings of man.

A telling confirmation of this comes from the analysis of Babylonian astronomy. Factual information about it is rather recent. The first steps in deciphering cuneiform tablets of astronomical content were taken only during the last decades of the nineteenth century, but it was not until Neugebauer joined the field in the 1930's that the pioneering work of Franz Xavier Kugler made a well deserved impact.[14] Kugler's achievement mainly consisted in shedding light on the true meaning of long lists of numbers accompanied with little if any explanatory texts. His masterful interpreting of some fifty such tablets forms now a part of the combined edition of about three hundred tablets deciphered and interpreted by Neugebauer.[15] Most of the tablets are from the Seleucid period, a circumstance of no small importance for our appraising of the basic mentality of Babylonian astronomy. In its most developed and valuable form it corresponded to establishing numerical relations of the succession of various celestial phenomena.

Unlike the great Greek astronomers, who aimed at theories broad enough to predict at any given time the geocentric longitude and in some cases the latitude of a planet, Babylonian astronomers showed little theoretical interest. They were preoccupied by the succession of specific phenomena, such as the heliacal risings and settings, zodiacal stations, oppositions, and the like.[16] Obviously, once these limited systematizations are established, intermediate positions can be calculated with more or less precision. It seems, however, that Babylonian astronomers expended no major efforts in this direction. The overwhelming majority of the tablets deal with the sequence of specific phenomena listed above. The reason for this should be clear to anyone mindful of the religious or rather magical background of Babylonian astronomy. Only "special phenomena" were of interest for the purposes of divination. Or as G. Sarton, hardly a scholar to play down the scientific achievements of ancient cultures, felt impelled to note, the planetary and lunar ephemerides of the Babylonians "were partly empirical and largely *a priori*; they suggest a complicated form of divination rather than a new branch of science."[17]

The non-scientific character of Babylonian astronomy is strikingly evident from the fact that it has never developed even tentatively a geometrical, or a mechanical model of the system of the planets. This is all the more strange as the scientifically best phase of Babylonian astronomy, the phase least encumbered by gross astronomical speculations, belongs to the same centuries during which ancient Greek astronomy made most extensive use of geometrical and mechanical models to explain the planetary system and the closed, spherical universe. During those centuries that witnessed the work of Calippus, Eudoxus, Apollonius, Hipparchus, and Aristarchus, contacts were numerous between Greece and Mesopotamia. Nevertheless, Babylonian astronomy shunned outside influences, especially those from Greece, with a tenacity which resulted in its complete isolation. While some recordings of eclipses by Babylonian priest-astrologers proved very valuable for Ptolemy, and while even Hipparchus was helped by Babylonian data in the discovery of the precession of equinoxes,[18] Babylonian astronomy, consisting largely in a mathematical formalism about the rise of planets and of the moon, made no impact on the Greeks. One would look in vain for a mention of it in the Greek scientific corpus or in Greek literature in general. As a matter of fact it remained unknown in the Western world until the late nineteenth century. References to it in Sanskrit and Tamil writings[19] illustrate, however, the actual passing of items of learning from Babylon to India. In the history of astronomy, Babylonian accomplishment during the Seleucid period is a largely isolated incident without which the history of astronomy would have practically remained the same.

In the isolation of Babylonian astronomy one is faced with more than a typical cultural phenomenon. True, cultures show time and again stubborn resistance to foreign elements. In referring to this feature of late Babylonian astronomy, Sarton could point to the present-day resistance to science among traditionalists in India, or more specifically to the opposition by Indian astrologers, whose number is anything but small, to scientific astronomy.[20] Yet, history provides more than one example of fruitful fusion of different cultures. Such a fusion was not possible between Greek and Babylonian astronomy. With the Greeks, astronomy as an intellectual pursuit achieved a fair measure of independence from astrology owing mainly to the emergence among them of a rational outlook on the world. With the Babylonians such an outlook was almost entirely missing. For them the observation of celestial phenomena remained in the service of a religious world view steeped in magic. This is shown, for instance, by the uneven attention given by Babylonian astronomers to the various planets. It was not Venus, the brightest planet, but Jupiter that commanded much of their interest, most likely because Jupiter was the star of Marduk, the chief deity in the Babylonian pantheon. They devised at least five different computational systems to come to grips with the apparent irregularities of Jupiter's motion across the sky.

It is also significant that for all its dependence on arithmetic and algebraic skill Babylonian astronomy was not creative from the viewpoint of mathematics. Babylonian algebra, from which Greek mathematics greatly profited, had reached its zenith a millenium and a half before the Seleucid era was ushered in. The lasting cultural contributions of Babylonians to astronomy, the division of the ecliptic into the stations of the zodiac, and the division of the circle into 360 parts, had sources other than the mathematical formalism of the Seleucid era. The undeniable barrenness of these notable feats seems to be the natural consequence of a chronic neglect of careful observational work. Neugebauer explicitly noted in connection with the Babylonian planetary theory "the minute role by direct observation in the computation of ephemerides." According to him "the real foundation of the theory is (a) relations between periods, obtainable from mere counting, and (b) some fixed arithmetical schemes . . . whose empirical and theoretical foundations to a large extent escape us but which are considered to be given and not to be interfered with by intermediate observations. It is a strictly mathematical theory which we are slowly learning to reconstruct from our fragments."[21]

In view of this it should be easy to recognize that the legendary precision of Babylonian astronomers, praised in so many histories of science, is one of those cliché details that must be taken with proper reservation. Babylonian astronomy was not interested in the facts of heaven as scientific data. Its interest lay exclusively in providing a fairly correct lunar calendar, completely subordinated to the service of religious rites. These in turn were expressions of an animistic and cyclic world view that provided no fertile soil for the development of the scientific enterprise. One should not, therefore, be surprised that the finest achievement of Seleucid astronomy, the arithmetic formalism of lunar theory, remained wholly unproductive from the scientific viewpoint. Though it clearly shows the recognition of five independent parameters determining the motion of the moon, it contains no traces of a geometric, let alone dynamic, model of the moon's motion. The concentration of Babylonian astronomers on the phases of the moon and of its eclipses reveals rather the bondage of Babylonian science to astrology and magic. In view of the oppressive role played by divination in every facet of Babylonian life, modern speculations about a "purely formalistic" school of mathematical astronomers in Seleucid Babylon should appear wholly arbitrary. Much less can statements be justified that refer to "an almost Comtean approach of early Babylonian astronomers."[22]

The principal reason for this lies in the fact that in ancient Mesopotamia astrology did not merely coexist with astronomy, but it wholly dominated the latter. The correlation of the two there has nothing in common with the simultaneous devotion of a Kepler or of a Tycho to astronomy and astrology. The Babylonian situation is worlds removed even from the attitude of Ptolemy who authored not only the *Almagest* but also the *Tetrabiblos*, the most influential astrological text ever written, of which he was more proud than of his most elaborate and exact planetary theory. The skilful calculators of Babylonian lunar theory were, it seems, primarily astrologers, magi, and soothsayers. Their principal compositions were incantations appropriate to any of the sundry phenomena of the heavens. Among them the most notable were, of course, the eclipses. Legion is the number of tablets on which all sorts of events on earth are connected with the moon's partial and total eclipses and with the various shapes of its horns. The invasion of locusts, the sickness of princes, the flourishing of market places, the peaceful reign of the king, the slaying of huge armies, general inundations, devastation of crops, eruption of fighting in the temple of Bel, the healing of the sick, are only a few of the countless events connected in ancient Mesopotamian omens with eclipses.[23]

Omens were also derived from other celestial phenomena, such as the relative position of the moon and sun, the rising of Venus and Mars, and from clouds, storms, thunders, and halos which also formed a part of the heaven in the old Babylonian world view. Needless to say one would look in vain for a fair measure of consistency in the enormous jumble of divinatory texts. Whatever regularity was observed in the heavens, it generated in the Babylonian mind a hapless surrender to patent contradictions in explanation. The low appearance of the moon could signal the clamour of the enemy in the land as well as the submission of a hostile country.[24] Again, the failure of the moon to appear at a time indicated by their calculations could mean the invasion of a powerful city or the counsel of the gods to bring happiness for the land.[25] The same exercise in inconsistencies characterizes the interpretative parts of Babylonian calendars. The only consistency there is the effort to describe each day a propitious one even when it was patently a harbinger of disaster.[26]

Underlying this morbid preoccupation with celestial omens was a conception of the cosmos in which celestial phenomena were the principal determining factors of what happened on earth in general and with human society and individuals in particular. This had to be so if the whole world constituted one huge organism or animal in which the small-scale patterns faithfully reflected those taking place on the cosmic scale. This parallelism between macrocosm and microcosm was grossly overemphasized in this century by the representatives of the Pan-Babylonian School.[27] Yet, whatever their exaggerations, the basic tone of Babylonian divinations brings out unmistakably the animistic features of the world conception of ancient Babylon. Its most telling evidence is the ancient Babylonians' addiction to haruspicy. Observations of physiological and pathological details of animal bodies certainly provided interesting details of biological information, and prompted perhaps some rudimentary systematization of it. The latter was merely a side effect. Clearly, no serious scientific information could be expected, for instance, from the anxious observation of the roaming of dogs of various colours around the temple.[28] Rationality was doomed in a land where the presence of a dog in a specific part of one's home was believed to entail its destruction by fire,[29] or where the future of the land was ascertained from the shape of the ears of lambkins.[30]

It should, therefore, appear most natural that the origins and development of the universe are accounted for in the *Enuma Elish*, the famed Babylonian cosmogonical poem, in heavily animistic terms.[31] According to the poem, only the divine parents, Apsu and Tiamat, and their son, Mummu, existed in the beginning. They represented in turn the sweet water, the salt water, and the mist hovering over both. The origin was, therefore, a chaotic mixture of waters and vapours with no definite entity or principle standing out in the amorphous vastness. As the poem pointedly notes, nothing could yet be seen, not even "pasture land or reed marsh." The process of primeval differentiation is described in the poem as the result of the generative power of Apsu and of the universal motherhood of Tiamat. The two brother-sister pairs born to them are Lahmu-Lahamu and Anashar-Knishar, but their actual cosmic role is not clarified. Anu, the son of the latter pair, is the one who becomes identified with the sky about which nothing further is said at this point. The roster of the original great gods closes with Ea, Anu's son, and this is basically all that is told about the origin or beginning of the world in the *Enuma Elish*.

Actually, it is almost a misnomer to call the *Enuma Elish* a cosmogonical poem. Its main concern is not the cosmos but the ultimate coming to power of Marduk, the chief deity of the Babylonian pantheon. His cosmogonical role is not about the real origins. He is merely attributed power and skill to let some transient order prevail in the unruly, violent, and chaotic state in which everything existing remains enveloped long after the cosmogonic process had run its first phase. It is, therefore, logical that there should follow a sequence of turbulent scenes which fill much of the remainder of the poem. Tiamat, the part-symbol of the original chaos, gives birth to a brood of boisterous lesser deities who turn against their own mother:

> They disturbed Tiamat and assaulted their keeper;
> Yea, they disturbed the inner parts of Tiamat,
> Moving (and) running about in the divine abode.[32]

The result is the beginning of a fateful clash between Apsu and Tiamat, the great mother. Apsu suggests the elimination of all the unruly gods so that he and Tiamat may sleep soundly. But Tiamat's motherly instincts revolt. The ensuing feud between the two and the ultimate defeat of Tiamat form the main body of the poem. The first phase of the struggle closes with Apsu's destruction. The gods fearful for their lives engage the good services of Ea who draws a magic circle around Apsu, puts him to sleep by incantations, tears off his crown, and slays him. Apsu's body becomes the foundation, the primeval mound, on which Ea then builds his abode. There Ea's wife, Damkina, gives birth to Marduk, a fearsome figure and a crudely anthropomorphic god:

> He sucked the breasts of goddesses.
> The nurse that cared for him filled (him) with awe-inspiring majesty.
> Enticing was his figure, flashing the look of his eyes,
> Manly was his going-forth, a leader from the beginning . . .
> Artfully arranged beyond comprehension were his members,
> Not fit for (human) understanding, hard to look upon.
> Four were his eyes, four were his ears . . .
> When his lips moved, fire blazed forth.
> Each of (his) four ears grew large,
> And likewise (his) eyes, to see everything . . .
> He was clothed with the rays of ten gods, exceedingly powerful was he;

The terror-inspiring majesty with its consuming brightness rested upon him . . .
He caused waves and disturbed Tiamat.[33]

Clearly, power, cosmic or otherwise, was the seed of further troubles in the
world as conceived in the *Enuma Elish*. Tiamat, disturbed by Marduk, the great-
grandson of her own son, begins to resent the loss of her husband, Apsu. She
connives a gigantic scheme to eliminate all those responsible for his death. Many
of the deities side with Tiamat:

They were angry, they plotted, not resting day or night;
They took up the fight, fuming (and) raging;
They held a meeting and planned the conflict.[34]

Tiamat herself prepares for battle by producing from her belly monster serpents,
"her irresistible weapons." The deities gathering around Ea finally decide on
casting their fortunes with the powerful Marduk. But he accepts the role of leader
only if he is given recognition by all gods as their supreme leader. In the bargaining
Marduk impresses them by showing the irresistible effectiveness of his word of
command. The gods around him burst into a hymn of worship:

Thy destiny, O lord, shall be supreme among the gods.
Command to destroy and to create, (and) they shall be!
By the word of thy mouth, let the garment be destroyed;
Command again, and let the garment be whole![35]

The poem states:

He commanded with his mouth, and the garment was destroyed.
He commanded again, and the garment was restored.[36]

But the poem also reveals that Marduk's word is not so powerful as to destroy
Tiamat. Nor does the "kingship over the totality of the universe" given to him
by the gods empower Marduk to create the heavens that are still to be fashioned.
He has to confront Tiamat in a violent encounter:

When Tiamat opened her mouth to devour him,
He drove in the evil wind, in order that (she should) not (be able) to close
  her lips.
The raging winds filled her belly;
Her belly became distended, and she opened wide her mouth.
He shot off an arrow, and it tore her interior;
It cut through her inward parts, it split (her) heart.
When he had subdued her, he destroyed her life;
He cast down her carcass (and) stood upon it.[37]

Deities that supported Tiamat are vanquished by Marduk in the same violent
manner. It is only after all this that the one having "kingship over the totality of the
universe" can turn to the task of producing an orderly heaven, symbol of the
order to be established below. But the heavens are not produced by his seemingly
all-powerful word of command. In the words of the poem:

The lord trod upon the hinder part of Tiamat,
And with his unsparing club he split (her) skull.
He cut the arteries of her blood
And caused the north wind to carry (it) to out-of-the-way places . . .
The lord rested, examining her dead body,
To divide the abortion (and) to create ingenious things (therewith).
He split her open like a mussel into two (parts);
Half of her he set in place and formed the sky (therewith) as a roof.[38]

The rest of Marduk's "creative" work is not much more inspiring or enlightening. His appointing the firmament with stars, with their constellations, and with
the signs of the zodiac are recounted in a style in which gory details are absent. The
poem pays distinct attention to Marduk's ordinances about the course and phases
of the moon. But the grossly animistic outlook is back in full force as the origin
of man comes up in the poem. Man is formed from clay mixed with a slain god's
blood. As another version of the Babylonian creation poem states, the production
of man is motivated by the desire of the gods that they could rest from any labour
"with man bearing the yoke."[39] In the Babylonian outlook man is, indeed, a slave
browbeaten by unpredictable gods who quickly punish any transgression, but whose
reward cannot be counted upon by a similar promptness and regularity. This
wilful atmosphere is also evident in the closing parts of the poem where the various
phenomena of nature and the manifold duties of man are assigned their guardian
deities well known for their connivance, capriciousness, and violence. Such gods
could certainly inspire fear and tremor in the puny slave, man, who in the closing
lines of the poem is urged, in a cruel twist of reasoning, to rejoice in the mocking
explanation of his destiny:

Let (man) rejoice in Marduk, the Enlil of the gods,
That his land be fruitful (and) it be well with him.
Reliable is his word, unalterable his command;
The utterance of his mouth no god whatever can change.
He looks on and does not turn his neck;
When he is wroth, no god can withstand his indignation.
Unsearchable is his heart, (all-)embracing his mind:
The sinner and the transgressor are an abomination before him.[40]

Obviously, a cosmos, and particularly a sky, with such origins could not function
as a paradigm of impersonal order but only as the personification of wilfulness.
Sumerians, Babylonians, and Assyrians were convinced that every part of nature
had a will of its own, often capricious and standing in continual conflict with one
another. Such parts of nature, or forces of nature, could in their belief be pacified
only by prayer and sacrifice. Needless to say, the outcome of such procedure was
ambiguous more often than not. All this was in line with the picture of the world
as a huge animal with apparently no beginning and end, subject to the various
periodic changes evident in the life of the animal kingdom. Changes in human
life, in society, and in the immediate physical surroundings of man were naturally
pictured as the effects of the periodic clashes of large-scale forces and phenomena
in nature. Most of these, the wind, the rain, the clouds, the daylight, and the night

were readily connected with the heavens. The observation of the heavens seemed, therefore, to be the logical clue for learning something about the course of events on earth. Those events were believed to have a recurring pattern because the phenomena of the sky followed one another in such a manner. The period of solar year and the periods of the moon were obviously the two principal ones of the heavenly cycles, and the correlation of these two formed the principal burden of the Babylonian observers of the sky. It was not, however, a scientific burden for them in the modern sense of the world. The ultimate motivation of their pre-occupation with the phenomena of the heavens came from that animistic, cyclic conception of the world in the same way as the observation of eclipses and the investigation of the entrails of animals were as many methods for them to divine ways and means for assimilating themselves with the cosmic life repeating itself for eternity.

A principal repetitive feature of the cosmos, the succession of solar years, dominated completely the Babylonian interpretation of the world. The year formed for them the principal paradigm of the animistic roots and cyclic patterns of existence. It also set the measure of order which the universe could have in their estimate. That measure was meagre indeed. A telling proof of this can be found in the various details of the great New Year festival, the importance of which in the entire texture of Babylonian life cannot be overestimated. Sumerian in its origins, many of its parts were faithfully acted out in the Akitu festival in Babylon. The culminating point of the ritual consisted in the sacred marriage between the king and the high priestess of the temple or ziggurat. The purpose of the rite was to secure for the king and his people the unimpeded flow of the forces of fertility for another round of the year-cycle. All this was spelled out most explicitly in the "fixing of destiny" which the high priestess read to the king following the comple-tion of their sexual union.[41]

In all this, a most ardent desire evidenced itself to influence, or at least to prog-nosticate, the course which history was to take until the rite was repeated again with the coming of a new year. The rite's specific aim did not consist in spelling out in advance details of future events. Thus, when particular predictions turned out to be wrong, the allegedly faulty observation of omens could readily take the blame, while the correctness of the principle was firmly upheld. The main concern rested with a general desire to re-experience once in every year the transition from chaos, represented by the yearly floods, to a more orderly and secure phase of life. But the ascendancy of order over chaos was not thought to be final. The rites of the Akitu festival are permeated with a fear that order, cosmic and social, may fall prey at any time to chaos. The faithful performance of the Akitu rites could secure the orderly course of cosmic and historic events for only a year. Beyond the one-year limit there lay a fearsome uncertainty with no truly hopeful perspectives.

Babylonian mentality was in a sense caught in a discontinuous succession of time units, a year long each, separated from one another by a ritual actualization of the original chaos. The fourth day of the week-long Akitu festival was a day of public wailing and mourning over the various manifestations of chaos, ethical and social, and it readily developed into actual anarchy on the streets of Babylon. Such studiously planned experiences involving the whole city-state could hardly generate a confident look at a more distant future. Babylonian culture seems to have been trapped in the treadmill of basically ominous yearly cycles. Perhaps this is the reason why in Babylon, prior to the Seleucid era, there had never developed a continuous chronology starting from some outstanding event, such as the counting of years in Greece from the first Olympiad. In Babylon the succession of years did not mean the accumulation of building blocks with continuous constructive potential. They rather meant the transition from one world period into another separated by the discontinuity of chaotic conditions.

Prior to the Hellenistic era there is no evidence that the Babylonian priest-astrologers have given specific cosmic interpretation to time cycles longer than a year, such as the cycles of the planets and the cycles of the moon's eclipses which they calculated with fair precision. In the long run, however, their fundamentally cyclic concept of cosmos and history was readily grafted on the longer celestial cycles. The classic illustration of this is the most important single work ever produced by an individual Babylonian scribe, the *Babyloniaka* (Babylonian history) of Berossos (c. 350 B.C.–c. 270 B.C.), a priest of Bel in Babylon. The work today is known only from fragments, though in its main lines is reliably reconstructed.[42] It seems to have consisted of three parts the first of which dealt with the origins of the world.

The cosmogonical fragment of Berossos' work is rather short,[43] but it nevertheless abounds in telling details. These not only prove the reliability of the fragment but also the measure in which Berossos' notion of the world was steeped in an organismic conception of the universe. At the very outset it is emphasized that out of the original darkness and water there came into being "strange and peculiarly shaped creatures." Among these were some men with two wings, others with four wings and two faces, again others having both male and female faces and genitals. The fragment recounts men with the legs and horns of goats, men with the hind parts of bulls, and still others with the tails of fish. Strange hybrids also characterize the animal world of the origins as described in the fragment. It mentions animals, part horse, part fish, and "other wondrous creatures, which had appearances derived from one another." According to the fragment all these could be seen depicted in the temple of Bel (Marduk) in Babylon and it was over this strange realm of all living that there "ruled a woman named Omorka. This in Chaldean is *thamte* [Tiamat], meaning in Greek 'the sea', but in numerical value it is equal to 'moon'."

The second part of the fragment starts with a reference to the formation of the heaven and earth from the two halves of the woman's body cut asunder by Bel. But far more interesting is the statement that the very same action of Bel which produced a certain structure in the universe also caused its eventual destruction. The separation by Bel of the darkness from light leads to the total destruction of all living, as they cannot endure light unmixed by darkness. The work of creation has, therefore, to start anew, foreshadowing its cyclic repetitions. This time the living (men and animals) are not produced from the chaotic waters but from the blood of gods mixed with earth. Bel (Marduk) in his typically cruel fashion orders the gods to cut off their heads to provide the needed blood. Most significantly, "the realm of the living" to be re-created includes the sun, the moon, the planets, and the stars as well. They all are refashioned out of earth and divine blood to possess special insight and understanding.

That such origins hardly led to an intellectually promising discussion of the cultural development of man can readily be surmised. In the fragments, Berossos attributes the emergence of culture in ancient Mesopotamia to a message to men from Oannes, a god-like sea monster, and from other gods symbolizing the animistic forces of nature. It was in that context that Berossos seemed to discuss the origins and scope of Babylonian astrology. The two other parts of the book are devoted to Babylonian chronology starting with the "ten kings before the Flood," and bringing it as far as the Persian era. The story of the Flood is, however, immediately followed by a reference of Berossos to future destructions and re-creations of the universe. The most detailed ancient recollection of Berossos' ideas on this point is due to Seneca, who as a Stoic was a firm advocate of the cyclic idea of the cosmos. He obviously found in Berossos a particularly useful source to quote because of the definiteness of the latter's idea in the matter:

> Some suppose that in the final catastrophe the earth, too, will be shaken, and through clefts in the ground will uncover sources of fresh rivers which will flow forth from their full source in larger volume. Berossos, the translator of [the records of] Belus, affirms that the whole issue is brought about by the course of the planets. So positive is he on the point that he assigns a definite date both for the conflagration and the deluge. All that the earth inherits will, he assures us, be consigned to flame when the planets, which now move in different orbits, all assemble in Cancer, so arranged in one row that a straight line may pass through their spheres. When the same gathering takes place in Capricorn, then we are in danger of the deluge. Midsummer is at present brought round by the former, midwinter by the latter. They are zodiacal signs of great power seeing that they are the determining influences in the two great changes of the year. I should myself quite admit causes of the kind.[44]

It is rather interesting that in discussing the period of cosmic conflagrations Seneca referred to Berossos and not to Plato. After all, the simultaneous, periodic return of all planets to the same point of the zodiac was considered by Plato a century or so prior to Berossos as having a major cosmic significance, and his view was received with fairly general acceptance. Yet, with Berossos the idea of periodic conflagration was not merely a part of speculative cosmology as it was with Plato, but formed part of a chronology encompassing the periods of both human and cosmic developments. Furthermore, Berossos' treatise shows particular features which clearly reveal its non-Platonic origin. His chronology covers the whole duration of the present world age divided by him into four phases that remind one of the successively shortening lengths of the Yugas. It is also remarkable that the lengths are multiples of 3600 years, which in the Babylon of Berossos was a perfect time unit called saros.[45] The first of the four phases of the world period is the age from the creation to the first of "ancient kings" that lasted, according to Berossos, 466 saros and 4 neros (4 × 600 years), or a total of 1,680,000 years. The second phase is the age of the ten "ancient kings" who ruled for 120 saros before the Flood. The third phase is the period between the Flood and the death of Alexander

H

the Great, with a length of 10 saros. From 323 B.C. on there would be, according to Berossos, 3 saros and 2 neros, or exactly 12,000 years. The total life-span of the cosmos between two destructions is given, therefore, by Berossos as exactly 600 saros, or 600 × 3600 years, a number of obviously special perfection in the sexagesimal system used by the Babylonians.

It may be argued that the support given by Berossos to the general concept of a cosmic Great Year also reflects Greek influences. Such is a possibility because of Berossos' close ties with Greek cultural centres. According to classical sources he spent his last years as the leader of a school of astronomy on the island of Cos. Whatever the extent of Berossos' indebtedness to Greek influence, his concept of cosmic cycles was more Babylonian than Greek. It fails to show, in close similarity with the Babylonian explanation of the cosmos, a striving for unitary explanation. This aspect of ancient Babylonian world view was already noted by Damascius (Nicolas of Damascus), one of the last of Neoplatonist philosophers (fl. 510). He was also the author of *Difficulties and Solutions of First Principles*, a treatise which contains a brief outline of the origin of the cosmos as conceived by the Babylonians. To the refined mind of a Neoplatonist the Babylonian approach to the problem had to seem basically defective. This is indeed the gist of Damascius' very first remark: "Of the barbarians the Babylonians seem to pass over in silence the one principle of the universe, and they assume two, Tauthe [Tiamat] and Apason [Apsu], making Apason the husband of Tauthe and calling her the mother of the gods."[46] Clearly, Damascius did not think highly of all this, and his disdain comes through as he remarks the number of generations of gods needed in Babylonian cosmology to produce the "fabricator of the world."

Basing an explanation on two principles, especially of the type implied in the Babylonian cosmogony, is not a procedure leading to great insights. Nor can the duality of principles do justice to the full meaning of the Greek word *arché* which means both principle and beginning. The most intimate relatedness of these two meanings had some appeal to the Greeks. Its crucial importance did not dawn on the Babylonians. Nor did they succeed in marshalling some of the intellectual encouragement which is ultimately anchored in one's firm holding to the indissoluble unity of final explanation (principle) and ultimate origin (beginning). Either of them can only be one, for it is as impossible to have two true explanations as it is to have two real beginnings for one and the same thing or process. Without the recognition of this elementary though monumental verity, reasoning can generate no lasting satisfaction but only despair and discouragement. The Babylonian cosmology with its heavily organismic colouring and with its cyclical overtones is a classic illustration of this. In the absence of a true *beginning* it lacked *explanation* and exuded pessimism. The cosmic process as depicted in *Enuma Elish* is an endless strife between warring deities symbolizing both cosmic and human destiny. Its complete, solemn reading at each year's Akitu festival impressed with renewed vividness a rather disconcerting state of affairs on the minds of each Babylonian generation. The "Dialogue of Pessimism," a Babylonian poem from the first millenium, reflects only too well the state of mind generated in such an ambience. True, pessimistic poems are not missing in any literature. But in the Babylonian literature the voice of pessimism sets the dominating tone.[47]

For such pessimism, cultural discouragement, and deep-seated intellectual inertia several explanations have been offered. The influence of climate on cultural prospects cannot be emphasized enough, and the Mesopotamian climate has gruesome aspects of its own. Yet, for many centuries the adversities of weather were successfully faced with relatively primitive techniques, and the fertility of the Tigris and Euphrates valley was well secured. This was done in spite of the many wars that time and again interrupted the peaceful accumulation of material and cultural acquisitions. The destructive role of wars in Mesopotamia should not, however, be exaggerated. Babylon and Assur enjoyed repeated periods of peace long enough to permit the emergence of a truly scientific endeavour in one or another direction at least. Again, the Sumerians and Assyro-Babylonians displayed more than enough intellectual acumen, industrial skill, curiosity, and patience in observations, that is, qualities indispensable for the creation of science. They have not come even moderately close to it. The promising creativity of Hammurabi's age was not followed up in later times either in literature, or in arts, or in legislation, let alone in matters of scientific learning.

The basic reason for this failure is neither geophysical, nor socio-economical, but has rather to do with Weltanschauung. A systematic investigation of the world and of its lawfulness presupposes a fair measure of confidence in the reasonability and likely success of such an enterprise. It was this confidence that the literate classes in ancient Mesopotamia were unable to cultivate in a sustained manner. They remained trapped in the disabling sterility of a world view in which not reason ruled but hostile wilfulness, the crushing blows of which threatened with repeated regularity. Believing as they did that they were part of a huge, animistic, cosmic struggle between chaos and order, the final outcome appeared to them unpredictable and basically dubious. All they could see was the endless alternation between the two. Not that they did not wish to contribute to a steady emergence of order. Not that they did not wish to influence nature, or rather its personalized forces, the gods. The animistic, cyclic world view made it, however, impossible for them to realize that to influence or to control nature one had to be able to predict accurately its future course. They lacked faith in the possibility of such a prediction as it implied the notion of an order free from the whims of animistic forces that inspired the vision of a collapse to occur time and again. As a result, the mastery of science could not become a proud feature of the culture of a land on which ziggurats cast their sombre omen.

NOTES

1. An informative discussion of the culture and history of ancient Mesopotamia can be found in H. W. F. Saggs, *The Greatness that Was Babylon: A Sketch of the Ancient Civilization of the Tigris-Euphrates Valley* (London: Sidgwick & Jackson, 1962) and Georges Roux, *Ancient Iraq* (London: Allen & Unwin, 1964). The best general summary of Babylonian science is the chapter, "Mesopotamia," by R. Labat and others in R. Taton (ed.) *History of Science*, Vol. I, *Ancient and Medieval Science*, translated by A. J. Pomerans (New York: Basic Books Inc., 1963), pp. 65–121.

2. *Herodotus* with an English translation by A. D. Godley (London: William Heinemann, 1926), vol. I, p. 281 (Book II, § 5).
3. See A. Parrot, *Ziggurats et Tour de Babel* (Paris: Albin Michel, 1949), p. 37.
4. Saggs, *The Greatness that Was Babylon*, p. 355.
5. On this point there is some controversy. But even if the sacred marriage had taken place in a temple adjoining the ziggurat, the latter still remained the focal point of the rites, and shared in their basic symbolism.
6. *Herodotus*, vol. I, p. 223 (Book I, § 178).
7. *Ibid.*
8. For a detailed discussion, see M. Levey, *Chemistry and Chemical Technology in Ancient Mesopotamia* (London: Elsevier Publishing Co., 1959).
9. Factual knowledge of Babylonian mathematics based on the deciphering of cuneiform texts and not on vague traditions and speculations, is relatively new. It is due to the pioneering investigations of F. Thureau-Dangin and O. Neugebauer. For a summary of the results now available the work to be consulted above all is Neugebauer's *The Exact Sciences in Antiquity* (2d ed.; New York; Harper & Brothers, 1962), chaps. 2, 3, and 5. Also very reliable is the chapter on Babylonian mathematics in B. L. van der Waerden, *Science Awakening*, translated by Arnold Dresden (New York: John Wiley & Sons, 1963).
10. The one major addition during the second period seems to be the explicit use of zero.
11. *The Exact Sciences in Antiquity*, p. 30.
12. For the most part Babylonian mathematicians did not solve simultaneous equations by the method of elimination through substitution. An exclusive reliance on this method came only with the Arabs. Nothing shows better the long-felt influence of Babylonian algebra than the fact that the methods of Diophantus (fl. 250), the great mathematician of the late Hellenistic era, often have more in common with the Babylonian than with the classical Greek algebraic tradition.
13. *Science Awakening*, p. 72.
14. Kugler's monumental works published between 1900 and 1924 were a sequel to studies initiated by two other Jesuits, J. Epping and J. N. Strassmaier.
15. *Astronomical Cuneiform Texts. Babylonian Ephemerides of the Seleucid Period for the Motion of the Sun, the Moon, and the Planets* (London: Lund Humphries, 1955).
16. See *ibid.*, vol. II, p. 273.
17. "Chaldean Astronomy of the Last Three Centuries B.C.," *Journal of the American Oriental Society*, 75 (1955):170.
18. See on this O. Neugebauer, "The Alleged Babylonian Discovery of the Precession of the Equinoxes," *Journal of the American Oriental Society*, 70 (1950):1–8.
19. For details on this, see O. Neugebauer's review of the sections on mathematics and astronomy in L. Renou and J. Filliozat, *L'Inde classique. Manuel des études indiennes* (Paris: Imprimerie Nationale, 1953) in *Archives internationales d'histoire des sciences*, 8 (1955): 168–73.
20. "Chaldean Astronomy of the Last Three Centuries B.C.," pp. 170–71.
21. *Astronomical Cuneiform Texts*, vol. II, p. 281.
22. The expression is that of M. R. Caratini in his account of Babylonian astronomy in R. Taton, *History of Science*, vol. I, p. 120.
23. For an extensive collection of such omens, see R. Campbell Thompson, *The Reports of Magicians and Astrologers of Nineveh and Babylon in the British Museum*, Original texts, with translation, notes, and introduction (London: Luzac & Co., 1900).
24. *Ibid.*, p. xlv, § 66 of text.
25. *Ibid.*, p. xlvii, § 82 of text.
26. See on this M. Jastrow, *The Religion of Babylonia and Assyria* (Boston: Ginn and Co., 1898), p. 376.
27. The founder of the School was Hugo Winckler, professor at the University of Berlin, but its short though virulent popularity was largely due to the writings of Alfred Jeremias.
28. See Jastrow, *The Religion of Babylonia and Assyria*, p. 398.
29. *Ibid.*, p. 399.

30. *Ibid.*, p. 394.
31. Of the many discussions and translations of this poem the one by Alexander Heidel, *The Babylonian Genesis: The Story of Creation* (Chicago: University of Chicago Press, 1942) stands out for its clarity and documentation. The poem consists of several parts, each on a separate clay tablet. Subsequent references indicate tables (parts) in Roman numerals, lines in Arabic numerals, together with the corresponding pages in Heidel's work. The passages are quoted with some minor orthographic and stylistic modifications.
32. I, 22–24, p. 19.
33. I, 85–88, 93–98, 103–04, 107, pp. 21–22.
34. I, 129–31, p. 23.
35. IV, 21–24, p. 37.
36. IV, 25–26, p. 37.
37. IV, 97–104, pp. 40–41.
38. IV, 129–32, 135–38, p. 42.
39. Tablet "Creation of Man," line 9; p. 67.
40. VI, 149–56, p. 60.
41. For further details, see E. D. van Buren, "The Sacred Marriage in Early Times in Mesopotamia," *Orientalia*, 13 (1944):1–72; Svend A. Pallis, *The Babylonian Akitu Festival* (Det Kgl. Danske Videnskabernes Selskab. Historisk-Filologiske Meddelelser, XII, 1; Copenhagen, 1926), especially pp. 292–97.
42. P. Schnabel, *Berossos und die babylonisch-hellenistische Literatur* (Leipzig: B. G. Teubner, 1923), especially pp. 176, 206–07.
43. It is given in English translation in Heidel, *The Babylonian Genesis*, pp. 77–78.
44. *Physical Science in the Time of Nero, Being a Translation of the Quaestiones naturales of Seneca*, by John Clarke (London: Macmillan, 1910), pp. 150–51 (Book III, chap. 29).
45. See on this Neugebauer, *The Exact Sciences in Antiquity*, p. 141.
46. This fragment is quoted in full in Heidel, *The Babylonian Genesis*, pp. 75–76.
47. The Gilgamesh epic illustrates this all too clearly. For a scholarly translation, see R. Campbell Thompson, *The Epic of Gilgamish* (London: Luzac & Co., 1928).

CHAPTER SIX

The Labyrinths of the Lonely Logos

When Berossos set forth his account of the cycles of cosmic and human history, Greek thought had already entered its Hellenistic period. The role Athens had played for almost two centuries shifted to Alexandria where the newly founded Museum with its Library offered unique facilities for scholarly work. The various branches of science owe a great deal to those facilities. They consisted not only of an unusually large number of books and various collections of scientific instruments and specimens, but also of lecture halls and, last but not least, of generous stipends which allowed scholars to devote themselves entirely to research and writing.

Some results of the scholarly work carried out in the Museum of Alexandria were truly spectacular. The very beginning of the Hellenistic period of Greek science[1] saw the publication of one of the greatest books written in history, the *Elements* of Euclid, containing propositions the content of which was fully grasped only by nineteenth-century geometry and mathematics. Two generations after Euclid, the custodian of the Library was Eratosthenes, the Father of mathematical geography, who made his name immortal by his surprisingly accurate measurement of the circumference of the Earth. About the same time the Museum also played host to the outstanding mathematician, Apollonius of Perga. Half a century later, the Museum of Alexandria was again the place where Hipparchus (fl. 150 B.C.) discovered the precession of equinoxes. In the first century B.C., an Alexandrian astronomer, Sosigenes, aided Julius Caesar in carrying out the historic calendar reform. To the very end of its existence the Museum of Alexandria could, indeed, boast of great names in astronomy and mathematics such as Ptolemy, Diophantus, Pappus, and Theon.

Alexandrian science excelled in other fields as well. Medicine received great impetus when in the early third century B.C., Herophilus and Erasistratus established medical schools there. The former did valuable discoveries in the vascular system, recognized the importance of the fourth ventricle, and was the first to point out the specific function of sensory nerves. Erasistratus in turn became the Father of comparative anatomy and of pathological anatomy, and discovered the difference between motor and sensory nerves. Next to medicine such leading figures of Hellenistic engineering as Ctesibius, Straton, and Heron worked in Alexandria.

Alexandria was the chief but not the sole place of learning in Hellenistic times. Scientific tradition remained alive to varying degrees in such older places of learning as Athens, Syracuse, and Cos, while fresh starts were made in Pella, Antiochia, Pergamos, and Rome. Also the movement of scholars between cities was vigorous. Apollonius of Perga worked both in his native town and in Alexandria. Galen, the greatest physician of antiquity, carried out his activities in Rome and in his native town, Pergamos. Those leaders of Hellenistic science, who like Archimedes of

Syracuse stayed in the same place much of their lives, saw to it that their findings received sufficient circulation as can be seen from Archimedes' famous letter on "The Method" to Eratosthenes in Alexandria.

For all the scientific achievements that took place during the six or seven Hellenistic centuries, this long period does not seem to equal in creative power the much shorter Hellenic phase of Greek scientific thought, or the period from Pericles to Alexander the Great. In more than one respect the feats of Hellenistic science were methodical though very valuable elaborations on themes, discoveries, and syntheses made in Hellenic times. Thus Euclid's *Elements*, published around 300 B.C., represented to a great extent the concluding phase of the spectacular emergence of scientific geometry to which no less than sixty known Greek geometers contributed during the previous four or five generations. What is true of geometry holds also of algebra, as its cultivation remained throughout the Hellenistic times distinctly geometrical in character.

In astronomy one may see the same great indebtedness of Hellenistic astronomers to their Hellenic predecessors. This is not to deprecate in any sense the brilliant contributions of scientists like Aristarchus, Hipparchus, Poseidonius, and Ptolemy. The working out of sound methods for estimating the distances of the moon, sun, and of their sizes in terms of the radius of the earth, the astoundingly precise measurement of the difference between the lengths of the solar and sidereal years with the simultaneous recognition of the precession of the equinoxes, the computation by Ptolemy of a long list of trigonometrical values and their skilful use in his great planetary system, were achievements of first rank. Ptolemy's analysis of the motion of the moon was in fact so exact as to lead him to the discovery of two recondite aspects of the moon's dynamics: its evection and nutation. Yet, the principal conceptual tools underlying the Ptolemaic theory of planets, the combination of circular paths, go back to Hellenic times, and in a rudimentary form were already hinted at by Plato himself.

This should be of no surprise. Plato held geometry in the highest esteem and it was in a sense a Platonic imprint on Greek astronomy, both Hellenic and Hellenistic, that its main achievements remained restricted to that part of astronomy which is largely geometrical: the investigation of the form and size of celestial bodies and of their paths across space. It was no coincidence that what happened in Greek astronomy took place in physics too. Its lasting advances were made in statics where a straightforward application of geometrical considerations came naturally. To advance from statics to dynamics required insights other than geometrical, and at that point even an Archimedes failed to break out from the conceptual confines of his scientific background, though he came tantalizingly close to formulating the basic propositions of infinitesimal calculus.

Development of calculus would have been a prime condition for making meaningful progress concerning the true dynamics of the inanimate physical world. On this point, however, ancient Greek scientific thought fell prey to the lure of sweeping generalizations that sidetracked the cultivation of physics for two thousand years. In short, the mistake consisted in the reduction of physics to biology for which there appeared several justifications. One, though not the foremost, was the undeniable success Hellenic science enjoyed in various branches of biological investigations, especially in medicine. In the Hippocratic corpus one finds not only marvellous insights into the causes and treatment of particular illnesses, but also the awareness of the need of both the rational and empirical procedures in medicine. It is rather difficult to improve on the sweep of diagnostic method as outlined, for instance, in the First Book of *Epidemics* where the physician is urged to be attentive to a long list of relevant details when treating his patient. In modern parlance the text expects him to bring to his treatment of the patient pertinent information not only from physiology, pharmacology, anatomy, and pathology, but also from psychology, meteorology, geography, and even sociology.[2]

To see the most mature form achieved during the Hellenic times in the field of biological sciences one must turn to the writings of Aristotle. It was he who turned zoology into a scientific discipline in his *History of Animals*, and laid lasting foundations for comparative anatomy in his *On the Parts of Animals*. His acumen as a biologist is perhaps even more brilliantly displayed in his *On the Generation of Animals* which remained until modern times the authoritative compendium on fertilization, embryology, and the birth and raising of the offspring. Nothing shows better the value of this book than that some of its inevitable errors went undetected until the nineteenth century. Darwin, for one, felt impelled to register his admiration for Aristotle, the biologist, following his reading of *On the Parts of Animals* in William Ogle's masterful translation: "I had not the most remote notion what a wonderful man he was. Linnaeus and Cuvier have been my two gods, though in very different ways, but they were mere schoolboys to old Aristotle."[3]

A gain is hardly ever without some loss. The extraordinary feats of Aristotle in biology were in a sense responsible for his failure in physics. The cultivation of the study of many aspects of the living organism invited a methodology which took its start from the purposiveness of biological systems. The emphasis on goals and purposes served biology only too well throughout its long history, and is staunchly defended by prominent biologists even in the present age of molecular biology and operational method. But in the days of Aristotle the espousal of final causes was far more than a methodological expedience. The realm of final causes stood then for the bedrock of intelligibility. The result was that investigation of any realm, living or not, was not considered satisfactory without attributing, rightly or wrongly, purposes to processes and phenomena of every kind, ranging from the fall of stones to the motion of stars.

Needless to say, Aristotle was not the originator of this approach. For all his criticism of his master, Plato, Aristotle remained a faithful disciple concerning the primacy of final causes in the art of explanation. As the most persuasive arguments in this respect rested with the conscious experience of man and with the behaviour of animals, the primacy of final causes meant in physics its reduction to a biological framework of thought. The whole edifice of physics as conceived by Aristotle is a monumental reply to a question raised by Plato in the *Timaeus:* "In the likeness of what animal did the creator make the world?" The question forms part of a paragraph where Plato probes into the ultimate nature of the cosmos, and he makes no secret of the fact that the intelligibility of the world depends on thinking about it as being the perfect animal. "The deity," he declares, "intending to make this world like the fairest and most perfect of intelligible beings, framed one visible animal comprehending within itself all other animals of a kindred nature."[4]

With his definition of intelligibility Plato merely echoed his master, Socrates. For it was Socrates who initiated that trend which produced along with most of the literary and philosophical gems of Greek humanism a wholly misguided form of physical science.[5] The motivations of Socrates could not have been more unobjectionable. He clearly saw that in the brilliant speculations of the Ionians and especially of the Atomists only inert matter existed, the configurations of which were ruled by chance. In such an account of the world, if taken consistently, there was no room for right or wrong, for moral responsibility and decision, or to recall Socrates' own problem, for obeying the voice of one's conscience even at the price of death. The details of Socrates' agonizing struggle with this issue are movingly told in the *Phaedo*, a work whose significance for the future course of science cannot be overestimated. To justify his own personal stance, wholly irrational from the viewpoint of "pure physics," Socrates outlined a new type of physics in which questions about purposes dominated. According to Socrates, the decisive scientific questions did not concern the size, shape, and location of the earth, but whether it was *best* for the earth to be of a given size and shape, and to be at a specific region of the cosmos. To save the meaning of what was typically human in the realm of existence one needed, so Socrates concluded, a reformulation of the methods and objectives of physics.[6]

The task of carrying out this programme fell to Aristotle, and he performed his chore with a singleminded devotion to the merits of the cause. To the modern reader of his *On the Heavens* and *Meteorologica* Aristotle's continual resorts to biological similes may appear tedious if not ridiculous. Earthquakes and animal digestion have so little in common as do the motion of the stars and the locomotion of quadrupeds. In fact the whole emancipation of science from the shackles of organismic or Aristotelian physics depended on achieving a depersonalized outlook on nature in which stones were not claimed to fall because of their innate love for the centre of the world. With Aristotle, however, the world, or at least its sublunary part, appears like a huge animal breathing, growing, decaying, subject to the cycles of birth and death for eternity. As the idea of cyclic returns in the universe had a distinctly biological or organismic hue, Aristotle found it much to his liking. In addition, Aristotle, always eager to arbitrate among the older philosophical dicta, could only relish the frequent occurrence of some form of a cyclic concept of the universe in the teachings of his forebears. Probably, all schools of pre-Socratic philosophy subscribed in one way or another to the belief that the universe was to perish and to resurge at regular intervals.

Among the Ionians Anaximander considered the infinity in space and time as a matrix that gave birth to an infinite succession of worlds. "From the infinite eternity comes the destruction, in the same way as does generation issued from it long before: all these generations and destructions reproduce themselves in a cyclic manner."[7] For Anaximander time itself was constituted by the succession of these cosmic destructions and regenerations.[8] According to another Ionian, Anaximenes, the world is eternal but "it is not the same world that exists forever; the one which exists is now one world and another later; its generation takes place anew after certain periods of time."[9] The wording of the opinions of Anaximander and Anaximenes, as is the case with many of the pre-Socratic fragments, comes from much later sources. The sources of Anaximander's dictum are works written by Eusebius and Hippolytus, both Christian Church Fathers. The opinion of Anaximenes is known from the writings of Simplicius, a most articulate defender of paganism in late antiquity. The unanimity of such disparate witnesses should dissipate doubts about the early popularity among the Greeks of the idea of a cyclic universe.

Actually, with each subsequent century the evidences in this respect become more direct, specific, and numerous. Among the Pythagoreans it was Philolaus, Alcmeon, Archytas, and Oenipodus who were recalled in later sources as proponents of the idea of the cyclic constitution of the cosmos. According to Philolaus the destruction of the universe would occur in two ways: either the fire of the skies spills over the earth, or the waters of the moon spread over the atmosphere.[10] Philolaus restricted, in line with his Pythagorean beliefs, the destruction and rejuvenation of the world to the sublunary regions. This distinction between the perishable sublunary world (cosmos) and the imperishable superlunary world (ouranos) is important to note as it became an article of faith almost universally accepted by subsequent generations of Greek philosophers. Another aspect of the Pythagorean doctrine of a cyclic cosmos, the reincarnation of souls, gave rise to vigorous controversies. The Pythagorean who may have derived the idea of reincarnation from the notion of cosmic cycles was Alcmeon. He argued that since the soul was divine, it had to share in the circular movement of the stars. The soul achieved this, so Alcmeon believed, by accompanying the cycles of bodily birth and death, and by producing the same bodily manifestations an infinite number of times.[11]

As will be pointed out later in detail, this numerically identical return of all beings, including humans, and for an infinite number of times, found no general favour among Greek thinkers. Other Pythagorean specifications of the cyclic concept of the universe were, however, eagerly followed up. One of these was Archytas' definition of the principal unit of time as determined by the fundamental cycle, or rather by the very soul of the universe. Eight or so centuries later Simplicius still tried to fathom the full meaning of the arcane statement of Archytas on the matter: "The time is the number of a certain motion, or rather in a general manner, the interval fitting the nature of the universe."[12] Simplicius was eager to point out that this definition was neither the one given by Aristotle, according to whom time was the number of movement, nor was it identical to the one offered by the Stoics who said that time was the interval of all movements. Archytas seems to have had in mind a most specific motion, that of the very soul of the world, which in turn caused all other motions and generations. "Of the number," wrote Simplicius, "which measures this movement Archytas says that it is already the producer of generation and that it goes on to the fabrication of the beings that are in the world; it is that number which determines without end the processes and transformations caused by the reasons which are born from it; it is that number which is the time fertile in works . . . But if Archytas declares that time, source of [all kinds of] generations, is the number which proceeds from the Soul of the World taken as a unity, it is clear that he also considered this movement as a unit of time; it seems that he regards time as constituted by the first movement which resides in the Soul [of the World] and also constituted by the movement that originates in that Soul; it is that second movement to which all other movements are referred and compared and by which they all are measured; it is indeed necessary that the measure may be superimposed on the subject to be measured and that at the same time it may enjoy with respect to the latter the role of a principle."[13]

What this lengthy passage certainly proves is that Simplicius was deeply impressed by Archytas' effort to make of cyclic recurrences a primordial principle underlying even the concept of number and time. A general effort along these lines is discernible in other Greek philosophers who shared Plato's admiration for the Pythagoreans. Plato's name comes naturally to mind in this connection as he added emphasis to another aspect of the Pythagorean speculations on cosmic cycles, namely, their relatedness to phenomena observed in the belt of the zodiac. The first to make explicit statements along these lines seems to have been Oenipodus,[14] but no astronomical considerations were implied in the biggest boost which the doctrine of the Great Year received in pre-Socratic times.

The source of that boost was Heraclitus who seems to have proposed the cyclic recurrence of all in every 10,800 years.[15] He very likely obtained that figure through Babylonian and Hindu influence. 10,800 is the product of 30 and 360, or the product of one "world-day" and of the number of days in an ideal year. Far more revealing is, however, the strikingly peculiar philosophy which Heraclitus built around the cyclic process as the fundamental pattern in nature. His was a baffling emphasis on the ultimate identity of darkness and light, and of such pairs of opposites as good and evil, pure and polluted, beginning and end, living and dead, waking and sleeping, young and old, whole and partial, connected and disconnected, in tune and out of tune. He earned, not without reason, the name "The Obscure," and this was perhaps the most flattering thing that could be said about one who got so hopelessly trapped in the cyclic identity of all that he came to doubt continuity itself within the cyclic process. His analogy of the ever fresh river is best remembered, but he also stated that the sun was new not only every day but in every moment. His thinking is indeed a classic example of the sad unbalance which is ultimately imposed on one's thinking by the acceptance of the endless, cyclic recurrence as the basic pattern of existence. Within that framework one was ultimately left with no consistency in reasoning and observation. It was no accident that, according to Heraclitus, the width of the sun was only one foot. Clearly, in a philosophy of nature steeped in the idea of perennial cycles there remained ultimately no room except for disconnected sense perceptions.

Few if any subsequent Greek philosophers followed Heraclitus in drawing all the implications of a thoroughly cyclic interpretation of nature. But in Greek philosophical tradition Heraclitus' name had certainly become connected with the idea of cyclic conflagrations and rejuvenations of the world which he probably did not discuss.[16] There is no reference to this among the dicta of Empedocles either who also imagined everything to oscillate between two extreme points or principles. According to him Love and Hate alternated in ruling over everything. The role of Love was to unite all bodies in one single, homogeneous sphere in which all differences disappeared. Discord, on the other hand, brought forth the distinctiveness of elements and produced the actual world. In this alternating process Empedocles discerned the only form of immortality "for the race of mortals," an expression in which he included men as well as things and their ultimate constituents, the elements. "In so far as they have the power to grow into One out of Many, and again, when the One grows apart and Many are formed, in this sense they come into being and have no stable life; but in so far as they never cease their continuous exchange, in this sense they remain always unmoved (*unaltered*) as they follow the cyclic process."[17] In addition to this general outline of cosmic cycles Empedocles' doctrine contained two particulars. One asserted the presence of a phase of rest between the actions of Love and Hate.[18] The other, a more important one, concerned the question, often pondered by subsequent Greek philosophers, whether things and humans would return in exact, numerical identity, or only the general classes of beings would see an endless rebirth.[19] Empedocles took the latter view which found greater favour among his successors than the doctrine of reincarnation which the Pythagoreans espoused so eagerly.

Great admirer as Plato was of Pythagoras and of his school, he did not take seriously their doctrine of reincarnation, possibly because he saw in it a degradation of the soul's God-like nature on which he laid so much emphasis. But Plato was most articulate on the great periods underlying all processes in the world. First of all he insisted that the circular motion of the heavens, the ultimate reason of the periodicity of the world, was the most perfect image in which the eternity of the divine essence could be made visible. Having tied time to eternity Plato went on to account for the various measures of time. They were given by the periodic returns of the sun and of the moon to the same point in the sky. Thus arose the day, the month, and the year as units of time, but what really interested Plato in this respect were not these commonplace details. He looked behind these obvious time units for a more recondite rationale, which resided, according to him, in the periods of the planets. He also complained that the admirable variety and combinations of planetary periods were not taken seriously enough by most of mankind: "Yet there is no difficulty in seeing that the perfect number of time fulfills the perfect year when all the eight revolutions, having their relative degrees of swiftness, are accomplished together and attain their completion at the same time, measured by the rotation of the same and equally moving. After this manner, and for these reasons, came into being such of the stars as in their heavenly progress received reversals of motion, to the end that the created heaven might be like as possible to the perfect and intelligible animal, by imitation of its eternal nature."[20]

This perfect number giving the length of the perfect year, or, as it was later called, the Great Year, had an importance in Plato's cosmological thought that cannot be stressed enough. He assigned to it a supreme role of causality, admitting that even his ideal state was no exception to the general law of periodic dissolutions and restorations. The inevitable decay of a well-governed state was due, according to him, to the fact that the revolutions of celestial orbs defined a period of fertility and barrenness, of healthy and defective breeding for each living species. "Not only for plants that grow from the earth but also for animals that live upon it there is a cycle of bearing and barrenness for soul and body as often as the revolutions of their orbs come full circles, in brief courses for the short-lived and oppositely for the opposite."[21] Thus, there would be a period for humans in which the begetting of their children would be ill-fated with the result that talents and abilities needed for the running of the state would drastically diminish. The immediate reason for this was that the guardians of the state could not help forgetting after a certain period of time the specific geometric ratio which determined the propitious season for begetting healthy and talented offspring. This geometric number, according to Plato, exerted supreme influence over the outcome of births, and derived from the same realm as did the "perfect number" which in turn determined "the period of divine begettings."

The shadow of a treadmill of inevitable ups and downs did not, however, represent the darkest aspect of a cyclic cosmology. The tone of approval which pervades the tale of the "Stranger" in the *Statesman*, suggests that Plato saw in that tale a logical exploitation of the idea of eternal cycles. After all, the tale resembled a reasoned discourse rather than a pleasant fable for which it was intended. The tale starts with the assertion that since it is a divine privilege to rotate always in the same direction, the motion of the material world had to be such as to shift periodically in reverse. This meant that the cosmic process alternated between two phases, one, in which divine and unfailing laws ruled bringing about a golden age, and another one, in which the divine guidance was withdrawn. But once the world was left on its own resources, it began to head towards a chaotic state. Its coming was signalled by the reverse of natural development, resulting in immaturity, feebleness, and in almost complete extinction.

The moment of reversal was marked according to the tale by cosmic convulsions and shudders reverberating across the globe. But what must have really made the listeners shudder was the ultimate rationale behind this ominous world picture. According to the tale, periodic reversals were inevitable because reason or God was only a co-ruler in the world. Opposite to God was the equally powerful factor represented by the "chaotic state." The universe could remember only for a while its "divine instructions" and proceed along progressive lines. Before long its material qualities let the evil principle unfold. The rest should be told in Plato's words:

This bodily element in its constitution was responsible for its failure. This bodily factor belonged to it in its most primeval condition, for before it came into its present order as a universe it was an utter chaos of disorder. It is from God's act when he set it in its order that it has received all the virtues it possesses, while it is from its primal chaotic condition that all the wrongs and evils arise in it – evils which it engenders in turn in the living creatures within it. When it is guided by the divine pilot, it produces much good and but little evil in the creatures it raises and sustains. When it must travel on without God, things go well enough in the years immediately after he abandons control, but as time goes on and forgetfulness of God arises in it, the ancient condition of chaos also begins to assert its sway. At last, as this cosmic era draws to its close, this disorder comes to a head. The few good things it produces it corrupts with so gross a taint of evil that it hovers on the very brink of destruction, both of itself and of the creatures in it.[22]

The structuring of the universe on the alternating rules of a rational and an irrational principle may not have been Plato's deepest cosmological preference. But his failure to turn his back most resolutely on this world picture should temper admiration for the "rationality" of the Platonic world view. At any rate, Plato saw everything, including human and social phenomena, as being under the ultimate influence of the cosmic law of cycles, the most fundamental of which had the length equivalent to the Great Year given by the perfect number. For him the manner in which the Great Cycle of the heavens produced the numerous smaller cycles on earth, including the periodicities of human society, was not physical in the obvious sense of the word. Underlying the physical, so Plato believed in a truly Pythagorean vein, was the realm of numbers and of geometrical proportions, on which he rested what is commonly called physical causality. This is why the Platonic scheme of the planets and fixed stars is a system of concentric, transparent shells that have no physical influence on one another. The disconnectedness of the heavenly spheres was retained in substance by Eudoxus (fl. 370 B.C.) who first supplied Plato's scheme with considerable geometrical sophistication to make it a more acceptable model of the actual motion of the planets. To account for their retrogressions, Eudoxus assigned to each planet several spherical shells whose total number amounted to twenty-seven in his system. While the shells of each individual planet were imagined to have some mechanical connection, the system of shells of one planet was in Eudoxus' system physically independent from the system of shells of each neighbouring planet.

Aristotle, whose main philosophical aim consisted in bringing the ideal world of Plato and the actual physical reality into a closely united system, did his best to endow the Eudoxian scheme of planetary motions with some physical characteristics. In the process Aristotle increased the number of shells to fifty-five and emphasized the physical connectedness between all of them.[23] This he obviously did to comply in full with his notion of physical causality, the ultimate source of which was the Prime Mover. To what extent Aristotle's Prime Mover was distinct from the uppermost heavens is a moot question. Nevertheless, it is clear that the daily rotation of the sphere of the fixed stars was in Aristotle's eyes a perfect image of the innermost activity of the Prime Mover. As all other motions were derivatives of the primary physical movement, the daily revolution of the fixed stars, it followed that throughout the cosmos this cyclic pattern of changes had to evidence itself.

Aristotle was eager to outline both the physical details and the philosophical justification of the transmission and transformation of immutable eternal cycles into variable terrestrial and atmospheric cycles. According to him this was possible because the periods and planes of the planets, sun, and moon differed from the rotation of the fixed stars.[24] It was in particular the sun's motion along the ecliptic that produced the most evident cycle on earth, the cycle of seasons. This the sun did by generating through friction a varying amount of heat depending on its position along the ecliptic.[25] The interplay of the process with a great number of physical factors, the countless variations of the surroundings on the earth, issued in a great number of cyclic changes of which the variations of the gestation period in animals received Aristotle's particular attention. He stated explicitly that these biological periods were governed by cosmic periods among which he singled out the periods of the moon and of the sun. Stressing the rather obvious role of the moon and the sun did not, however, distract Aristotle's attention from the ultimate cyclic process in the cosmos. Underlying the cycles of the moon and the sun, he noted with emphasis, there were other, more fundamental cycles.[26]

Another area of terrestrial phenomena where Aristotle found detailed evidence of the cyclic process was what today would be called geophysics. In the *Meteorologica* he analysed with great finesse the slow changes of the land and sea. He was, of course, aware of the fact that many of those transformations "escape our observation because the whole natural process of the earth's growth takes place by slow degrees and over periods of time which are vast compared to the length of our life, and whole peoples are destroyed and perish before they can record the process from beginning to end."[27] The lack of direct evidence did not, however, shake his confidence about the ultimate nature of those changes. "The process must be supposed to take place in an orderly cycle."[28] The lengths of biological cycles and the lengths of geological cycles formed, so to speak, the two extremes on the scale of cycles observable in nature. The former could be extremely short while the latter were extremely long. To intimate this great length Aristotle noted that "just as there is a winter among the yearly seasons, so at fixed intervals in some great period of time there is a great winter and excess of rains."[29]

It is uncertain whether in speaking about a "great winter" Aristotle had in mind the Great Year. Aristotle's reticence about the Great Year forms one of the features that distinguish his views on the cyclic nature of the happenings in the cosmos. Also Aristotle, unlike Plato, the Pythagoreans, and several pre-Socratics, exempted the superlunary world from the periodic cosmic cataclysms.[30] This was in line with his belief in the incorruptible nature of the heavens. Again, Aristotle differed from the Pythagoreans concerning the perennial rebirth of the same individuals. According to Aristotle, beings, such as man, animals, and all inert objects composed of corruptible matter, cannot return in a numerically identical manner. He expected them to reappear only as species. On the other hand, Aristotle considered the cyclic recurrence of celestial phenomena numerically identical in character.[31]

Aristotle's concern about such details illustrates only too well the extent to which he subscribed to the universal validity and fundamental importance of cyclic returns. He put himself on record in this respect in sweeping and explicit terms. The relevant and lengthy passage comes fittingly enough from his book *On Coming-To-Be and Passing-Away*. There he stated that order in the universe derived from the periodicity of processes, and that it was in that periodicity that the eternal character of the world came most tellingly to the fore. "God, therefore, following the course which still remained open, perfected the universe by making coming-to-be a perpetual process; for in this way 'being' would acquire the greatest possible coherence, because the continual coming-to-be of coming-to-be is the nearest approach to eternal being. The cause of this continuous process, as has been frequently remarked, is cyclical motion, the only motion which is continuous."[32] He noted with obvious satisfaction that even straight motion, which dominated the sublunary region, was also in some sense a cyclic motion, as fire, air, and water were in a continuous transformation due to the upward and downward replacement of these elements.[33] Aristotle, always intent to systematize, hastened to apply his favourite principle in this connection too: partial realizations of a phenomenon are only possible because there exists a supreme form of it. Thus, minor cyclic processes occur because they all are subject to the influence of the most fundamental cycle in nature, the daily revolution of the sphere of fixed stars.

It was from this fundamental cycle that Aristotle derived all necessary processes and formations in the world. He considered these processes perennial because the principal cycle was such. Eternity and absolute necessity were for Aristotle two aspects of one underlying principle: the cyclic process. In his own words: "Therefore, it is in cyclical movement and cyclical coming-to-be that absolute necessity is present, and if the process is cyclical, each member must necessarily come-to-be and have come-to-be, and, if this necessity exists, their coming-to-be is cyclical. This conclusion is only reasonable, since cyclical movement, that is, the movement of the heavens, has been shown on other grounds to be eternal, because its own movements and the movements which it causes come-to-be of necessity and will continue to do so; for if that which moves in a cycle is continually seeking something else in motion, the movement of those things which it moves must also be cyclical."[34]

To subject everything in the world with such inexorable logic to an ironclad wheel of necessity could be considered from the heights of abstract and detached reasoning a mental feat of exceptional sweep. When, however, considered from an existential immediacy, the feat implied some ominous consequences which in time manifested themselves in all their gravity. Fascinating as could be the unswerving loyalty to cycles it was subject to its inner logic that could not be ultimately escaped by any idea or attitude. Circular motion with its returning upon itself is diametrically opposed to a one-directional flow of things and events. It is in this latter form that the passing of time presents itself in human experience. Preoccupation with the idea of universal cyclic recurrences leads naturally to the weakening of a concept of time which gives to each human action a unique character and unequivocal meaning. More concretely, the meaning of historical succession and what is based on it, the concept of progress, would in such a framework lose their significance and, more specifically, their inspirational value. This is well illustrated by a section of *Problems*, a work of which Aristotle is not the author, but which was certainly composed in the Lyceum and which incorporates several chapters with an unmistakably Aristotelian ring. The particular problem in question is the meaning of the terms "before" and "after" and deserves to be quoted in full:

Should it be in the sense that the people of Troy are before us, and those previous to them before them and so on continuously? Or if it is true that the universe has a beginning, a middle and an end, and when a man grows old he reaches his limit and reverts again to the beginning, and those things which are nearer the beginning merit the term "before," what is there to prevent us from regarding ourselves as nearer the beginning? If that is true, then we should be "before." As therefore in the movement of the heavens and of each star there is a circle, what is there to prevent the birth and death of perishable things from being of this nature, so they are born and destroyed again? So they say that human affairs are a cycle. The suggestion that those who are continually being born are numerically identical is absurd, but we could accept that they are the same in "form." In that sense we should be "before," and we should assume the arrangement of the series to be of the type which returns to the starting-point and produces continuity and is always acting in the same way. For Alcmaeon tells that men die because they cannot connect the beginning with the end – a clever saying, if one supposes him to be speaking in metaphor, and not to wish his words to be taken literally. If then, there is a circle, and a circle has neither beginning nor end, men would not be "before" because they are nearer the beginning, nor should we be "before" them, nor they "before" us.[35]

The weakening of historical consciousness which exudes from the passage is logical consequence of a systematic insistence on the cyclic pattern of all happenings. What the disciples did was merely a variation on a theme which their master articulated several times while making short references to intellectual history. In four different works of Aristotle one comes across a statement that should give pause to the careful reader. He should reflect on the full bearing of words that on the surface may appear entertaining asides. The context in each case indicates that Aristotle was eager to draw the logical consequences of his belief in a cyclic patterning of all forms of existence and experience. In the third chapter of the First Book of the *Meteorologica*, where he traced out the atmospheric circulation of water, he referred approvingly to a dictum of Anaxagoras on the ether and added: "We maintain that the same opinions recur in rotation among men, not once or twice or occasionally, but infinitely often."[36] In *On the Heavens*, while referring to the fact that throughout all past time men repeatedly recognized the incorruptible character of the heavens, Aristotle hastened to note: "We cannot help believing that the same ideas recur to men not once or twice but over and over again."[37] In the *Metaphysics*, after describing the system of fifty-five spherical shells needed for the motion of planets, he noted again with satisfaction the ancient character of man's belief in the divinity of the upper heavens. In the same breath he also voiced his firm conviction that the age-long tradition about the divine or imperishable nature of the sky was merely the evidence of a perennial re-emergence of the same ideas through a cyclic process.

The passage also gives a glimpse of the inhibitory impact of the belief in cyclic recurrences. The treadmill of perennial returns not only generates pessimism by its spectre of the inevitable decay of man's achievements but it also invites the setting of a complacently low ceiling on attainable goals. For this is what is revealed in Aristotle's words: "probably every art and every philosophy has often reached a stage of development as far as it could and then again has perished."[38] Obviously, what was true in art and philosophy had also to be valid for the patterns of social and political life. In *The Politics* Aristotle noted in connection with the caste system and with the custom of using a common table by an entire village community: "It is true indeed that these and many other things have been invented several times over in the course of ages, or rather times without number; for necessity may be supposed to have taught men the inventions which were absolutely required,

I

and when these were provided, it was natural that other things which would adorn and enrich life should grow up by degrees. And we may infer that in political institutions the same rule holds.''[39]

The word "necessity" was very aptly used by Aristotle in the context, provided its full Aristotelian meaning is kept in mind. The necessity of which Aristotle spoke is more than the accidentally occurring constraint posed by the environment. What may seem accidental for the cursory analyst of events and circumstances was in Aristotle's eyes an inevitable and periodically recurring consequence of the great cycles that ruled the cosmos and everything and everybody in it. It was not long before Greek thought began to strain under the oppressive weight of such an outlook on existence and human endeavour.

As a matter of fact a principal if not the primordial preoccupation of the Stoics,[40] who from the third century on eclipsed the Peripatetics, consisted in an agonizing effort to escape the debilitating clutches of the prospect of eternal recurrences. Already the founder of the Stoics, Zeno, who came from the town of Citium in Cyprus, was imbued with the belief, perhaps under Oriental or more specifically Babylonian influences, that after each world destruction the same people would attend to the same needs. Anytus and Melitus (politicians of ancient times) would once more make requisitions, Bousiris (a legendary king of Egypt) would start again killing a foreigner each year, and Hercules would display anew his physical prowess by vanquishing Bousiris and his sons. This picturesque account of Zeno's reflections on the recurrence of all is due to the second-century Christian author, Tatian, who turned a gnostic.[41] His younger contemporary, Alexander of Aphrodisias, the noted commentator of Aristotle and certainly most familiar with the philosophical tradition, reported much the same about Chrysippus who fused Stoic teachings into a great system of thought. Alexander referred to a book of Chrysippus, now lost, entitled "On the Cosmos," in which the Stoic belief in the numerical recurrence of individuals was stated most explicitly: "Difference between former and actual existences of the same people will be only extrinsic and accidental; such differences do not produce another man as contrasted with his counterpart from a previous world-age."[42] Two centuries later the Christian bishop, Nemesius, provided a lucid account of the details of the doctrine of Chrysippus and of his teacher Cleanthes: "When all of the planets return with respect to both latitude and longitude exactly to the same point where they were located in the beginning when the World was formed for the first time, thay all will become the cause of the extinction and destruction of all beings. Then, as the planets retrace exactly the same route which they had already traversed, each being that had already been produced during the previous period will re-emerge once more in exactly the same manner. Socrates will exist again, and Plato as well, and also each man with his friends and fellow citizens; each of them will suffer the same trials, will manage the same affairs; each city, each village, each camp will be restored. This reconstitution of the Universe will occur not once, but in a great number of times; or rather the same things will reoccur indefinitely to no end. As to the gods, who are not subject to the [cosmic] destruction, it is enough for them to witness only one of these periods to know henceforth everything that should occur in subsequent periods; nothing shall indeed occur that may differ from what had already been produced; all things reappear in the same manner, with no differences whatever, not even the slightest ones."[43]

The agreement of the gnostic Tatian, of the pagan Peripatetic, Alexander, and of the Christian bishop, Nemesius, guarantees not only the reliability of their reports about the thinking of the Stoics. It also indicates the depth and vividness with which the Stoic belief in eternal recurrences remained alive through the Hellenistic centuries. An equally relevant point in the reflections of the Stoics in this connection is the desperate effort of some of them to surmount the plain absurdity of such an outlook, an absurdity easily felt by anyone who had not thrown to the winds all common sense. For as Origen noted,[44] most Stoics came so much under the sway of the fateful prospects of eternal recurrences that they no longer saw any reason to exempt from it even the highest heavens and the gods themselves. But once in the grips of inexorable returns the acumen of leading Stoics could only offer inept verbalizations in place of explanation. "In attempting to remedy the absurdities in some way the Stoics say that in every cycle all men will be in some unknown way indistinguishable from those of former cycles. To avoid supposing that Socrates will live again, they say that it will be some one indistinguishable from Socrates, who will marry some one indistinguishable from Xanthippe, and will be accused by men indistinguishable from Anytus and Meletus. But I do not know," Origen remarked, "how the world can always be the same, and one world not merely indistinguishable from another, while the things in it are not the same but are indistinguishable."[45]

Origen was a Christian apologist and a most influential one. The sundry absurdities of the doctrines of pagan philosophers were most welcome grist to his mill. Still, one would make a mistake in assuming that his apologetic purpose exaggerated the perplexities of the Stoics. Plutarch, the famed historian and for years the chief priest at Delphi, reported how they struggled to make room for that invincible human aspiration for *one* providence, *one* destiny, and *one* purpose. But as Plutarch aptly put it, the really preposterous idea was not that there should be an uncounted number of supreme gods bearing the name Zeus. What he found inadmissible was the cosmology of those who "make an infinite number of suns and moons and Apollos and Artemises and Poseidons in the infinite cycle of worlds."[46] Several centuries later, when antiquity flickered its last, Simplicius still could speak about this problem of the Stoics so as to indicate that its existential impact long outlasted the philosophical fortunes of the Stoa. Having embraced the doctrine of palingenesis, wrote Simplicius, the Stoics stated "that a man will be reborn who is the same as I am; they also ask themselves and rightly so, if I shall be numerically the same as I am now, or if I shall be the same by essential identity, or if I shall be different by the fact of my insertion into a Universe other than this one?"[47]

For a Stoic the inevitable emergence of a new universe, and the endless succession of such universes constituted a fundamental tenet. A telling indication of this is the care by which the leaders of the Stoa elaborated on the physical characteristics of the succession of worlds. A detailed consideration of the physics of the Stoics has, therefore, interest of its own, not only because the Stoics represented in classical antiquity the last important philosophical school to formulate a physics (and cosmology) in line with their basic philosophical postulates. Because of its close ties to the doctrine of eternal recurrences, an analysis of the physics of the Stoics[48] should afford insights about the intellectual feedback of the belief in eternal returns on the fortunes of scientific endeavour. In particular, it should be rewarding to search for some clues to the disappointing failure of the Stoics to develop valuable ideas about physical processes into real physics.

Some of their initial steps in this respect were truly promising. Trying to avoid the soulless mechanism of atomism and the splitting of the realm of being into two by Platonic and Peripatetic dualism, the Stoics were led toward a world picture in which process, not being, dominated. Theirs was a physics in which foremost attention was paid to processes that today come under the heading of thermodynamics. As the basic Stoic aspiration was vitalistic and monistic, they had to assume a concept of being in which everything fused into one all-encompassing living continuum. Thus, they pictured matter, inert in itself, as wholly permeated by a force of vital tension, the pneuma. The identification by the Stoics of the pneuma with a subtle form of fire was rather natural. Most physical processes are accompanied by some tension in the form of resistance, expansion, pressure, and so forth, all of which bring about temperature changes. Also, in man's own experience, and this was of crucial importance in the Stoic framework of thought which set so much emphasis on ethical effort, both noble aspirations and unwholesome passions cause tensions with a rise in the body's temperature. One of the last Stoics, Marcus Aurelius, noted explicitly that man was faithful to himself as long as he carefully mastered this dynamic tension within himself.[49] Finally, there was the ancient doctrine of Heraclitus that made the tension of opposites into a fundamental law of nature. Accordingly, the Stoics declared that of the four main forms of matter as defined by Empedocles, fire and air had binding qualities, while water and earth needed binding factors to remain together. In other words, while fire and air produced tension and compression, water and earth were factors of decompression and expansion.

From this it followed that all actual forms of all beings were determined by the special ratio of hot and cold, tensioning and de-tensioning, compressing and decompressing factors. With a remarkable consistency the Stoics tried to accommodate within this framework of explanation all processes in nature from the most minute to the most gigantic ones. As to the latter end of the gamut they pictured the world as a finite sphere that oscillated through cycles of expansion and contraction in the infinite void surrounding it. Cleomedes was most explicit on this point as he replied to critical Peripatetics for whom the idea of an infinite void was anathema. According to them matter was bound to pour out into the infinite void and to be scattered there with no chance of coalescing again. "Our reply is," reads Cleomedes' answer, "that this could never happen, because the cosmos has a *hexis* (tension) which holds it together and protects it, and the surrounding void cannot affect it. It maintains itself by the rule of an immense force, contracting and again expanding into the void according to its natural transmutation, alternately dissolving into fire and starting creation again."[50]

It seems to have been a secondary question to the Stoics whether this dissolution into fire (*ekpyrosis*) was a violent conflagration or a slow, almost imperceptible combustion. One Stoic text describes the conflagration of the universe with a reference to the gradual transformation of swamps into dry grounds, and vice versa, due to the alternation of dry and humid climate.[51] The all-important point is that the world goes through endless cycles, the period of which is determined by two extreme thermal points, namely, when the cosmic fire (pneuma) is at its maximum, and when the humid element (water) is reaching its dominating role. At the latter point the whole cosmos turns into sluggish matter. Between these two points the relative amount of pneuma and sluggish matter is in constant transformation.

By giving a distinctly thermodynamic account about cosmic processes the Stoics achieved a first, and they were also first in making bold guesses about wave propagation in the pneuma as the main factor in the production of audio and visual effects. They referred to the analogy of water waves produced by a stone and argued that the propagation of sound and light (waves) in the pneuma was not circular but spherical.[52] Such waves would be produced in the pneuma because it had, as had stretched strings, the quality of tension. A statement attributed to Chrysippus even shows that the Stoics groped for making their physical analogies more amenable to quantitative formulations. Chrysippus is reported to have explained vision as being due to a cone whose base is at the object and whose apex is at the eye. Within this cone there is air (pneuma) in a state of special stress which activates the eye as a stick would activate any part of the skin.[53] A favourite Stoic expression for this stress was *tonike kinesis* (tensional motion), which according to the Stoics moved in the various substances simultaneously inward and outward. The outward movement was believed to produce quantities and qualities while the opposite movement was credited with the unity of any particular substance.[54]

That this was somewhat paradoxical had become apparent in due course. The best the Stoics and their sympathizers could do was to point out further analogies derived from cases of dynamic balance. Thus, there was the case of a man trying to swim up-river but held in the same place by the contrary current, or the case of a bird hanging in midair. Even more appealing was the reference to the hanging of a grain in the centre of a bladder being blown up.[55] Here, clearly, the symmetrical "tensional motion" of the air streaming into the bladder kept the grain in equilibrium, and they used this as an explanation for the equilibrium position of the world in the centre of the infinite void. This was about as far as they could go. A quantitative analysis of the various phenomena, which they explicitly ascribed to tensional motion acting in both directions, was beyond their ken. It would have involved familiarity with the mathematics of standing waves and with the basic laws of dynamics.

Thus, the Stoics kept applying their basic tenets to natural phenomena in a highly interesting but basically qualitative manner. A case in point is Poseidonius who was obviously helped by the concept of the tensional motion of the pneuma to offer some mechanical explanation of the correlation of oceanic tides with the motion of the moon. If the ocean has a daily, monthly, and yearly period as the heavens have, it is, according to him, because of the ocean's sympathy (or kinetic tension) with the moon and possibly the sun.[56] But beyond this promising though very generic physics of the tides Poseidonius was unable to go. His attention was conditioned by his satisfaction to sight in earthly processes replicas of heavenly cycles. Of these the most fundamental was the period determined by the successive and simultaneous return of all planets to the same point of the zodiac. For the Stoics this was of major importance. Their definition of time was inseparably tied to the idea of cycles, and cycles always represented a whole. To break such a notion of time into atomic units and overcome thereby the paradox of the arrow that very much agitated the Stoics appeared to them a most unnatural procedure. It was contrary to that continuum in the wholeness which according to their basic preferences was the supreme feature of the perennial life of the cosmos endlessly going through its life-cycles. Time for the Stoics was a manifestation of that living wholeness as stated explicitly by the Stoic Apollodorus of Seleuca (fl. 120 B.C.): "Time is the interval of movement of the cosmos . . . and the whole time is passing just as we say that the year passes, on a *larger circuit*"[57] (Italics added).

This larger circuit, the Great Year, was for the Greek mind a circular barrier that deprived it of insights and aspirations without which science could not reach a self-sustaining stage. Even in the case of that proverbial braveness and calmness that characterized the attitude of the best Stoics in the face of grave existential perplexities, the idea of eternal recurrences acted as a mesmerizing prospect. Their desperate efforts to explain it away illustrates this only too well. On the level of scientific speculations the belief of the Stoics in eternal recurrences could not, therefore, fail to exert an inhibitory influence. The impasse into which Stoic thought was brought by a definition of time based on the "larger circuit," or Great Year, contrasts sharply with the fine progress of the Stoics toward the understanding of infinitesimals along the parameter of extension. The Stoics were particularly intent to reflect on the method of convergence, the scientific value of which was shown by Eudoxus in his proving that the volume of a cone was one-third of the volume of a cylinder of the same base and height. Eudoxus' proof consisted in approximating the volume of the cone by many flat cylinders that formed a series of steps. It was this procedure in geometry that became fully generalized in that famous statement ascribed by Plutarch to the Stoics: "There is no extreme body in nature, neither first nor last, with which the size of a body comes to an end. But every given body contains something beyond itself and the substratum is inserted infinitely and without end."[58] The full scientific usefulness of such a statement could not, of course, be exploited without the mathematical techniques of calculus. But similar speculations about the infinitesimals of time were also indispensable for the development of calculus and of what rests on it, exact physical science. This could only be formulated when time was no longer conceived as a mirror image of eternal recurrences, but rather as an uninterrupted one-directional flow of events.

If the predicament of the Stoics is already instructive about the inhibitory impact of eternal recurrences, the evidence provided by Epicurus, the leader of the Stoics' chief competitors for philosophical hegemony, is nothing short of dramatic. In the history of physical science Epicurus is best remembered as one who not only refocused attention on the merits of Democritus' atomism, a favourite target of the Peripatetics and the Stoics, but also added an important modification to it, the molecular theory. Much less frequently is, however, recalled the truly schizophrenic attitude of Epicurus toward science. While Democritus' atomism knew no bounds as regards the application of the principle, "atoms in eternal motion," Epicurus frantically warned against attempts to demonstrate the validity of strict, unequivocal physical laws in heavenly phenomena. In all three of his extant philosophical letters Epicurus returned to this point with noticeable animation. Needless to say, the chance collision of atoms which according to him underlies all phenomena, especially the successive production and dissolution of worlds, left Epicurus with ample room for "learned ignorance" about the particular manner of physical processes. He did his best to emphasize the principle of "several possible explanations" in all cases. He went to the extremes in noting that ascribing a strictly unique and specific form of physical causality to celestial phenomena "would cause the greatest disturbance in men's souls;"[59] that only by admitting more than one cause in explaining the motion of heaven can one "live free from trouble;"[60] "that to assign a single cause for these occurrences . . . is madness;"[61] that "it were better to follow the myths about the gods than to become a slave to the destiny of the natural philosophers: for the former suggests a hope of placating the gods by worship, whereas the latter involves a necessity which knows no placation."[62]

This last passage brings to the surface two extremely important aspects of the broader ramifications of Epicurus' grappling with the inexorability of eternal returns. The first is theological; the second is scientific. Obviously, the gods of Epicurus were not made of such strong fibre as to elicit man's trust whatever the vicissitudes. Epicurus presented a pathetically tragic figure while trying desperately to save at least the minimum of religious comfort his gods could provide. He kept insisting in all his three letters that the realm of the divine should not be considered the cause of the motion or rather cycles of the heavens, because this would only make the gods subject to the same ironclad cyclic fate. This in turn would destroy man's belief that somehow the fate of cycles might be circumvented by recourse to the gods. For this is what Epicurus emphasized when he pleaded: "do not let the divine nature be introduced at any point into these considerations, but let it be preserved free from burdensome duties and in entire blessedness."[63] His letter to Herodotus stated exactly the same: "the motions of the heavenly bodies and their turnings and eclipses and risings and settings, and kindred phenomena to these, must not be thought to be due to any being who controls and ordains or has ordained them and at the same time enjoys perfect bliss together with immortality."[64]

It is not only the impotency and disarming confusion of a certain religious or theological outlook which is revealed in passages like these. They should also convey with an almost brute force the inhibitory influence of the belief in eternal returns. It was under such influence that scientific enterprise could appear to Epicurus as a road leading to an enslavement by "the destiny of natural philosophers." He clearly indicated what he meant by this as he singled out the astrologers who more than any other group of scientists were given to what he called the maddening business of assigning one exclusive cause to celestial phenomena. He could just as well have referred to astronomers. Within the realm of Greek science there had never been more than a tenuous distinction between the practice of astronomy and astrology. At any rate, by Epicurus' time astrology was already on its way to becoming a dominating and intellectually most respectable preoccupation of the Hellenistic world with the inevitable result that a Ptolemy could, without fear of criticism, consider his treatise of astrological divination, the *Tetrabiblos*, a work of far greater importance than his compendium of mathematical astronomy, the *Almagest*.

Astrology tried its best, of course, to make the most of the strictly astronomical study of the motion of planets against the background of the fixed stars. This motion was cyclic and to attribute to it a single, strictly causal factor meant for Epicurus the acceptance of that inexorable cyclic fate which, as he rightly sensed it, robbed man of his peace of mind and even of a modicum of his sense of purpose. Herein lies the ultimate root of the dichotomy if not schizophrenia of Epicurus' interpretation of existence and understanding. On the one hand he claims science, as he speaks boldly of the endless formation and dissolution of universes. Yet, at the same time he turns against science, because he sees in it, if taken consistently, a potential advocate of the oppressive cyclic fate which should upset all normalcy in man's reflection on himself, on society, and on the universe.

One cannot be but sympathetic to Epicurus' agony, and sense at the same time the antiscientific force emanating from the ominous spectre of eternal recurrences. To save the possibility of a modest measure of human happiness in the face of an inexorable cosmic treadmill, Epicurus saw only one escape hatch. It consisted in jettisoning consistent, scientific thinking. Against the demands of science Epicurus made an almost pitiful recourse to a primitive, philistine common sense which prescribed among other things the rejection of even the most reasonable conclusions about the true size of celestial bodies: "The size of the sun (and of the moon) and the other stars," he stated, "is for us what it appears to be; and in reality it is either (slightly) greater than what we see or slightly less of the same size. And every objection on this point will easily be dissipated, if we pay attention to the clear vision, as I show in my books about nature."[65] In those books, now lost, Epicurus took pains to give alternate explanations of celestial phenomena. In doing so he aimed at achieving one of his major objectives, the discrediting of what he called "the slavish artifices of astronomers."[66] Little did he guess that what he actually advocated was a debacle of scientific thinking.

This is not to suggest that Epicurus had much success in shaking the confidence of Greek astronomers in the value of their geometrical devices which enabled them to "save the phenomena," that is, to calculate with remarkable exactitude the motions of the sun, the moon, and the planets. Greek astronomers revelled in finding a great number of cycles defined by these motions. Geminus of Rhodes gave in his *Isagoge* (Introduction to astronomy), written during the first century B.C., a detailed account of how these various cycles were arrived at.[67] First there were the cycles needed to co-ordinate the seasons with the lengths of solar and lunar years. In this class Geminus mentioned the octaeteris or eight-year cycle consisting of 2922 days, the 16-year and 160-year cycles, and Meton's cycle of 19 years. Meton excelled as an astronomer in Pericles' time, and in his cycle the phases of the moon occur on the same date in every 19 years. Among later refinements of Meton's cycle can be mentioned the 76-year cycle introduced by Calippus (fl. 340 B.C.) and the 304-year cycle worked out by Hipparchus (fl. 130 B.C.). The luni-solar cycle was further refined by Aristarchus who gave its length as 2434 years, a figure which Censorinus mistakenly interpreted as Aristarchus' value for the Great Year.[68]

About the length of the Great Year a variety of estimates were put forth during the whole duration of classical antiquity. Apparently, the widespread belief in the doctrine of the four ages of the world, as stated with poetic beauty by Hesiod and by many poets after him,[69] inspired renewed efforts to determine its exact length. Astronomers here could not be of much help as the revolutions of all planets were not known with sufficient accuracy to allow a convincing computation of their least common multiple. Taking rough values one could assign the period of 59 years to the Great Year, as did Oenopidus (fl. 520 B.C.), who was later followed by Philolaus (fl. 480 B.C.), while Democritus seems to have preferred the figure of 77 years. Obviously such short periods could not do justice to a concept of the Great Year which implied the conflagration of the universe. It was, therefore, inevitable that determinations of the Great Year relied not so much on exact astronomy as on fanciful analogies. This was the case with Heraclitus who, as was already mentioned, claimed that the Great Year consisted of 10,800 years. If it is true that Plato in his turn proposed 36,000 years as the length of the Great Year, he too was led by poetic analogies and not by astronomy.[70] At any rate, considerations about the precession of the equinoxes could have no part in Plato's estimates. Moreover, neither Hipparchus nor Ptolemy connected the Great Year with the precession of the equinoxes, the discovery and precise calculation of which constituted a chief glory of Hellenistic science.[71]

Ptolemy's great astrological treatise, the *Tetrabiblos*, refers explicitly to the great cycle determined by the return of all heavenly bodies to the same position.[72] Not that Ptolemy considered the cycle of the Great Year as particularly useful for prognostication. For this there were several reasons. First, for Ptolemy the mathematical apparatus of astronomy represented a mere formalism which could be changed if it did not yield fairly good quantitative results. The orbits, periods, epicycles, eccentrics, and deferents did not correspond, according to him, to real pieces of celestial mechanism. If they were used by astronomers it was only so because such devices helped effectively "to save the phenomena," or the calculation of the future position of planets. This "positivism" of Ptolemy entailed a moderate skepticism concerning the value of astronomical theory, and kept him from subscribing to the doctrine of identical returns which seemed to imply a perfect and definitive knowledge of the heavens. Another reason for Ptolemy's reluctance to support the doctrine of identical returns derived from the obvious impossibility to keep exact records of the details accompanying the actual coming together of all planets. Such an event reached so far back in time as to fall outside the experience of man. At the same time no judicious reader of the *Tetrabiblos* can fail to note that all events and human characteristics can only be cyclically repetitive in a system where the motion of planets ultimately determines everything.

By the early part of the second century when the *Tetrabiblos* was written, astrology, or the art of prognostication through astronomy, to recall Ptolemy's definition of it, had ruled supreme in the intellectual climate of the Hellenistic world and so did the belief in the inexorability of eternal returns. It is no wonder that the works of the poets and philosophers of Roman times are full of revealing references to the Great Year. As to the poets, the cyclic world picture is implied in their allusions to the idea of the Four Ages of the World. From the time of Hesiod this represented a nostalgic theme, a backward look to the idyllic simplicity of a Golden Age, and a rather despondent acceptance of the ills of the present as part of an irreversible fate. It is only on rare occasion that the poets looked forward as Virgil did in the famous fourth *Eclogue*.[73]

Among the philosophers it was also the pessimistic component of the concept of cyclic returns that prevailed. Cicero, for one, clearly betrayed an awareness of the despondency exuding from the doctrine of the Great Year, though as a true Stoic he tried to face the inevitable prospect with saddened courage.[74] In the closing sections of his *De republica*,[75] which contains the famous "dream of Scipio," Cicero took note of the Great Year to warn that no statesman should expect perennial fame as his reward. The world is subject to periodic catastrophes which reduce to ashes all cultures and obliterate all memory and fame. For Cicero the basic unit of time is the period separating the conflagrations and he calls it emphatically *the* year. He is also very clear on the role of stars in determining the length of the cosmic period: "Men commonly reckon a year solely by the return of the sun, which is just one star; but in truth when all the stars have returned to the same places from which they started out and have restored the same configurations over the great distances of the whole sky, then alone can the returning cycle truly be called a year; how many generations of men are contained in a great year I scarcely dare say."[76] Yet, he volunteers an estimate which is interesting for two reasons. First, according to Cicero the time which had elapsed from Romulus' death to Scipio's victory over Carthage did constitute less than one-twentieth of the Great Year, an estimate which puts the length of the Great Year as somewhat more than 20×573, or 11,460 years.[77] Such a figure may suggest either Cicero's unfamiliarity with the true period of the precession of the equinoxes, or one may perhaps assume that in speaking of the stars Cicero meant only the planets and borrowed some earlier opinion about the length of time needed for the return of all planets to the same position at the same time. However that may be, Cicero's belief in the eternal recurrence of all is strongly conveyed in his concluding allusion to the coincidence of the sun's eclipse and of Romulus' death. The allusion also refers to a future eclipse of the sun in the same section of the sky, at the same season, and with all the stars being again in the same position.

A hundred years later the same acceptance of the Great Year is in evidence in the writings of Pliny the Elder and Seneca, two Romans who paid special attention to scientific topics in the early phase of the Empire. Pliny's mention of the Great Year is short but highly expressive of the bondage in which the idea of a closed and cyclically repetitive universe held the best minds. In the opening pages of the Second Book of his *Natural History* Pliny rejected the "madness" of the plurality of the worlds and voiced the truth of the Great Year, which constituted in his eyes the fundamental form of the universal influence of planetary motion on all events on earth.[78] In his turn, Seneca was far more prolific on the matter. His interest in the topic is evidenced not only by his quoting, as was shown in the previous chapter, a statement of Berossos on the cyclic destructions and rejuvenations of the world. Seneca was also eager to emphasize that not only fire but all forces of nature, such as floods and earthquakes, would play a part in the cosmic catastrophes. Nevertheless he gave to water the principal role in bringing about the periodic dissolution of the world. He depicted in detail the gradual erosion of all soil by water, the splitting of rocks and mountains by the sudden eruption of enormous springs, and the rapid rise of flood waters everywhere, to such an extent as to obliterate with fearful rapidity the distinctness of seas. "There will be no Adriatic any longer, no strait in the Sicilian Sea, no Charybdis, no Scylla . . . All these names will be obliterated – Caspian and Red Sea, Ambracian and Cretan Gulfs, the Pontus and the Propontis. All distinctions will disappear. All will be mixed up which nature has now arranged in its several parts."[79]

Needless to say, all this beckoned the end of mankind, and Seneca gave a graphic description of its gloomy details. Against the final floods there will be no defence, "nor will walls and battlements afford protection to any. Temples will not save their worshippers, nor citadels their refugees. The waves will anticipate the fugitives, and sweep them down from their very stronghold. Some enemies will hasten from the west, others from the east. A single day will see the burial of all mankind. All that the long forebearance of fortune has produced, all that has been reared to eminence, all that is famous and all that is beautiful, great thrones, great nations – all will descend into the one abyss, will be overthrown in one hour."[80]

Following the lengthy account of the terse fate that "there will one day come an end to all human life and interests,"[81] Seneca referred in a few words to the emergence of a new world from the muddy remnants of the old. For a moment optimism seemed to gain the upper hand: "Ocean will be banished from our abodes into its own secret dwelling-place. The ancient order of things will be recalled. Every living creature will be created afresh. The earth will receive a new man ignorant of sin, born under happier stars."[82] Yet, Seneca almost immediately sank into something akin to despair. He could not help foreseeing that the newly emerging man would retain his innocence but for a short while: "Vice quickly creeps in; virtue is difficult to find; she requires ruler and guide. But vice can be acquired even without a tutor."[83]

The despondency which Seneca felt over the inevitable prospects of the dissolution of all human achievements and over the unavoidable dominance of base instincts is to be kept in mind for a proper evaluation of his often quoted words about scientific progress. They are contained in the last section of his work where he faces the baffling questions posed by the comets. Obviously, the ancient world picture, or rather the system of concentric shells composing the sky, could hardly account for the "irregular" properties of the comets and especially for their rare, haphazard appearances. True, Seneca held out the hope that one day man may learn about the recondite secrets of nature, among them answers concerning the comets. At the same time he cautioned against sanguine expectations and impatience, and recalled that some beautiful discoveries about eclipses and other celestial phenomena were only of recent origin. Thus, one could look forward to penetrating even farther into the mysteries of nature:

The day will come, when the progress of research through long ages will reveal to sight the mysteries of nature that are now concealed. A single lifetime, though it were wholly devoted to the study of the sky, does not suffice for the investigation of problems of such complexity. And then we never make a fair division of the few brief years of life as between study and vice. It must, therefore, require long successive ages to unfold all. The day will yet come when posterity will be amazed that we remained ignorant of things that will to them seem so plain.[84]

One should not, however, conclude from this that Seneca was painting a rosy picture about the ultimate prospects of scientific advance. He noted that for all the scientific search for reasons underlying the phenomena, there was a total confusion as to the nature of the reasoning mind: "We shall all admit that we have a mind, by whose behest we are urged forward and called back; but what that mind is which directs and rules us, no one can explain any more than he can tell where it resides. One will say that it is breath; another, a kind of harmony; another, a divine force and part of God; another, subtlest air; another, disembodied power. Some will even be found to call it blood, or heat. So far is the mind from being clear on all other subjects that it is still in search of itself."[85] With these words, Seneca did not wish, in all likelihood, to shake man's confidence in rational inquiry, yet the frankness of his words conveys a less than unshakable trust in the rationality of the universe as far as the inquiring mind is concerned. He also lays great emphasis on the deeply paradoxical nature of the universe. Nature is such, he claims, that she "is made up of contrarities," that, "the whole concord of the universe is a harmony of discords," that nature "does not turn out her work according to a single pattern," and that "he has little conception of nature's power who thinks that she may not do exceptionally what she does not do repeatedly."[86] But even more emphatically Seneca speaks in the same context about man's proclivity for unreasonable and unproductive conduct. He reminds those who are impatient with an only partial knowledge of the world that one should not entertain high expectations for the very reason that the extent of vice has not yet revealed itself in its full strength. Still, for all its infancy, vice (and one may recall the increasingly downward trend implied in the succession of the Four Ages) is already effectively blocking the progress of science:

[Vice] is still in its infancy, and yet on it we bestow all our efforts: our eyes and our hands are its slaves. Who attends the school of wisdom now? Who thinks it worth while to have more than a bowing acquaintance with her? Who has regard for philosophy or any liberal pursuit, except when a rainy day comes round to interrupt the games, and it may be wasted without loss? And so the many sects of philosophers are all dying out for lack of successors. The Academy, both old and new, has left no disciple. Who is there to hand down the precepts of Pyrrho? That famous school of Pythagoras, despised of the rabble, can find no master. The new sect of the Sextii, which contained the vigour of Rome, started with great enthusiasm, but on the very threshold of its career is also dead.[87]

No wonder that Seneca's concluding words on the subject of progress are markedly despondent. The growing lure of vice prevents, according to him, not only further advances on the basis of discoveries already made, but also their bare preservations.[88] Seneca is not at all sure whether this trend can ever be overcome. In any case, the final outcome would show in his estimate relatively little difference. The bottom of the well of knowledge would not be reached even with the concerted efforts and good will of all. Since it is patently utopistic to expect a dramatic rise of constructive attitudes, the prospects of scientific quest remain very poor indeed. Such is the gist of the final sentence of Seneca's work: "We search for her [scientific knowledge] on the surface, and with a slack hand."[89]

For all the intellectual humility, noble vision, and diligence of Seneca, the expression "slack hand" applies in a sense to his compendium of scientific knowledge. His *Quaestiones naturales* and also Pliny's *Natural History*, both composed around the middle part of the first century A.D., evidence the lack of perspective and confusion into which the Greek scientific heritage was gradually sinking. Both works are vast and often uncritical collections of facts, opinions, and legends, in addition to undigested excerpts from earlier writings on scientific subjects. They are a far cry from the much clearer air which Greek science exuded a century or two earlier. Their prose freely mixes fancy with valuable insights in much the same way as does the poetry of Lucretius in his *De rerum natura*[90] which also pays due homage to the ancient dogma of the Great Year.

In a good display of consistency, Lucretius was not reluctant to state that the atomic doctrine entailed the continual agglomeration and dissolution of worlds: "Whatever increases and nourishes other things from itself must be diminished, and remade when it receives things back."[91] The present world appeared to Lucretius to be in its youth, otherwise poets would have sung of battles much older than the Theban war and the siege of Troy. More importantly, the youth of the actual world was indicated to Lucretius by the very fact that not all arts and crafts had already reached their greatest perfection. Great improvements were still being made in ship-building, musical tones were invented "yesterday," the true scientific theory of nature and of the system of the world was only discovered "recently" and, added Lucretius with no small pride: "I am the first to describe it in our own mother tongue."[92] Yet, this outburst of optimism and evocation of progress is immediately dampened by the ominous spectre of an inevitable, disastrous end of all. "If you believe by any chance," Lucretius turned to his reader, and he indicated no doubt about the correctness of his assumption, "that all these things have been the same before, but that the generations of men have perished in scorching heat, or that their cities have been cast down by some great upheaval of the world, or that after incessant rains, rivers have issued out to sweep over the earth and overwhelm their towns, so much the more you must own yourself worsted, and agree that destruction will come to earth and sky."[93] Clearly, as Lucretius perceived it, the Great Year was a most powerful reminder of the ultimate futility of all efforts.

A most revealing insight into the deeper aspects of that sense of futility is provided in the famous Meditations of Marcus Aurelius, the great philosopher-emperor of late Roman times. His memorable advocacy of Stoic resignation rested on the consideration that only the fleeting moment of the actual present counted. "No man can part with either the past or the future,"[94] he emphasized, a point which was not without merit but which actually reflected on his part a deep-seated despondency about the meaning of time. If only the actually lived moment mattered, then it made no difference whether one's life-span was short, long, or stretching over, as he put it, three thousand or thirty thousand years. While the actual moment carried with itself the immediate experience of conscious participation, lengthier spans of time were in his eyes of little if any existential value. The reason for this was that when the ever recurring character of time was considered against a typical human life-span, the disconcerting impact of endless repetitions came forcefully to the fore: 'These two things, then, must needs be remembered: the one, that all things from time everlasting have been cast in the same mould and repeated cycle after cycle, and so it makes no difference whether a man see the same things recur through a hundred years or two hundred, or through eternity: the other, that the longest liver and he whose time to die comes soonest part with no more the one than the other."[95] And he warned in the same breath indicating the total futility of trying to contravene the mesmerizing prospects of a cosmic treadmill: "For to gruntle at anything that happens is a rebellion against Nature, in some part of which are bound up the natures of all other things."[96]

If the cosmos of Marcus Aurelius was a treadmill it was ultimately so because it had no absolute origin and absolute end. He denied an absolute beginning to the formal part of man or his soul because it was a minute part of the universe in which everything changed into everything throughout eternity. Such a view, he added, lost nothing from its validity because the universe had been "arranged according to completed cycles."[97] Actually, his references to the cyclic framework of the universe were meant to make the point stronger. The highest aim of human efforts consisted, according to him, in one's conscious acceptance of being a pebble in the gigantic, cyclic torrent of the universe. That torrent rushed through all humans, now producing them, now burning them to ashes and in an uncounted number of times: "How many a Chrysippus, how many a Socrates, how many an Epictetus hath Time already devoured."[98] To keep this in mind was the acme of understanding: "Whatsoever man thou hast to do with and whatsoever thing, let the same thought struck thee."[99] This had to be so if it were true what Marcus Aurelius laid down as a fundamental tenet: "The same upwards, downwards, from cycle to cycle are the revolutions of the Universe."[100] Whether the succession of cycles was brought about by conflagration or by "everlasting permutations"[101] could only be a question of secondary importance. Although Marcus Aurelius urged against being overcome by the disarming spectre of a universe of haphazards, the vision of a universe burning into ashes again and again was equally discouraging.

The sombre mental atmosphere cast by the belief in the Great Year persisted to the very end of Roman times. One of the last non-Christian Roman writers on scientific matters, Macrobius (fl. 410), provides a lengthy illustration of the implications of the idea of the Great Year on man's cultural endeavour. The detailed account by Macrobius of the physical processes leading to the periodic global cataclysms is already indicative of the wider importance he attached to the whole question. Due to the influence of the ethereal fire filling the superlunary regions, so Macrobius believed, first there develops an excess heat in the upper atmosphere and this ultimately results in the scorching of much of the earth. General conflagration produces in turn an excess of moisture which brings about a universal flood leaving only small portions of the dry land intact. It is these small portions which serve as "seedbeds for replenishing the human race, and so it happens that on a world that is not young there are young populations having no culture, whose traditions were swept away in a debacle."[102]

Macrobius then goes on to describe the perennially recurring cultural pattern in a markedly dispiriting tone: "They [the small remnant of mankind] wander over the earth and gradually put aside the roughness of a nomadic existence and by natural inclination submit to communities and associations; their mode of living is at first simple, knowing no guile and strange to cunning, called in its early stage the Golden Age. The more these populations progress in civilization and employment of the arts, the more easily does the spirit of rivalry creep in, at first commendable but imperceptibly changing to envy: this, then, is responsible for all the tribulations that the race suffers in subsequent ages. So much for the vicissitudes that civilizations experience, of perishing and arising again, as the world goes on unchanged."[103]

Macrobius seemed to suggest that the conflagrations and floods may occur several times during the span of one Great Year which he specified as 15,000 years. At any rate, he did not compute the Great Year from one conflagration to the next. "The world-year begins," he stated, "when anyone chooses to have it begin, as Cicero did when he marked the beginning of his world-year by the sun's eclipse at the hour of Romulus' death."[104] But he was very emphatic on the point that the time corresponding to the period of the Great Year is completed when "the sun is again eclipsed in the same position, and finds all the stars and constellations of the sky in the same positions that they held when it was eclipsed at Romulus' death."[105] Macrobius did not specify his sources about the 15,000 years as the length of the Great Year. His main concern was to justify Cicero's estimate already discussed in this chapter. That Macrobius expressed his own thought in the form of a commentary on Cicero's "Dream of Scipio," shows the popularity which Cicero's sober moralizing enjoyed throughout Roman antiquity. The same fact also casts some light on the discouraging status of scientific knowledge in late Roman times. Thus, Macrobius used Cicero's short reference to the smallness of the earth in the universe as an occasion to give a detailed summary of man's scientific knowledge about the universe. While his account is precious because it served for the early Middle Ages as a chief information about antique science, it also shows the gradual deterioration of the scientific tradition as the closing centuries of classical antiquity ran their course.[106]

There remained some bright spots in the Eastern part of the Empire. In the latter half of the fifth century the Academy in Athens still could boast of teachers like Proclus, who worked heroically to keep alive and summarize the Greek philosophical and scientific heritage.[107] The Prologue of his commentary on Euclid is an incisive summary of the main philosophical problems that arose within ancient Greek mathematics. The methodological and philosophical aspects of ancient Greek astronomy received a classic summary in his *Hypotyposis* in which he stressed the use of excentric orbits in explaining planetary motions. He also collected in a

logical sequence all the propositions of Aristotle about the world that lent themselves to geometrical demonstrations.

But Proclus was also an enthusiastic student of Ptolemy's *Tetrabiblos*. He wrote
both a paraphrase of and a commentary on it. His addiction to astrology was so
great that he composed an original interpretation of the influence of eclipses on
human affairs in terms of their occurrence along the zodiac. Not surprisingly,
Proclus was an ardent advocate of eternal recurrences. He found it perfectly appropriate to refer to the periodic recurrence of every great thought and discovery in
his famous Prologue to his commentary on Euclid. In speaking about the origins
of geometry, Proclus took pains to emphasize that what he said on the topic was
true only of the "present age of the world." The divine Aristotle, Proclus recalled,
had already observed "that the same opinions often subsist among men, according
to certain orderly revolutions of the world: and that sciences did not receive their
first constitution in our times, nor in those periods which are known to us from
historical tradition, but have appeared and vanished again in other revolutions of
the universe; nor is it possible to say how often this has happened in past ages, and
will again take place in the future circulations of time."[108] Consequently, he felt
it his scholarly duty to note that crediting the Egyptians with the invention of
geometry could only be done insofar as the "present age or revolution" of the
world was considered.

Proclus displayed thorough consistency when in a book now largely lost he
defended the eternity of the world as advocated by Plato in the *Timaeus*. Clearly,
a world in which everything, even human history and culture, recurred cyclically
had to be eternal. It was also a world in which no startlingly new discoveries could
be expected. The whole tone of Proclus' scientific, literary, and philosophical work
illustrates this. He was a diligent and acute compiler who did his work in the belief
that all important advances had already been made. When he criticized now and
then the accepted world view, this was not done in the manner of a drastic break
with tradition. His insistence that the fifth substance, or the ether, did not differ
from ordinary fire was not construed with the aim to assert the uniformity of the
whole cosmos. He emphasized with equal vigour that light coming from the sun
differed in essence from light emitted from ordinary fire. No basically new scientific insight could indeed emerge in an interpretation of the world that had already
been moulded for centuries on the pattern of eternal recurrences.

As one would expect it, there were only a few and somewhat inept attempts
in late pagan antiquity to expose the futility of the belief in the Great Year. One
of them was Plutarch's account of the Great Year in his *Obsolescence of Oracles*,[109]
a motley collection of topics grouped around the contention that, as the number of
people had diminished during the previous hundred years or so, there was no need
of as many prophecies as before. In such a context a sarcastic criticism of the Great
Year, as offered by Plutarch, could not carry much weight. Plutarch most likely
only served evidence that whenever he could slight the Stoics he eagerly seized
the opportunity. Needless to say, the general belief in the Great Year was too deep
and too strong to yield to occasional literary lampooning. To overcome its fateful
and ubiquitous grip on intellectual pursuit more enduring considerations and convictions were needed than the ones which Hellenistic antiquity could provide. As a

result many a promising scientific start made during that period came sooner or later to an abortive stop.

Those who insist that in classical antiquity there existed a genuine belief in cultural progress[110] are constructing arguments that actually aggravate the puzzle of the factual absence of a self-sustaining progress in the field of physical sciences. Moreover, the claim about the existence of such a belief is rather difficult to reconcile with other attitudes that were widely shared in classical antiquity. Among these was a certain smug satisfaction with the material commodities already secured. More importantly, evidences of this come from contexts in which the theme of intellectual progress is analyzed. A classic case is the opening of Aristotle's *Metaphysics* which deals with the respective merits of manual labour, expert craftsmanship, and speculative thought. What is surprising is not that Aristotle extolled the latter over the other two, but rather his insistence that philosophy emerged only when most of the conceivable material benefits had already become available. For Aristotle the convincing evidence of this was provided by what happened in Egypt and Greece. As to Egypt, he stated that after all useful crafts had been developed nothing stood in the way of cultivating sciences. As a result, "sciences concerned neither with giving pleasure to others nor with the necessities of life were discovered, and first in such places where men had leisure. Accordingly, it was in Egypt that the mathematical arts were first formed, for there the priestly class was allowed leisure."[111] As to Greece, Aristotle emphasized the same. It is evident, he stated, that the pre-Socratics "pursued science in order to understand and not in order to use it for something else. This is confirmed by what happened; for it was when almost all the necessities of life were supplied, both for comfort and activity, that such thinking began to be sought."[112]

These opinions of Aristotle expressed the general consensus of classical times. As one would expect, those who carried on the work of the Stagirite professed similar convictions. Theophrastus, Aristotle's successor as director of the Lyceum, noted with satisfaction the material abundance of his own time as contrasted with the times of the Trojan war: "Life in earlier times was lacking in equipment . . . and the arts [crafts] had not been brought to perfection; the life of our own day on the other hand is equipped with everything conducive to ease, enjoyment, and amusements in general."[113] The same view was voiced by opponents of the Aristotelian way of thinking. Democritus, the leader of atomists and the chief protagonist of the void, saw no justification in trying to improve existing material conditions. His precept was that "One must keep one's mind on what is attainable, and be content with what one has, paying little heed to things envied and admired, and not dwelling on them in one's mind."[114] This passive attitude toward the external world was also voiced in his dictum that one should not take risks but rather accept the ordinary course of events: "Chance is generous but unreliable. Nature, however, is self-sufficient. Therefore it is victorious, by means of its smaller but reliable (power) over the greater promise of hope."[115] In line with this he did not consider life an opportunity to change man's material conditions: "One should realize that human life is weak and brief and mixed with many cares and difficulties, in order that one may care only for moderate possessions, and that hardship may be measured by the standard of one's needs."[116]

K

What Democritus, the natural philosopher, stated was echoed by Xenophon, the historian. Moreover, Xenophon's deprecating of the value of manual crafts, or "illiberal arts," was put forward as a genuine expression of the Socratic tradition. In his *Oeconomicus*, a discussion on estate management, Xenophon put the following words in Socrates' mouth: "The illiberal arts, as they are called, are spoken against, and are, naturally enough, held in utter disdain in our states. For they spoil the bodies of the workmen and the foremen, forcing them to sit still and live indoors, and in some cases to spend the day at the fire. The softening of the body involves a serious weakening of the mind. Moreover, these so-called illiberal arts leave no spare time for attention to one's friends and city, so that those who follow them are reputed bad at dealing with friends and bad defenders of their country. In fact, in some of the states, and especially in those reputed warlike, it is not even lawful for any of the citizens to work at illiberal arts."[117] Centuries later the same conviction animated Archimedes, the greatest engineer in antiquity, who refused to write a handbook on engineering. As Plutarch reported in his biography of Marcellus, Archimedes looked upon "the work of an engineer and every art that ministers to the needs of life as ignoble and vulgar."[118]

The fact that the best results of ancient engineering were used mostly for purposes of warfare or as devices of deception and magic in temples also provides a vivid illustration of a passive attitude toward nature. One could speculate about nature in order to understand it, but one was not supposed to supplement his speculations about nature by submitting them to tests consisting in changes imposed systematically on nature. This entailed the barring of repeated experiments together with the creation of isolated, ideal situations and processes within the realm of matter. While nature was thought to be repetitive, that is cyclic, artificially produced recurrences of events (or systematic experimentation) were considered contrary to nature and the dictates of reason.

To account for the complacency with regard to existing material commodities, to explain the timorous, diffident approach towards nature, one should look for factors in addition to the customary list of socioeconomical ones.[119] Those, for whom thought is not merely the inevitable outcome of material and social conditions, let alone conditionings, will readily recognize the inhibitory influence of the belief in eternal recurrences on the prospects of the Greek scientific enterprise. In that belief man was merely a bubble on the inexorable sea of events whose ebb and flow followed one another with fateful regularity. If such a belief inspired anything it was not confidence. It hardly encouraged conviction in the rationality of nature, nor did it enhance man's readiness to dominate nature. It did not generate intellectual curiosity or appreciation of experiments aimed at controlling nature. In particular, the belief in eternal cycles imposed on thinking a concept of time which could only be cyclic, therefore fundamentally repetitious and ultimately meaningless.

Aristotle's seemingly innocuous dictum that "time itself is regarded as a circle"[120] had indeed been pregnant with momentous implications for the ultimate fortunes of the Greek Logos. For all its brilliance, for all its spectacular initiatives, it remained trapped within a spacious labyrinth where every move and enterprise led in the final analysis back to the same starting point. The possibilities within such a framework had become exhausted within a relatively few generations, and in the end nothing remained except to write commentaries on the great classics of the truly creative phase of the Greek intellectual endeavour. Originality has never been the main hallmark of commentaries, and this turned out to be the case as the closing phases of Greek intellectual history had run their course. Opinions all too often repeated had readily turned into hallowed tenets.

Among the basic points of the Aristotelian codification of the Greek world view that were steadily reasserted by the commentators was the eternity of the cosmos. Its opposite, the creation or absolute beginning of the cosmos, was rejected with a persistent resolve throughout Greek antiquity. The only logical alternative to that denial was the acceptance of perennial returns and the Greek mind made the choice with a consistency worthy of its logical powers. But the choice could not escape its own logic either. Once the possibility of an absolute beginning was abandoned, there remained little encouragement to think in terms of real starting points. There was not left enough trust either to explore the potentialities of a one-directional flow of time and events. But without these and related conceptual categories, no meaningful advance could be made toward the intellectual stage embodied in classical physics.

A striking evidence of this is provided by the main philosophical concerns, perplexities, indignations, and stubborn traditionalism of Simplicius, the last great figure of ancient Greek philosophical tradition. His commentaries on Aristotle's cosmology, physics, and metaphysics reveal in stark directness the predicament of the Greek Logos, its lonely wanderings, and its strange shunning of a new light which unexpectedly came to diffuse over the confines of its mighty labyrinth. Simplicius had only scorn for well argued suggestions that terrestrial and heavenly matter were the same, that the motion of bodies did not require a continuous contact between the moved and the mover, that processes were not ultimately circular but one-directional, that the motion of stars did not really rule man's mind and his relation to nature.

On the face of it such claims were purely philosophical or scientific. But Simplicius knew only too well about their ultimate, non-philosophical wellspring, the very first article of faith in the Christian creed, the belief in a personal, rational Creator. With this tenet Simplicius had no patience, and precisely because of this he was unable to sense its portentous bearing on natural philosophy. Absolute beginning could hardly appear reasonable to one who insisted that the Pythagorean concept of time resting on the idea of eternal returns represented the most satisfactory synthesis and analysis of time.[121] He took an even more glaringly wrong tack as he held up to ridicule the mind that "could conceive of such a strange God who first does not act at all, then in a moment becomes the creator of the elements alone, and then again, ceases from acting and hands over to nature the generation of the elements one out of another and the generation of all the rest out of the elements."[122]

The passage should speak for itself to anyone sufficiently familiar with the main conceptual foundations of modern science. In the passage, Simplicius denounced in advance and unwittingly the very considerations that helped science to a viable birth a thousand or so years later. The passage is also an implicit reference to the notion of a universe created *out of nothing* and *in time*. Such a universe could no longer loom as a fateful omen and threaten the puny man with its inexorable cycles. In the passage also there was included the notion of a general impetus given to matter, and the notion of laws imbedded in created matter to govern its further differentiation. What this implied was the all-important concept of an autonomy which could be accorded to nature only by a rational Creator, remaining forever consistent with His creative plans and decisions. This autonomy in turn meant that the investigation of nature's laws was no longer a risky attempt to pry into God's secrets, but a laborious enterprise to which the principal handiwork of God, man, was entitled and equipped.

The thinker held up for ridicule by Simplicius was John Philoponus, a Christian, who was fully aware of the theological inspiration of his insights which, if cultural conditions had been favourable, might have meant a new starting point for physics.[123] His startling propositions and the great opportunity missed by science in the middle of the sixth century will be discussed in a later chapter. It is, however, even more important to note that John Philoponus' scientific ideas were the product of the long maturation of a unique world view, whose origin, endurance, and mighty growth kept defying the laws of probability. To understand the ultimate emergence of science in the modern sense, a close and detailed look at that most improbable development forms an indispensable condition.

NOTES

1. Here only a few titles can be mentioned of the enormously large literature dealing with ancient Greek science. In addition to the sections on Greek, Hellenistic, and Roman science in R. Taton (ed.), *History of Science*, Vol. I, *Ancient and Medieval Science*, translated by A. J. Pomerans (New York: Basic Books, 1963), special mention should be made of the works of S. Sambursky, *The Physical World of the Greeks*, (London: Routledge and Kegan Paul, 1956) and *The Physical World of Late Antiquity* (New York: Basic Books, 1962). Although Sambursky often draws far-fetched parallelisms between ancient Greek and twentieth-century physics, his works are valuable because of the careful documentation of the philosophical motivations underlying ancient Greek and Hellenistic physics and astronomy. The problem of the failure of Greek science to become a self-sustained enterprise is usually referred to in a more or less cursory manner in any major work on the subject. The only modern analysis of ancient Greek science in which careful attention is paid to the influence of the belief in the Great Year on the fortunes of Greek scientific thought is Pierre Duhem's *Le système du monde: histoire des doctrines cosmologiques de Platon à Copernic*, a ten-volume work, of which the first five volumes were published between 1913–15 and the remaining five, posthumously, between 1954–59 by Hermann, Paris. Strangely enough, only four brief references to Duhem are given in the monograph on the subject by Charles Mugler, *Deux thèmes de la cosmologie grecque: devenir cyclique et pluralité des mondes* (Paris: Librairie C. Klincksieck, 1953).
2. "Declare the past, diagnose the present, foretell the future," such is the soundly scientific motto given in the "Second Constitution" of the *Epidemics* (Book I, sec. 11). See *Hippocrates* with an English translation by W. H. S. Jones (London: Heinemann, 1923), vol. I, p. 165.
3. Quoted in F. Darwin, *The Life and Letters of Charles Darwin* (London: Murray, 1888), vol. III, p. 252.

4. *Timaeus*, 30–31; see *The Collected Dialogues of Plato*, edited by E. Hamilton and H. Cairns (New York: Pantheon Books, 1963), p. 1163.

5. This new type of physics is aptly named "organismic" physics. Its origins, claims, fallacies, and tenacious hold on human thought are discussed in detail in chap. i of my *The Relevance of Physics* (Chicago: The University of Chicago Press, 1966).

6. See sections 46–47 of the *Phaedo*.

7. *Die Fragmente der Vorsokratiker griechisch und deutsch*, by H. Diels, edited by W. Kranz (Dublin/Zurich: Weidmann, 1968), 12A–10. This work will hereafter be referred to as *Diels*.

8. *Diels*, 12A–11.

9. *Diels*, 13A–11.

10. *Diels*, 44A–18.

11. *Diels*, 24A–12.

12. As related by Simplicius, in his *In Aristotelis categorias commentarium*, edited by K. Kalbfleisch (Berlin: G. Reimer, 1907), p. 350; also quoted in Duhem, *Le système du monde*, vol. I, pp. 80–81.

13. *Ibid.*, pp. 350–51; in Duhem, p. 82.

14. *Diels*, 41–47.

15. See on this, *Heraclitus, The Cosmic Fragments*, edited with an introduction and commentary by G. S. Kirk (Cambridge: The University Press, 1954), pp. 300–01. On Heraclitus' dicta on cyclic returns and the Great Year, see especially pp. 294–302. The figure 10,800 was wrongly remembered as 18,000 in late antiquity. A case in point is the report of Joannes Stobaeus (fl. A.D. 500) in his *Eclogarum physicarum et ethicarum libri duo*, edited by A. Meineke (Leipzig: B. G. Teubner, 1860), p. 66 (Lib. I, cap. 8).

16. See Kirk, *Heraclitus*, pp. 335–38.

17. *Diels*, 31B–17. The passage is quoted in the translation of K. Freeman, *Ancilla to the Pre-Socratic Philosophers* (Cambridge, Mass.: Harvard University Press, 1966), p. 56.

18. As reported in Aristotle's *Physics*, 250b–251a.

19. See on this point the discussion by Simplicius in his *In Aristotelis de Caelo commentaria*, ed. I. L. Heiberg (Berlin: G. Reimer, 1894), p. 294.

20. *Timaeus*, 39d; see also 31–34. Quoted from *The Collected Dialogues of Plato*, p. 1168. At this point mention should be made of the detailed discussion of cosmic cycles in Plato's and Aristotle's thought in *Primitivism and Related Ideas in Antiquity*, by A. O. Lovejoy and G. Boas, with supplementary essays by W. F. Albright and P. E. Dumont (New York: Octagon Books, 1965). The problem is, however, broached by them as an example of ancient longing for a Golden Age.

21. *Republic*, 546; quoted from *The Collected Dialogues of Plato*, p. 775.

22. *The Statesman*, 273b–d; *ibid.*, pp. 1038–39.

23. See *Metaphysics*, 1073b–1074a.

24. *On the Heavens*, 286a.

25. *Meteorologica*, 341a.

26. *Generation of Animals*, 778a.

27. *Meteorologica*, 351b. Quotation is from *Aristotle, Meteorologica*, with an English translation by H. D. P. Lee (Loeb Classical Library; Cambridge, Mass.: Harvard University Press, 1952), p. 109.

28. *Ibid.*, 351a; see *transl. cit.*, p. 107.

29. *Ibid.*, 352a; see *transl. cit.*, p. 115.

30. See on this, *Meteorologica*, 352a; *On the Heavens*, 280b.

31. In his work, *On Coming-to-be and Passing-away*, 338b; see E. S. Forster's translation *Aristotle, On Sophistical Refutations; On Coming-to-be and Passing-away* (Cambridge, Mass.: Harvard University Press, 1955), p. 329.

32. *Ibid.*, 336b–337a; quoted in Forster's translation, pp. 317–18.

33. *Ibid.*, 338a.

34. *Ibid.*, 338a–b; quoted in Forster's translation, p. 327.

35. *Aristotle, Problems I*, Books I–XXI, with an English translation by W. S. Hett (rev. ed.; London: W. Heinemann, 1953), p. 367.

36. *Meteorologica*, 339b.

37. *On the Heavens*, 270b.

38. *Aristotle's Metaphysics*, translated with commentaries and glossary by H. G. Apostle (Bloomington: Indiana University Press, 1966), p. 209 (1074b).

39. *The Politics of Aristotle*, translated by B. Jowett (Oxford: Clarendon Press, 1885), vol. I, pp. 223–24 (1329b).

40. On the views of the various early representatives of Stoic thought, the primary source is the collection of Stoic fragments edited by J. von Arnim, *Stoicorum veterum fragmenta* (Leipzig: B. G. Teubner, 1903–24).

41. In his *Adversus Graecos*; see *Stoicorum veterum fragmenta*, vol. I, p. 32.

42. *Stoicorum veterum fragmenta*, vol. II, pp. 189–90.

43. *Ibid.*, vol. II, p. 190.

44. *Contra Celsum*, Book IV, § 68; in *Origen: Contra Celsum*, translated with an introduction and notes by H. Chadwick (Cambridge: The University Press, 1965), p. 238.

45. *Ibid.*; the embarrassment of the Stoics in this connection is noted again by Origen in the same work, Book V, § 20.

46. In his *The Obsolescence of Oracles*, chap. 29; quoted from *Plutarch's Moralia* with an English translation by F. C. Babbitt (London: W. Heinemann, 1936), vol. V, p. 435.

47. *Simplicii in Aristotelis Physicorum libros quattuor posteriores commentaria*, ed. by H. Diels (Berlin: G. Reimer, 1895), p. 886 (Lib. V, cap. iv).

48. As impressively given in S. Sambursky's *Physics of the Stoics* (New York: The Macmillan Co., 1959). In the work no adequate attention is, however, given to the preoccupation of the Stoics with cyclic recurrences.

49. See *The Communings with Himself of Marcus Aurelius Emperor of Rome together with His Speeches and Sayings*, revised text and a translation into English by C. R. Haines (London: W. Heinemann, 1916), p. 219.

50. Quoted in Sambursky, *Physics of the Stoics*, p. 113.

51. Quoted in Sambursky, *The Physical World of the Greeks*, p. 200.

52. See text *ibid.*, p. 138.

53. *Ibid.*

54. *Ibid.*, p. 139.

55. See *Physics of the Stoics*, p. 109.

56. See *The Physical World of the Greeks*, p. 142.

57. Quoted in *Physics of the Stoics*, p. 106.

58. Quoted from *The Physical World of the Greeks*, p. 155. The passage is from the "Common Conceptions against the Stoics," one of Plutarch's two works against Stoic philosophy. The other is "Contradictions of the Stoics."

59. "Letter to Herodotus" in *Epicurus: The Extant Remains*, with short critical apparatus, translation and notes by Cyril Bailey (Oxford: Clarendon Press, 1926), p. 49.

60. "Letter to Pythocles," *ibid.*, p. 59.

61. *Ibid.*, p. 79.

62. "Letter to Menoeceus," *ibid.*, p. 91.

63. "Letter to Pythocles," *ibid.*, p. 65.

64. "Letter to Herodotus," *ibid.*, p. 49.

65. "Letter to Pythocles," *ibid.*, p. 61.

66. *Ibid.*, p. 63.

67. *Gemini Elementa astronomiae*, Greek text with German translation and notes by C. Manitius (Leipzig: B. G. Teubner, 1898), chap. viii, "On the Months," pp. 100–23.

68. See on this Sir Thomas Heath, *Aristarchus of Samos the Ancient Copernicus* (Oxford: Clarendon Press, 1913), pp. 314–16.

69. See on this chapter ii, in A. O. Lovejoy and G. Boas, *Primitivism and Related Ideas in Antiquity*.

70. The question is discussed in Heath, *Aristarchus of Samos*, pp. 171–73.

71. This holds true, of course, only of Ptolemy's *Almagest* where the precession of the equinoxes is discussed in Book VII, chap. 3, in a truly operational spirit which characterizes the whole book.

72. In Book I, chap. 2. See pp. 15–17 in *Ptolemy, Tetrabiblos*, edited and translated into English by F. E. Robbins (Cambridge, Mass.: Harvard University Press, 1964).

73. Such passages, collected by Lovejoy and Boas in their *Primitivism* (chap. 12, pp. 368–88), constitute a trickle as compared with classical texts conveying the contrary outlook.

74. Cicero's ideas about the Great Year are fully discussed in an article by P. R. Coleman-Norton, "Cicero's Doctrine of the Great Year," *Laval Théologique et Philosophique*, 3 (1947): 293–302.

75. A full translation of the section of Cicero's *De republica* containing "The Dream of Scipio" is given in *Macrobius, Commentary on the Dream of Scipio*, translated with an Introduction and Notes by William Harris Stahl (New York: Columbia University Press, 1952).

76. *Ibid.*, p. 75.

77. *Ibid.* In his *Hortensius*, extant only in fragments, Cicero was reputed to have assigned 12,954 years as the length of the Great Year. See on this Coleman-Norton, "Cicero's Doctrine of the Great Year," p. 297.

78. See *Pliny, Natural History* with an English translation by H. Rackham (Cambridge, Mass.: Harvard University Press, 1938), p. 193 (Book II, chap. 6).

79. *Physical Science in the Times of Nero: Being a Translation of the Quaestiones Naturales of Seneca* by John Clarke (London: Macmillan, 1910), p. 153 (Book III, chap. 29).

80. *Ibid.*

81. *Ibid.*, p. 152.

82. *Ibid.*, pp. 155–56 (Book III, chap. 30).

83. *Ibid.*, p. 156.

84. *Ibid.*, p. 298 (Book VII, chap. 25).

85. *Ibid.*, pp. 297–98 (Book VII, chap. 24).

86. *Ibid.*, p. 301 (Book VII, chap. 27).

87. *Ibid.*, p. 307 (Book VII, chap. 32).

88. *Ibid.*, p. 308.

89. *Ibid.* Seneca's admiration for Berossos seems to have been widely shared during the golden age of the Roman Empire. Vitruvius, who held an official position as an architect during the reign of Augustus, made several enthusiastic references to Berossos in his *De architectura*. See *Vitruvius, On Architecture*, text and English translation by F. Granger (Cambridge, Mass.: Harvard University Press, 1962), vol. II, pp. 227, 245, 255 (Book IX, chaps. 2, 6, and 8).

90. Subsequent quotations will be to *Lucretius, De rerum natura* with an English translation by W. H. D. Rouse (Cambridge, Mass.: Harvard University Press, 1937).

91. *Ibid.*, p. 363 (Book V, lines 322–23).

92. *Ibid.*, p. 363 (Book V, lines 336–38).

93. *Ibid.*, pp. 363–65 (Book V, lines 337–44).

94. *The Communings with Himself of Marcus Aurelius* (see note 49), p. 39 (Book II, § 14).

95. *Ibid.*

96. *Ibid.* (Book II, § 16).

97. *Ibid.*, p. 115. (Book V, § 13).

98. *Ibid.*, p. 173. (Book VII, § 19).

99. *Ibid.*

100. *Ibid.*, p. 247. (Book IX, § 28).

101. *Ibid.*, p. 267. (Book X, § 7).

102. *Macrobius, Commentary on the Dream of Scipio* (see note 75), p. 219.

103. *Ibid.*

104. *Ibid.*, p. 221.

105. *Ibid.*

106. Thomas Whittaker's *Macrobius or Philosophy, Science and Letters in the Year 400* (Cambridge:

Cambridge University Press, 1923) is a valiant but unconvincing effort to put Macrobius' account of antique science under the best possible light. It is clear that history of science was not Whittaker's speciality.

107. On Proclus' life and work a most informative discussion is Laurence Jay Rosan's *The Philosophy of Proclus: The Final Phase of Ancient Thought* (New York: Cosmos, 1949).

108. The quotation is from the still very readable translation made by Thomas Taylor, *The Philosophical and Mathematical Commentaries of Proclus on the First Book of Euclid's Elements* (London: Printed for the Author, 1792), vol. I, p. 98.

109. See edition cited in note 46, p. 435 (chap. 29).

110. For a recent and most articulate presentation of the claim that classical antiquity was sufficiently permeated with, and motivated by the idea of progress, see the work by the late Ludwig Edelstein, *The Idea of Progress in Classical Antiquity* (Baltimore, Md.: The Johns Hopkins Press, 1967). As early as 1955 Edelstein devoted a paper to the question, and was much preoccupied with it during the last years of his life. He presented his case in a brilliant lecture delivered at the Tenth International Congress of Science, held in Ithaca, N.Y., in 1962. His work, published posthumously, contains the four completed chapters of his systematic discussion of the question. Unfortunately, the crucial issue of *scientific* progress is only touched upon lightly in Edelstein's work which is more erudite than convincing. It will hardly make much dent on the general consensus which prevailed for the past fifty years, largely under the impact of J. B. Bury's work, *The Idea of Progress: An Inquiry into Its Origin and Growth* (London: Macmillan, 1920). In spite of the political unity created by the Roman Empire, classical antiquity witnessed no systematic approach to learning and general education without which progress, literary and especially scientific, could not develop. Nor was classical antiquity animated by that measure of optimism which is needed for a sustained effort on behalf of progress.

111. *Aristotle's Metaphysics*, 981b (pp. 13–14).

112. *Ibid.*, 982b (p. 15).

113. This passage from Theophrastus' *On Pleasure* survived in *Athenaeus, The Deipnosophists*, with an English translation by Charles Burton Gulick (London: W. Heinemann, 1955), vol. V, p. 299.

114. Quoted in the translation of Kathleen Freeman from her *Ancilla to the Pre-Socratic Philosophers*, p. 109.

115. *Ibid.*, p. 108.

116. *Ibid.*, p. 117.

117. See *Xenophon, Memorabilia and Oeconomicus*, with an English translation by E. C. Marchant (London: W. Heinemann, 1923), p. 391.

118. *Plutarch's Lives* with an English translation by B. Perrin (London: W. Heinemann, 1955), vol. V, p. 479. This disdain for the practical *and* dynamical largely derived from a lack of motivation toward the real world, and helped prevent Archimedes from formulating the fundamentals of infinitesimal calculus.

119. Such a list is B. Farrington's work, *Greek Science: Its Meaning for Us* (Baltimore: Penguin Books, 1961). The debilitating weaknesses of Farrington's Marxist interpretation of Greek science were laid bare by L. Edelstein, "Recent Trends in the Interpretation of Ancient Science," *Journal of the History of Ideas*, 13 (1952): 573–604. Edelstein, of course, discussed the first edition of Farrington's work published in two parts, in 1944 and 1949, respectively.

120. See the concluding section of Book IV of the *Physics* (223b), where Aristotle speaks approvingly of those who derive the cyclic pattern of terrestrial and historical events from the cyclic motion of the heavens: "This is also the reason why human affairs and all things that come and go in the natural course of events are commonly said to move in cycles: all affairs are judged in terms of time; and they come to a stop and then start afresh in cyclical fashion. Some even look upon time itself as in some sense a cycle: time measures and is measured by circular motion; and the cyclical theory of events implies a circle of time in the sense of time measured by circular motion." Quoted in the translation of Richard Hope, *Aristotle's Physics* (Lincoln: University of Nebraska Press, 1961), p. 88.

121. See *Simplicii in Aristotelis categorias commentarium*, edited by K. Kalbfleisch (Berlin: G. Reimer, 1907), p. 351.
122. *Simplicii in Aristotelis physicorum libros quattuor posteriores commentaria*, p. 1151; quoted in the translation of S. Sambursky, *The Physical World of Late Antiquity*, p. 168.
123. Details of this will be given in Chapter Eight.

CHAPTER SEVEN

The Beacon of the Covenant

In a history of science, which is mainly a meticulous listing of particular discoveries (and most histories of science are still of this type), there can be no room for a discussion of the culture of the Old Testament. After all, the scientific attainments of the Hebrews are practically non-existent when set beside the achievements of the Greeks, and are insignificant even compared to the much less developed science of ancient Babylon and Egypt. The historian of science who notes the nomadic character of the ancient Hebrews and their heavy dependence on the culture of their more powerful neighbours, may feel justified in limiting his remarks on the people of the Bible to a few pages. If, in addition, he recalls the religious genius of the Hebrews, he implicitly offers a convenient excuse for their inferiority in matters scientific. The widely shared conviction about the fundamental opposition between religious and scientific orientation then readily cloaks such a procedure in the aura of objectivity.[1]

For an analysis of the history of science which dares to extend its search beyond the glitter of easy listings of facts and data, the foregoing procedure is doubly inadequate. This holds true particularly of an inquiry into such tantalizing questions of the history of science as to why science failed to develop in half a dozen great cultures, why it came to a standstill in Greece after a splendid start, and why it finally emerged more than a thousand years later in clearly identifiable circumstances. Such a probing into the history of science cannot easily brush aside factors, however extraneous they may appear to the mentality in which science is today cultivated. Needless to say, the historian must first reconstruct before tackling the task of interpretation. Reconstruction is in turn an exercise in judgment. Selections ought to be made between relevant and irrelevant data, and here the historian cannot help relying on the wisdom of hindsight. The importance of the scientific thinking of a John Philoponus, for instance, is evident only when viewed from a much later vantage point in scientific history. And so is the decisive role of his Christian monotheism which he brought for the first time to play upon scientific considerations.

Unless such points are firmly kept in mind one can hardly grasp the full portent of the brave stand taken by the seers of ancient Israel against the powerful gods of Egypt, Babylon, and Assur. In their actual settings the utterances of Isaiah and other prophets must have appeared highly presumptuous. The foreign gods that symbolized the cultural superiority and military strength of Mesopotamia and the Nile valley could inflict their deadly blows at will. Puny Israel was no match for its mighty neighbours. Yet, the brave front put up by the prophets was not without some remarkable aspects of its own. Their audacity was not the result of momentary impulsiveness. An Isaiah, a Jeremiah, and the rest of the prophets were not first in proclaiming the absolute superiority of Yahweh, as the one and only God. They merely rephrased a conviction that stubbornly had kept itself alive for centuries within a small group of nomadic tribes known as the Israelites. Against heavy odds they not only survived on a hazardous crossroad of great civilizations. They also succeeded in handing down from generation to generation an outlook on the world that set them radically apart from all their neighbours.

This was all the more remarkable as the Israelites shared their neighbours' belief in a flat earth floating on water, and in a firmament resting on columns located at the extremities of dry land. The Hebrew way of thinking was also moulded, as was the case with their great and small neighbours, by the same huge geographic unit stretching from the Euphrates to the Nile valley. In the case of Babylonians and Egyptians the impact of landscape was such that they came to accept nature as the supreme, perennial reality, which as a living being goes forever through the processes of birth, growth, death, and rebirth. The Babylonian origin of Abraham's progeny and their early sojourn in Egypt could have all too readily locked their minds in a similar concept of nature. One would look in vain, however, for a deification of nature in the Hebrew interpretation of the external world. The most ancient parts of the Bible already show that for the Hebrews external nature was an irrefragable evidence of a supreme, absolute, wholly transcendental Person, the Lord of all.

The word Creator was deliberately omitted in the foregoing context. The literary codification of the concept of a Creator, let alone that of a creation out of nothing, becomes evident only at a relatively late phase of a development, the origins of which hardly go beyond the tenth century B.C., or the time of Solomon. The first documents which formed the beginnings of what later became known as the books of the Bible, merely emphasized the idea of the utter dependence of everything on one single Being, who stood absolutely alone and could not be challenged by anything or anybody, either on the level of cosmic or of historical processes. Of these two categories it was clearly the latter that dominated the Hebrew mind. In the biblical view God is primarily and ultimately a person, whose most unique characteristic is to reveal His unspeakable transcendence in His most immediate concern for the children of Abraham.

The God of the Bible is the God of Abraham, Isaac, and Jacob; that is, the God of the Covenant, or a God who freely binds Himself to the welfare of mankind through the mediation of Abraham's progeny. He is a God who is ultimately predicated on that mysterious prompting which took Abraham out of the land of Ur and set him on a course in which the sole beacon was the unflinching confidence generated in Abraham that God would not fail to accomplish His part of the Covenant. The terms of the Covenant had some cosmic perspective: Abraham's progeny would reach a number comparable only to the myriads of stars and to the immensity of the grains of sand on the seashore. In that prospect the plethora of blessings was implied, as in primitive, nomadic society the large number of offspring was the decisive condition of enduring prosperity. It was in such a perspective that the external nature or cosmos was reflected upon in the Bible. There the universe is a dwelling place for man, a persuasive evidence of God's loving care. The second and by far the earlier of the two accounts of creation in the book of Genesis is hardly a cosmogony in the usual sense of the word. Its central theme is not about the main parts and forces of nature, but about man, and this is made absolutely clear at the outset: "At the time when Yahweh God made earth and heaven there was as yet no wild bush on the earth nor had any wild plant yet sprung up, for Yahweh God had not sent rain on the earth, nor was there any man to till the soil."[2] The author of the account is clearly skipping over nature to plunge into the primary topic of his narrative: the making of man by God, an act which includes the preparation of the whole of nature for him. This is what is emphasized in the detailed description of the Garden of Eden, in the rule about the use of fruits from the various trees, in the naming of all animals by man, and in the formation by God of a helpmate for man.

In all this there is a total lack of even a rudimentary piece of scientific detail. The account is, however, replete with a highly elevated mentality which constitutes the very climate of scientific thinking. Primitive as some details may appear in Genesis 2, it is animated by an uncompromising consistency of explanation which is the hallmark of scientific reasoning. In Genesis 2 there is only one effective cause, the power of God, through which heaven and earth and everything on earth has been formed. Yahweh God is an exclusive source of effectiveness. He is not challenged or complemented by any force or principle. He is the sole and supreme Lord of all. For all the primitiveness of the world picture of Genesis 2, it exudes a clear atmosphere undisturbed by what turns all other ancient cosmogonies into dark and dispirited confusion: the infighting among the gods and the lurking in the background of an irreconcilable antagonism between spirit and matter, good and evil.

This is not to suggest that Genesis 2 ignores evil in individual and collective history. Yet the snake, the symbolic instigator of evil, is itself utterly dependent for its existence upon Yahweh God, and so is man who is seduced into defying God's dictate. Evil, unlike in most other cosmogonies, is here strictly circumscribed in its power and extent by God's sovereignty and goodness. All this does not exhaust the uniqueness present in the second account of the creation in Genesis. In it one also finds intimated with sufficient clarity the formation of the universe once and for all, and an equally singular destiny for man and mankind. Or as the biblical text states, the head of the snake will be crushed for good by the progeny of the woman. Already at its earliest phase nothing could be more alien to the biblical outlook than the prospect of an endless tug of war between opposite cosmic and moral forces. The absolute sovereignty, rationality, and benevolence of God leaves no room for a senseless replay of the greatest of all happenings, the formation of heaven and earth and of man's privileged endeavour in the world.

The story of Yahweh making the world and preparing it for man, who is God's special handiwork, was obviously told many times before receiving its final written form and was certainly repeated on countless occasions following its definitive phrasing. It was only natural that such a unique story should become a unique source of inspiration. The documentary evidence of this lies in the earliest Psalms. Some of them undoubtedly have a Davidic authorship and they form a touching poetical counterpart of the world view of Genesis 2. Thus, in Psalm 8 Yahweh's ability to overcome His (and His people's) enemies is illustrated by His power which set the heavens, the stars, and the moon in place *and* by His astonishing care which made man a ruler over the works of God's hands. In another of the earliest Psalms, the nineteenth, one finds the same perspective. The lawfulness of the heavens which declares the glory of God, and the vault of heaven which proclaims His handiwork, form a spectacular backdrop to the highest form of law, the law of Yahweh showing the path of happiness to man.

It should not, therefore, be surprising that one finds in the earliest Psalms a most confident vision of nature. When almost three millennia after the composition by David of Psalm 23, Kant spoke of it as the most comforting page in the Bible, he merely echoed an already hallowed sentiment expressed in the style of disarming simplicity and charm:

> Yahweh is my shepherd,
>   I lack nothing.
> In the meadows of green grass he lets me lie.
> To the waters of repose he leads me;
>   there he revives my soul.
>
> He guides me by paths of virtue
>   for the sake of his name.
> Though I pass through a gloomy valley,
>   I fear no harm;
> beside me your rod and your staff
>   are there, to hearten me.
>
> You prepare a table before me
>   under the eyes of my enemies;
> you anoint my head with oil,
>   my cup brims over.
>
> Ah, how goodness and kindness pursue me,
>   every day of my life;
> my home, the house of Yahweh,
>   as long as I live!

The reference to the "house of Yahweh" represents a point of utmost importance. If the vastness of Yahweh's care for His people is measured only by the breadth and width of the world, it is also a care which is channelled through one specific point on earth, the seat of His Covenant, the temple in Jerusalem. This fusion of Cosmos and Covenant, this sameness of the Lord of nature and of the Lord of a chosen people, forms a primordial feature of biblical religiosity. Psalm 24 goes spontaneously from Yahweh, to whom belong the world and all it holds and all who live in it, to the question: "Who has the right to climb the mountain of Yahweh, who the right to stand in his holy place?" The reply could hardly be more concise and revealing at the same time. It emphasizes the indispensability of an impeccable attitude toward God and man. As to the latter, a particular duty, the avoidance of lies, stands for the whole range of man's obligations toward his neighbours. The former point is dealt with in a similar manner. Only one, though a most concrete and most tangible obligation is mentioned, as representative of the broad range of man's service of God: the man who pleases Him is the one "whose soul does not pay homage to worthless things."

The worthless things are the idols of the gods of the heathen. Consorting with them in any form constituted the worst possible offense in Old Testament religiosity as the sovereignty of Yahweh was patently incompatible with any pantheon of gods. Such is at least the clear implication of what belief in Yahweh stood for. Belief, however, is also a subjective conviction and not only a more or less doctrinaire proposition. But man's mind is hardly a stable structure and even more volatile is his adherence to the once recognized truth. The foregoing and other Psalms, when sung by the Israelites, echoed not only the happy sentiments of those in safe possession of priceless truths. They were at times muttered with the agony of the believer engulfed by an atmosphere of unbelief.

Belief in the Lord of all and in man's total dependence on Him could often appear as a pleasant illusion in a cultural ambience such as the one enveloping Yahweh's people. As compared with the religion of various Canaanite tribes, to say nothing of the religions of mighty Egypt and Assur, Yahweh's religion was a rather abstract and humble affair. The chief attractiveness of idols lay in their concreteness and in the natural expectation that such gods could be swayed by the sweetness and glitter of sacrificial gifts. With Yahweh nothing even remotely similar was possible. To portray Him in any form constituted a capital crime. He far transcended all, nature and man; He was not to fall captive to man's wishes and scheming. On the other hand man's nature, being what it is, gravitated not toward Yahweh but toward the idols and their far more "human" world.

How far that trend could go is worth recalling. The pre-exilic prophets' greatest concern derived from the ubiquitous and persistent presence of idolatry on the hills of Judah and Israel. Their tireless denunciation of Jewish idolatry is well exemplified in Isaiah's thundering:

> The land is full of soothsayers,
> full of sorcerers like the Philistines;
> they clap foreigners by the hand.
> His [Yahweh's] land is full of silver and gold
> and treasures beyond counting;
> his land is full of horses
> and chariots without number;
> his land is full of idols . . .
> They bow down before the work of their hands,
> before the thing their fingers have made.[3]

Idolatry was in the prophetic perspective the source of all upheavals, internal and external, rocking the House of Jacob, and conversely, the final and secure establishment of Israel depended on the total and definite cessation of all idolatry. Or as Isaiah put it: "That day, man will look to his creator and his eyes will turn to the Holy One of Israel. He will no longer look after the altars, his own handiwork, nor gaze at what his hands have made: the sacred poles and the solar pillars."[4]

For a historian of ideas, few topics could be as suspenseful as the ultimate fate and fortune of a conviction which a relatively few tried to impose in its pristine purity on the wavering minds and loyalties of the vast majority, as one Jewish generation followed another. Would their uphill fight maintain its momentum, or would it in the long run be stopped and reversed for good by the downhill trend of idolatrous, polytheistic, and naturalistic proclivities? The Second Book of Chronicles is a graphic record of that see-saw battle in which the forces of idolatry seemed to gain further ground in each turn. In fact, with the ascent in 687 of the twelve-year-old Manasseh to the throne of Judah, the final victory of idolatry seemed to be at hand:

> Manasseh was twelve years old when he came to the throne and he reigned for fifty-five years in Jerusalem. He did what is displeasing to Yahweh, copying the shameful practices of the nations whom Yahweh had dispossessed for the sons of Israel. He rebuilt the high places that his father Hezekiah had demolished, he set up altars to the Baals and made sacred poles, he worshipped the whole array of heaven and served it . . . He built altars to the whole array of heaven in the two courts of the Temple of Yahweh. He caused his sons to pass through the fire in the Valley of Benhinnom. He practised soothsaying, magic and witchcraft, and introduced necromancers and wizards. He did very many more things displeasing to Yahweh . . . He placed the image of the idol he had made in the Temple . . .[5]

Coloured and exaggerated as may appear this account given by a spokesman of strict monotheism, the future for the survival of the faith in Yahweh, the sole and supreme Lord of all, could not have been darker. Nothing shows this more forcefully than the fact that for decades the Book of the Law had remained hidden in an obscure corner of the Temple. The restoration of the Mosaic Covenant to its pre-eminence in 622, in the eighteenth year of the reign of Josiah, was done with extraordinary determination, but it could not forestall further reversals of fortune between monotheistic and polytheistic trends. The concise summary of what followed after Josiah is also given in 2 Chronicles, whose author viewed the problem not only in retrospect but also through the lenses of one of the most cataclysmic and catalytic events in Jewish history, the destruction of Jerusalem in 585 and its tragic aftermath, the forty-year long captivity in Babylon. By all counts of probability, the faith in Yahweh, the Lord of all, should have completely disintegrated through those traumatic events and trials. What else could have more effectively proved the illusory character of a God than His apparent inability to protect His own city and nation from total devastation? What else could have more efficiently undermined the isolated faith of a few tens of thousands in the sole Maker of Heaven and Earth than the lengthy and oppressive exposure to the cult of the gods of Babylon, the very embodiment of power, success, and refinement?

What actually happened constitutes a most baffling chapter in cultural history. Jewish monotheism emerged from the cauldron of captivity in a far more robust and in a far more incisive form. The confrontation with the pantheon of Babylon channelled startlingly new vigour into the faith in Yahweh as the sole God of all. This can best be gauged from that cutting satire with which Deutero-Isaiah puts to ridicule the making and worship of idols. The passage is iconoclasm at its best. The biting remarks on the nothingness of idols seem to reduce them literally to ashes together with the carvers and blacksmiths who make them. Their toil and exhaustion is made to evoke the haplessness of their artifacts. But the sharpest stricture is reserved for their blindness which makes them insensitive to what should be most obvious: "They never think, they lack the knowledge and wit to say, 'I burned half of it [timber] on the fire, I baked bread on the live embers, I roasted meat and ate it, and am I to make some abomination of what remains? Am I to bow down before a block of wood?' A man who hankers after ashes has a deluded heart and is led astray. He will never free his soul, or say, 'What I have in my hand is nothing but a lie!'"[6]

It is in that exposure of the mental blocks produced by idolatry that lies the uniqueness of that satire. A comparable counterpart of it is simply non-existent in any other ancient literature, religious or philosophical. The same holds true of that soaring monotheism which generated the courage to decry idolatry, this most pervasive, most hallowed, but also most detrimental practice of all ancient cultures. The style of Deutero-Isaiah rises to its highest level as he describes Yahweh's power beside which all forms and symbols of strength fade into nothingness:

> Go up on a high mountain,
> joyful messenger to Zion.
> Shout with a loud voice,
> joyful messenger to Jerusalem.
> Shout without fear,
> say to the towns of Judah,
> 'Here is your God'.
>
> Here is the Lord Yahweh coming with power,
> his arm subduing all things to him.
> The prize of his victory is with him,
> his trophies all go before him.
> He is like a shepherd feeding his flock,
> gathering lambs in his arms,
> holding them against his breast
> and leading to their rest the mother ewes.
>
> Who was it who measured the water of the sea in the
>   hollow of his hand
> and calculated the dimensions of the heavens,
> gauged the whole earth to the bushel,
> weighed the mountains in scales,
> the hills in a balance?
>
> Who could have advised the spirit of Yahweh,
> what counsellor could have instructed him?
> Whom has he consulted to enlighten him,
> and to learn the path of justice
> and discover the most skilful ways?
>
> See, the nations are like a drop on the pail's rim,
> they count as a grain of dust on the scales.
> See, the islands weigh no more than fine powder.
> Lebanon is not enough for the fires

nor its beasts for the holocaust.
All the nations are as nothing in his presence,
for him they count as nothingness and emptiness.

To whom could you liken God?
What image could you contrive of him?

A craftsman casts the figure,
a goldsmith plates it with gold
and casts silver chains for it.
For it a clever sculptor seeks
precious palm wood,
selects wood that will not decay
to set up a sturdy image.

Did you not know,
had you not heard?
Was it not told you from the beginning?
Have you not understood how the earth was founded?
He lives above the circle of the earth,
its inhabitants look like grasshoppers.
He has stretched out the heavens like a cloth,
spread them like a tent for men to live in.
He reduces princes to nothing,
he annihilates the rulers of the world.
Scarcely are they planted, scarcely sown,
Scarcely has their stem taken root in the earth,
than he blows on them. Then they wither
and the storm carries them off like straw.

'To whom could you liken me
and who could be my equal?' says the Holy One.
Lift your eyes and look.
Who made these stars
if not he who drills them like an army,
calling each one by name?
So mighty is his power, so great his strength,
that not one fails to answer.

How can you say, Jacob,
how can you insist, Israel,
'My destiny is hidden from Yahweh,
my rights are ignored by my God'?
Did you not know?
Had you not heard?

Yahweh is an everlasting God.
he created the boundaries of the earth.

He does not grow tired or weary,
his understanding is beyond fathoming.
He gives strength to the wearied,
he strengthens the powerless.
Young men may grow tired and weary
youths may stumble,
but those who hope in Yahweh renew their strength,
they put out wings like eagles.
They run and do not grow weary,
walk and never tire.[7]

The justification for quoting such a lengthy passage lies in its obviously inimitable conceptual and emotional richness that would defy any effort to condense and paraphrase it. The passage throbs with a living unity making the ancient tree of the monotheism of the Covenant blossom into a magnificent new foliage. The old part was the emphasis on the unity of cosmic and human history based on the sameness of the Maker of the world and of the Shepherd of His people. The new foliage was the triumphant confrontation with the idolatrous great cultures and the ensuing bold assertion that idolatry deprived man of basic insights into the fundamental characteristics of the world. That new understanding was in a sense an old possession, but the confrontation in Babylon articulated it in a novel manner. The chief evidence of this is the creation story in the first chapter of Genesis. A classic target of sophomoric rationalism and vicious bias which often parade in the cloak of science, the first chapter of Genesis is, in fact, a most lucid expression of that faith in the rationality of the universe without which the scientific quest in man could not turn itself into a self-sustaining enterprise.

The story of Genesis 1 has, of course, been often presented as a feeble derivative of Babylonian and other creation myths. The bias implied in such an evaluation of the biblical account of creation is characteristic of that rationalism which cannot go beyond the immediate wording of the story and catch sight of its monumentally simple structure and highly elevated message. The structure is patterned after the logical procedure followed in every construction work: first comes the framework, and afterwards the details of each major section. Whereas the sober compliance with this straight pattern sets Genesis 1 markedly apart from other ancient cosmogonies, there is nothing original in the list of the principal parts of the world and their embellishments as given in Genesis 1. The list would have been much the same in any part of the land stretching from the Nile to the Euphrates and beyond. The second and third days witness the formation of the structure in the analogy of the erection of a nomadic shelter which consists of a roof stretched over a chosen piece of ground. Thus, we see the cosmic roof, or the vaults of heaven, appear to separate the upper from the lower waters, and the dry land emerge to set limits to the oceans.

Such is certainly a terse account of the structure of the whole world. The author of Genesis 1 could very well have been more prolific. He might have mentioned the columns supporting the heavenly vault at the extremities of the dry land, and the tooth-like protrusions of the dry land reaching far into the depths of the ocean, as if to indicate the stability of the dry land floating on the waters. But his main interest lies not in details of cosmography. He rushes through with the business of the fourth, fifth, and sixth days, the embellishment of the heavenly vault with celestial bodies, and the replenishing of the lower regions with their respective populations of birds, fish, and animals of all kinds.[8] Clearly, the chief message of Genesis 1 is not so much about the world as about God and man.

As to God, He is described in the analogy of the typical working man whose first action is to provide light at the very start (the first day) and whose activity is organized, in conformity with the accepted Mesopotamian tradition, around a seven-day period. With that, all strict similarity ceases between Genesis 1 and its Near-Eastern counterparts. For all the Mesopotamian flavour of Genesis 1, its author uses the common lore with unusual skill to drive home some very uncommon points. These are the absolute sovereignty and precedence of God over any and all parts of the world, the infinite power of God who brings things into existence with sheer command, and his overflowing goodness which can only produce an intrinsically good world of matter, both in its entirety and in its parts. None of these assertions are, of course, made by the author of Genesis 1 in the sophisticated language of philosophy. Around 500 B.C., philosophical discourse was even in Greece only at its beginning, and its twenty-five centuries of subsequent vicissitudes has only served proof of the indispensability of ordinary language as the primordial vehicle of understanding.

There is nothing abstract in the assertion repeated after the formation of each part and portion of the cosmos that "it is good." There is no obvious trace of the abstractness of the idea of a "creatio ex nihilo" in the grand opening sentence: "In the beginning God created the heaven and the earth."[9] But when Genesis 1 is read, as it should be, with an eye on the *Enumah Elish*, the Babylonian creation myth, one cannot help being overawed by the diametrically opposite thrust of the two accounts about the origin of the universe. It is through that implicit contrast that Genesis 1 offers its unique message: it is not the chaos, but God who is primordial, and to such a limitless extent that the unsavoury details of the emergence of gods from the chaos are not even considered worthy of rebuttal. Nor is any word wasted on arguing over the Babylonian dualism of matter and spirit, evil and good. The goodness of God, the maker of all, is simply reasserted with the air of a matter-of-fact certainty that stands far above any questioning. His all-powerfulness is conveyed in the same concrete style which uses the idiomatic expression "heaven and earth" to designate the whole universe. To make this point crystal clear God's sovereign action is reasserted as the sole source of existence with respect to each section of the structure of the universe, and to the embellishment of each section. If there is any philosophical or rather linguistic sophistication in Genesis 1 it concerns the specificity of "making" when God is the maker. The word *bara* does not mean "to create out of nothing," though is reserved in the Bible for that most specific type of making which is God's exclusive privilege.[10] It is not a word which achieved a special significance through philosophical encounters. It is rather a word which took on a special hue by the instinctive realization of the demands when man's parlance was about things divine. The word *bara* forms, therefore, a part, a most incisive part, of that uphill trend in conceptualization toward ever more refined and categorical statements about God, the Maker of heaven and earth.

Before this further development within the Covenant is taken up, the second main message of Genesis 1, the message about man, should be considered in some detail. In line with the earlier phases of the Covenant's spirit the universe is depicted in Genesis 1 as principally an abode where man can develop his unique potentialities. These derive from the fact that only man is said to be the image of God, in sharp contrast to other creation stories imbued with pantheistic tendencies. Man and his wife are, therefore, given in Genesis 1 the mandate to "multiply, fill the earth and conquer it."[11] They are commanded to "be masters of the fish of the sea, the birds of heaven and all living animals on the earth."[12] The full measure of the vitality and depth of such a command and of the belief in it could unfold only in a long, historical process of which more will be said in a later chapter. Attention at this point should rather be centred on the subordination of the cosmic view to the destiny of mankind. That destiny is unique, coming as it does from God, the unique source of existence. Consequently, the universe too, being in its ultimate meaning an abode for man, is not an agglomerate of capricious events and processes but something which is complete because of the co-ordination of all its parts to its wholeness. The deeper meaning of that completeness was, however, to be grasped along the parameter of time, as mankind's completeness could only be achieved in time. Therefore, the phrase, "the heaven and earth were completed with all their array," is immediately followed by the description of God's rest on the seventh day, an image which henceforth would serve as the classic evocation of the final spiritual and cosmic completion of the Covenant in a new heaven and in a new earth.[13]

About Genesis 1, one should also note its literary structure which puts it in the class of didactic poetry in prose. Whether it was recited as such is rather conjectural. There can, however, be no doubt in this respect about the Psalms composed in the post-exilic period. In several of them one finds an account of the work of Creation which could only come from a conviction about its crucial relevance for everything man thinks and does. In turn, these Psalms, when echoed on countless lips on count-less occasions, could not fail to give renewed impetus to that uphill trend of the Covenant's monotheism and outlook on the world in general, the endurance of which kept defying all considerations of historical probability. History is, to be sure, greatly influenced by outbursts of enthusiasm. But the enthusiasm exuding from Psalm 136 is not the enthusiasm of conquerors and of masses of men coaxed into lurid celebrations and pursuits. The enthusiasm of Psalm 136 is of a distinctly spiritual character anchored in the belief in the transcendence of God and of spiritual values. A most striking of these values is the confident outlook on existence and nature inspired by the Covenant's doctrine about Yahweh as the Maker of heaven and earth:

> He alone performs great marvels,
>   his love is everlasting!
> His wisdom made the heavens,
>   his love is everlasting!
> He set the earth on the waters,
>   his love is everlasting!

> He made the great lights,
>   his love is everlasting!
> The sun to govern the day,
>   his love is everlasting!
> Moon and stars to govern the night,
>   his love is everlasting![14]

No nation, no sequence of generations could produce songs like this outside the group of those who cast their lot with the faith of their fathers in the God of the Covenant. He is a God, who as Psalm 18 tells it, intervenes irresistibly on behalf of his people. His power is illustrated by his utter superiority over the main elements of the nascent world as listed in the *Enumah Elish*. As Yahweh sweeps down from high above, the heaven and earth bend, the watery darkness quivers, the bed of the seas shakes, the foundations of the mountains are laid bare. As a result, the rescue of those who trust in Him becomes a foregone conclusion. They are drawn up by Him from chaotic deep waters, the standard symbol of confusion, of absence of purpose, and of the merciless dissolution of all.

It should not be surprising that this unconditional and firm trust in Yahweh produced a warm, confident, optimistic appraisal of nature which once more sets apart the realm of the Covenant from the surrounding cultures. Psalms 65 and 107 speak eloquently of the confidence of seafarers buffeted by stormy seas. Doubts disappear about the ultimate calm of nature when it is recalled that His deeds "bring shouts of joy to the portals of morning and evening," that is, to the whole world stretching between the rising and setting points of the sun.[15] Only robust confidence could issue from an outlook which pictured the deep places, or the ultimate foundations of the world, as being in the hands of the Lord. The same outlook ascribed to the Lord in a memorable phrase even the strength of hills.[16] There is, indeed, a logical development between the early, glowing description of the Promised Land in Deuteronomy, and the subsequent extension of that warm appraisal to the whole of nature. Behind both there lay the same unshaken belief in a reasonable Lord of all who can be spoken to in words overflowing with joy:

> You visit the earth and water it,
>   you load it with riches;
> God's rivers brim with water
>   to provide their grain.
>
> This is how you provide it:
> by drenching its furrows, by levelling its ridges,
> by softening it with showers, by blessing the first-fruits.
>
> You crown the year with your bounty,
> abundance flows wherever you pass;
> the desert pastures overflow,
> the hillsides are wrapped in joy,
> the meadows are dressed in flocks,
> the valleys are clothed in wheat,
> what shouts of joy, what singing![17]

That song was the natural echo of the theme which sets the tone of the first pages of the Bible about the ultimate characteristic of external nature: It is good. The exclusion of an evil principle of equal rank with God entails in turn that the biblical world view has no room for a concept of nature in which capricious, dark forces dominate. Again, within the context of the Covenant, the world is not an all-encompassing entity containing the source of all life, human and divine, and unfolding that life through an endless chain of inexorable, blind cycles. The world, being the handiwork of a supremely reasonable Person, is endowed with lawfulness and purpose. These are the direct result of the never failing and benevolent surveillance by Yahweh over the entire world. The regular return of seasons, the unfailing course of stars,[18] the music of the spheres,[19] the movement of the forces of nature according to fixed ordinances,[20] are all the results of the One who alone can be trusted unconditionally. Thus, the prophet Jeremiah praises the faithful recurrence of harvests as the sign of Yahweh's goodness.[21] Moreover, he establishes a remarkable parallel between Yahweh's unfailing love and the eternal ordinances by which Yahweh set the course of stars and the tides of the sea.[22]

The coupling of the reasonability of the Creator and the constancy of nature is worth noting because it is there that lie the beginnings of the idea of the autonomy of nature and of its laws. But this can be seen only in retrospect. Long before the biblical heritage about the Creator could inspire reflections most germane to science, it had to prove repeatedly its vitality in crises that either threatened individuals or the body of believers as such. The former case can be studied in that giant of literary masterpieces, the Book of Job.[23] Faced with the quick series of heavy blows at his fortune, health, and family, Job also had the added trouble of being importuned by friends who press him for an explanation of his tragic predicament. Needless to say, Job remains short of an answer that would dissipate all the darkness of the mystery of suffering. But while a long list of reflections about the problem reveals the frustrating volatility of even the seemingly best established answers, the waves of scepticism do not prevail over one specific consideration. It is anchored in the justice and faithfulness of God as firmly displayed in the great work of creation. Against this monumental evidence of reason and good will no defiance is valid. The picture Job paints about the might of the Maker of heaven and earth emphasizes the sovereign will of God who seemingly can play havoc with mountains, trample the sea waves, shake the pillars of the vault of heaven. Nevertheless Job admits that "his [Yahweh's] heart is wise, and his strength is great . . . his works are great, beyond all reckoning, his marvels past all counting."[24] Bowing to Yahweh's infinite and unfathomable wisdom manifest in the created world is, in fact, the ultimate solution to Job's predicament. In a most dramatic list of questions Yahweh himself spells this out "from the heart of the tempest":

> He said:
> Who is this obscuring my designs
>   with his empty-headed words?
> Brace yourself like a fighter;
>   now it is my turn to ask questions and yours to inform me.

Where were you when I laid the earth's foundations?
    Tell me, since you are so well-informed!
Who decided the dimensions of it, do you know?
    Or who stretched the measuring line across it?
What supports its pillars at their bases?
    Who laid its cornerstone
when all the stars of the morning were singing with joy,
    and the Sons of God in chorus were chanting praise?
   . . .

Can you fasten the harness of the Pleiades,
    or untie Orion's bands?
Can you guide the morning star season by season
    and show the Bear and its cubs which way to go?
Have you grasped the celestial laws?
    Could you make their writ run on the earth?
Can your voice carry as far as the clouds
    and make the pent-up waters do your bidding?
Will lightning flashes come at your command
    and answer, 'Here we are'?[25]

These two short sections represent only a small portion of Yahweh's questions unfolding the great and small details of the marvellous arrangement present everywhere in the created world. Job's answer to Yahweh's questions should, however, be quoted in full:

I know that you are all-powerful:
    what you conceive, you can perform.
I am the man who obscured your designs
    with my empty-headed words.
I have been holding forth on matters I cannot understand,
    on marvels beyond me and my knowledge.
I knew you then only by hearsay;
    but now, having seen you with my own eyes,
I retract all I have said,
    and in dust and ashes I repent.[26]

Such a finale may appear as a dejected capitulation on Job's part in the face of an overpowering, inexorable, if not despotic deity. But the passage is rather the evidence of the intellectual honesty which the Covenant was capable of generating, in which the sombre aspects of man's puniness were not in a cowardly way swept under the rug. This courage came all the more naturally as the tone of optimism would readily reassert itself within the Covenant, a feat that could not be achieved millenia later by evolutionary naturalism, the chief spokesmen of which looked in vain for an escape hatch from the clutches of a most pessimistic and nihilistic finale in store for all. Within their system of reasoning there could be no logical reason for hoping against hope, an attitude so magnificently displayed by those living from the vitality of the Covenant when the hour of confrontation came with the Hellenistic world.

The confrontation took place on two fronts. One centred on the temple of Jerusalem and the homeland of the people of the Covenant, the other on Alexandria, the cultural hub of late antiquity, which had a sizeable Jewish population by the middle of the second century B.C. On the home front the confrontation meant martyrdom or apostasy. The two Books of Maccabees[27] contain the shocking details of a full-scale religious persecution during which every observance of the ritual laws of the Covenant was savagely punished by death. The trials, as one would expect, prompted both apostasies and heroic resistence. The defectors rationalized that understanding should be reached with the pagan world and its culture at any price, as they saw the source of Jewish misfortunes in the claim about the absolute uniqueness of the Covenant. Those rather ready to die based their course of action precisely on the consideration that the doctrinal and spiritual heritage of the Covenant was the highest form of understanding available for man which had to be maintained in its pristine purity whatever the price. The touchstone of that purity was the faith in the Maker of heaven and earth.

The immense bearing of that faith on single human destinies comes through with overpowering force and clarity in the heroic fate of seven brothers arrested with their mother. The story is told in great detail in the Second Book of Maccabees obviously because of its extraordinary instructiveness. The words spoken by the brothers and their mother, as the turn of each came to be put to death, are carefully recorded,[28] so that one can now follow step by step the unfolding of the conceptual riches of the faith in Yahweh, the Creator of all. The last words of the oldest of the brothers recall God's Covenant with Moses, as the sure token of His ultimate mercy. The final statements of the second, third, and fourth brothers refer to the foundation of the Covenant, to the creating God, the King of the world, the Lord of heaven, whose purpose in creating will not be defeated by evil designs. Consequently, the final resurrection of the faithful is mentioned by them as the deepest consequence of the creation of man by God. In the words of the second, "The King of the world will raise us up, since it is for his laws that we die, to live again for ever." The third is even more specific: "It was heaven that gave me these limbs; for the sake of his laws I disdain them; from him I hope to receive them again." The fourth then boldly turns the table on the persecutor king: "Ours is the better choice, to meet death at men's hands, yet relying on God's promise that we shall be raised up by him; whereas for you there can be no resurrection, no new life." The fifth and sixth brothers also speak as those who really are in a position to make accusations, revealing thereby a courage which astonishes the king and his attendants.

The most astonishing part of the story is the attitude of the mother who keeps encouraging each of her sons in the span of one single terrible day with the words: "I do not know how you appeared in my womb; it was not I who endowed you with breath and life, I had not the shaping of your every part. It is the creator of the world, ordaining the process of man's birth and presiding over the origin of all things, who in his mercy will most surely give you back both breath and life, seeing that you now despise your own existence for the sake of his laws." She did not waver when the king offered to save her youngest son if he showed readiness to reject the faith. The king even tried to engage the help of the mother to achieve his aim but the mother had her own brand of persuasion. Bending over her son she uttered words never before registered: "My son, have pity on me; I carried you nine months in my womb and suckled you three years, fed you and reared you to the age you are now (and cherished you). I implore you, my child, observe heaven and earth, consider all that is in them, and acknowledge that God made them out of what did not exist, and that mankind comes into being in the same way. Do not fear this executioner, but prove yourself worthy of your brothers, and make death welcome, so that in the day of mercy I may receive you back in your brothers' company."

It would be wholly missing the point to argue whether the enunciation of the idea of "creation out of nothing" comes from the very words of the mother or reflects the conviction of the unknown author of 2 Maccabees. What is of crucial significance is the appearance of the phrase at a dramatic juncture in the history of the Covenant. The mother's words are reported with that fullest approval and admiration which sets the tone of the narration of her own and her sons' heroic stand for the Covenant's God, whose sovereignty is of such extent that all beings owe to His command their emergence from sheer non-existence. As was formerly the case with the confrontation in Babylon, the pressure of trial produced priceless benefits both with respect to doctrine and to the impressiveness of existential testimony.

The sudden emergence of a most specific formulation of what ultimately is meant by creation was, however, conditioned by the spread of a new influence from the outside. Hellenistic trends certainly brought to the attention of Jewish theologians the struggling of Greek philosophers with the concept of coming into being out of nothing. What Aristotle so explicitly rejected was undoubtedly embraced by Jewish theologians as a most appropriate formulation of the great proposition about the Creator of all, on which everything, Covenant and Cosmos alike, rested. It is in that spirit that the "Wisdom of Jesus Ben Sira,"[29] probably written on the eve of the Maccabean war, emphasized the glory of God as revealed both in nature and in the history of the Covenant. The lengthy passage on nature begins with the famous declaration: "By the words of the Lord his works came into being and all creation obeys his will," and contains phrases like "He has imposed an order on the magnificent works of his wisdom" and "All things go in pairs, by opposites . . . the one consolidates the excellence of the other . . ." After this comes a detailed account of the sun, moon, stars, rainbow, and of the rest of nature as evidence of God's glory with emphasis on the creative power of God's word, the source of all order and motion in nature. The concluding consideration is the admission of man's inability to fathom all marvels of God's visible works.[30]

The closer contact of Jewish thinkers with Hellenistic learning in Alexandria had, as one would expect, rather original reflections on the world and its Maker. There, it was not political oppression but cultural refinement steeped in polytheism and nature worship that could gradually obscure the unique world view of the Covenant. This uniqueness did not rest on some special technical information about the world. The author of the Book of Wisdom[31] recalls with appreciation the various lores of knowledge which Hellenistic culture could offer. But for him it is because of the perspective provided by the belief in the Creator of all, the source of all wisdom, that one should rejoice for having been taught:

> the structure of the world and the properties of the elements,
> the beginning, end and middle of the times,
> the alternation of the solstices and the succession of the seasons,
> the revolution of the year and the positions of the stars,
> the natures of animals and the instincts of wild beasts,
> the power of spirits and the mental processes of men,
> the varieties of plants and the medical properties of roots.[32]

Moreover, the wisdom derived from the contemplation of the attributes of Yahweh provides one with insights that reach far beyond the purely intellectual realm. These insights instruct man in the wholeness of his attitudes, sharpen his judgments about the intangibles and imponderables of life, and strengthen his will to follow the path of righteousness and virtue. This is why true wisdom is described as the "breath of the power of God," the "emanation of the glory of Almighty," the "reflection of the eternal light," and the "untarnished mirror of God's active power."[33]

Most importantly, the wisdom in question is a chief benefit of being a loyal part of the Covenant, because only in the perspective provided by the Covenant can man grasp the uppermost of all knowledge, which is about the ultimate destiny of mankind and cosmos alike. This is why the second part of the Book of Wisdom forms a summary of the Covenant's history from the first man to the occupation of the Promised Land. Entering the land of promise is seen, through its antecedent, the sojourn of Israel in Egypt. Egypt in turn provides the classic example about the disastrous consequences of idolatry. One of these is the reluctance of men steeped in idolatry to recognize and carry out God's will. Consequently, the plagues which hit Egypt shortly before the exodus of the Jews are presented as a punishment inflicted upon the Egyptians for their idolatry.

In the context several points are emphasized which are of great importance. As to the punishments, it is pointed out that God's power could have inflicted the damage in one single devastating blow, but such would not have been in line with the procedure typical of God who interacts in history in the same manner as He rules the cosmos: "They could have dropped dead at a single breath . . . But no, you ordered all things by measure, number, weight."[34] The passage is worth noting not only because it instances the ready assimilation of a typically Greek philosophical or scientific idea. The real importance of the passage is in its subsequent impact. The ordering of all things by measure, number, and weight served as inspiration and assurance for those who in late antiquity assumed the role of champions of the rationality of the universe. A thousand years later the expression was gladly seized upon by those who daringly started out on the road to unfold the marvels of God's handiwork along the lines of quantitative inquiry.

The context of the foregoing passage also contains a statement about God's powerful hand which "from formless matter created the world."[35] The expression may be an indirect suggestion that the shaping of the world by God out of chaotic matter is not wholly alien to the "creation" of the world as described in Plato's *Timaeus*. What is, however, wholly alien to the Book of Wisdom is the pantheistic, emanationist outlook of Plato's cosmology. Nothing shows this better than the detailed discussion there of another tragic consequence of idolatry. It consists, according to the Book of Wisdom, in man's folly which can obscure for him what should be most obvious: the ultimate dependence of everything in heaven and on earth on the power and wisdom of the Creator. The famous passage, which categorically states that man is guilty if from the contemplation of things visible he fails to recognize their invisible Author, deserves to be quoted in full:

> Yes, naturally stupid are all men who have not known God
> and who, from the good things that are seen, have not been
>   able to discover Him-who-is,
> or, by studying the works, have failed to recognize the Artificer.
> Fire, however, or wind, or the swift air,
> the sphere of the stars, impetuous water, heaven's lamps,
> are what they have held to be the gods who govern the world.
>
> If, charmed by their beauty, they have taken things for gods,
> let them know how much the Lord of these excels them,
> since the very Author of beauty has created them.
> And if they have been impressed by their power and energy,
> let them deduce from these how much mightier is he that has
>   formed them,
> since through the grandeur and beauty of the creatures
> we may, by analogy, contemplate their Author.[36]

While laying bare the foolishness of those investigators of nature who "fall victim to appearances seeing so much beauty," the author of the Book of Wisdom treats them with distinct compassion and understanding. He attaches "small blame" to them as he sees in the sedulous investigation of nature the evidence of an eager though covert search for God. Still, their guilt remains in virtue of the decisive consideration:

> if they are capable of acquiring enough knowledge
> to be able to investigate the world,
> how have they been so slow to find its Master?[37]

No alleviating circumstance is, however, found for idolators and for makers of idols. The toil of the latter is held up to ridicule in a passage which constituted a fine variation on a theme already explored by Deutero-Isaiah. What is strikingly new in this respect in the Book of Wisdom is the penetrating analysis of the social and cultural ills generated by idolatry. Thus, idolatrous honours accorded to kings dim the critical sense of the governed for the abuses of those who govern. Idolatry turns upside down the scale of values to the extent that what patently constitutes an external and internal turmoil would appear as the normal course of events with the result that idolators give "to massive ills [such as follow] the name of peace":

> With their child-murdering initiations, their secret mysteries,
> their orgies with outlandish ceremonies,
> they no longer retain any purity in their lives or their marriages,
> one treacherously murdering the next or doing him injury
>   by adultery.
> Everywhere a welter of blood and murder, theft and fraud,
> corruption, treachery, riots, perjury,
> disturbance of decent people, forgetfulness of favours,
> pollution of souls, sins against nature,
> disorder in marriage, adultery, debauchery.[38]

This description will appear as hopelessly biased to those who keep reconstructing Hellenistic antiquity from a very narrow segment of it represented by the best in its literature and scientific achievements. They then seek an answer to the puzzling decay of scientific and cultural efforts of late antiquity in causes that have relatively little to do with it. They strangely overlook in that particular historical context the general truth that the effective cultivation of science needs an atmosphere if not of actual honesty and virtue, at least an atmosphere in which crime, falsehood, vices of all kinds are clearly recognized for what they are. For it is the very soul of science to call a fact a fact in all truth and honesty. Such an attitude cannot emerge in the relatively narrow field of scientific pursuit if parodies of facts, norms, and values are taken for genuine along much of the gamut of human experience. Historians of science would do well to meditate at length on this. From the terse statement, "The worship of unnamed idols is the beginning, cause and end of every evil,"[39] there is much to be learned even about that evil which caused science to wither away in late antiquity.

Scientific breakthroughs, or new scientific instruments, are never easy to make. But they should seem to appear child's play when compared with the task of bringing about a never before experienced cultural or rather moral climate in which the good, right, and truthful are accorded, in principle at least, unconditional respect. Those who achieved this enormous feat did so in the deep conviction that they were the true, spiritual heirs to the Covenant. The principal aim of the Master from Nazareth consisted in bringing out the basic feature, love, in the image of the Father and Maker of all. If what he said about the love of God and neighbour was already extraordinary, no less astonishing was the effectiveness of his words and of his matchless remarks, governs the whole of creation, keeps clothing the lilies own life as the utmost compliance with the Father's will. That will, to recall some of his inimitable remarks, governs the whole of creation, keeps clothing the lilies of the field, prevents sparrows from falling to the ground haphazard, and instituted the human race, as male and female, from the creation. Deep-seated consciousness of the unique importance of the creation sets the tone of one of his few recorded utterances of prayer that starts with the exclamation: "I bless you, Father, Lord of heaven and of earth."[40] It was the time passed since the days of creation which he used as the backdrop to indicate the enormity of the trials which would signal the approach of the final phase of cosmos and history. His emphasis on the redemptive restoration of all was anchored in the infinite love of a Father who would certainly secure the ultimate triumph of the purpose of creation, the union of all in God.

One may or may not agree with the extraordinary claim which the Master from Nazareth made about himself. Many millions have been deeply touched by his humaneness, uncounted others still find distasteful his emphatic other-worldliness. Yet, only the councils of rigid rationalists and dogmatic Marxists would deny that anyone ever changed the world as much as he did, and for the better. His immediate disciples were impressed by him to the point of according him divine honours. But their worship of him was only matched by the magnitude of their and their master's concern, lest any crack, however slight, be made in the edifice of the jealously strict monotheism of the Covenant. The Master from Nazareth most emphatically endorsed the hallowed phrase of the Covenant as standing for the first of all commandments: "Listen, Israel, the Lord our God is the one Lord, and you must love your Lord your God with all your heart, with all your mind, and with all your strength."[41] The Israel he had in mind was not to stay confined to its national self. Its spiritual endowment had the destiny to spread across a world lost in polytheism and idolatry.

It was, therefore, most logical that the clash between Christianity and the Roman Empire came to a head on the issue of monotheism. The preaching of the Gospel had to be done in an ambience filled with the smoke of food sacrificed to idols. Paul's instructions to the Corinthians is most instructive in this respect. Speaking of the daily use of leftover sacrificial food he noted that no food was forbidden for those who "know that idols do not really exist in the world and that there is no god but the One. And even if there were things called gods, either in the sky or on earth – where there certainly seem to be 'gods' and 'lords' in plenty – still for us there is one God, the Father, from whom all things come and for whom we exist."[42]

The same unity between the old and new phase of the Covenant comes through Paul's speech on the Aeropagus to a purely pagan audience.[43] Unquestionably, Paul spoke to pagans on more than one occasion. Still, Athens and its Acropolis were the symbol of the best in heathen culture, and so the speech was recorded as most typical of the Christian message to the pagan world. The message was anchored in God, the Creator, "who made the world and everything in it," "who gives everything – including life and breath – to everyone," with an abiding purpose, to be sure. This was recalled in Paul's paraphrase of the words of Genesis: "From one single stock he not only created the human race so that they could occupy the entire earth, but he decreed how long each nation should flourish and what the boundaries of its territory should be. And he did this so that all nations might seek the deity and, by feeling their way towards him, succeed in finding him."[44]

About the ways of finding God, Paul made statements harking back to the Wisdom Books. Men being the children of God "have no excuse for thinking that the deity looks like anything in gold, silver or stone that has been carved and designed by a man."[45] This affirmation of guilt is both tempered and emphasized by Paul in a reasoning which sheds full light on what has ever since been the very core of Christian thinking. Prior to the appearance among men of that most unique phenomenon, Christ, there was some excuse for ignoring God. But with Christ having lived an incomparable life, God now "is telling everyone everywhere that they must repent, because he has fixed a day when the whole world will be judged, and judged in righteousness, and he has appointed a man to be the judge. And God has publicly proved this by raising this man from the dead."[46]

The rest is well known. Reference to resurrection evoked laughter on the part of some present. Many others discovered in the risen Christ the supreme evidence of God's love, mercy, and the proof of the reasonableness of all He made both in space and in time, that is, in cosmic and human history. To keep this in mind is of crucial importance if one is to grasp the impact of Christianity on human culture. The insistence of Christianity on the rationality of nature, and on the ability of the human mind to recognize the Creator of nature, was never meant to be an isolated philosophical proposition. The persistence by which Christianity maintained, reaffirmed, and elaborated further these propositions derived from the unparalleled influence that Christ's figure was able to exercise not only from the distance of one, but of five, fifteen, and even fifty generations. What still animates Christianity is the conviction spelled out by Paul that "Christ is the image of the unseen God"[47] in a most unique sense, and that only in Christ are deposited the ultimate rationality and purposefulness of everything in heaven and on earth. For to use Paul's words again, "It is the same God that said, 'Let there be light shining out of darkness,' who has shone in our minds to radiate the light of the knowledge of God's glory, the glory on the face of Christ."[48] This is why Paul felt entitled to declare that by God's doing, Christ "has become our wisdom, and our virtue, and our holiness, and our freedom."[49] Thus, for the Christian the ideal of perfection is tied to the ideal of the "perfect man in Christ,"[50] that is, a man who searches not for narrow logic but for understanding in its broadest sense which gives justice to the facts of nature as well as to the facts of history, and which satisfies man's senses as well as his innermost aspirations.

The Christian certitude about the rationality of nature, about man's ability to investigate its laws, owes its vigour to the concreteness by which Christ radiated the features of God creating through that fulness of rationality which is love. "I am certain of this," wrote Paul, that "neither death nor life, no angel, no prince, nothing that exists, nothing still to come, not any power, or height or depth, nor any created thing, can ever come between us and the love of God made visible in Christ Jesus our Lord."[51] This passage is the capstone of a lengthy effort on Paul's part to convince his own flesh and blood about Christ. The argument begins on grounds commonly shared, the pity of Jews and Christians alike for men "who keep truth imprisoned in their wickedness."[52] In a tone recalling the Wisdom Books Paul reasserts man's ability to recognize the Author of nature: "For what can be known about God is perfectly plain to them since God himself has made it plain. Ever since God created the world his everlasting power and deity – however invisible – have been there for the mind to see in the things he had made."[53] Again the Wisdom Books are echoed when Paul unfolds the sad consequences of man's refusal to recognize God from the created world. Such a refusal is tantamount to making "nonsense out of logic"[54] and to the further darkening of the mind: "The more they called themselves philosophers, the more stupid they grew."[55] The final stage of the process was the wholesale erosion of public and private honesty and morality.

A world like that could not be transformed by reasoning alone, however persuasive. Only a memorable progression of martyrs could tilt the balance decisively. The immediate issue was monotheism. The larger implications were distinctly cultural. The example and teachings of Christ conjured up the picture of a more humane society, based on a more honest use of reason and on a substantially greater measure of willingness to follow its dictates. The relatively little which is available in the form of written records about the thinking of the first generation of Christian martyrs shows clearly their full awareness of being the true representatives of rationality precisely in virtue of their faith in the Creator of all who revealed Himself in a most special way in Christ. The famous letters of Ignatius, the venerable bishop of Antioch, are a case in point, and so is the first Epistle of Clement to the Corinthians, which deals with the urgency of peace among Christians, especially in times of "sudden and repeated misfortunes and calamities which have befallen us."[56]

The attitude exuding from the earliest documents of persecuted Christianity shows also that writings destined to form the canonical books of the Christian Church were written in an atmosphere in which the official or canonical readily grew out from the communal. Thus, the last canonical book of the New Testament, the Apocalypse of Saint John,[57] reveals a traditional pathos and perspective of faith. It culminates in a vision which contrasts the ultimate downfall of the symbolic Babylon, or human culture, victimized by idolatry, with the ultimate triumph of the symbolic Jerusalem, or the community of those faithful to the Covenant. Of the former the voice of the angel declares: "Babylon has fallen, Babylon the Great has fallen, and has become the haunt of devils and a lodging for every foul spirit and dirty loathsome bird. All the nations have been intoxicated by the wine of her prostitution; every kind in the earth has committed fornication with her, and every merchant grown rich through her debauchery."[58]

This downfall is as final as is its counterpart, the emergence of a new heaven and a new earth. Or as John tells his own vision: "Then I saw a new heaven and a new earth; the first heaven and the first earth had disappeared now, and there was no longer any sea. I saw the holy city, and the new Jerusalem, coming down from God out of heaven, as beautiful as a bride all dressed for her husband."[59] This new Jerusalem is, therefore, the completion of the Covenant which started with the creation of the world in the beginning. The very last statement made about God in the canonical books fittingly recalls His very first aspect stated in the very same books: "He is God, the Lord of all, whose name is "God-with-them." And just as the very first words of God created the first heaven and earth, the last words of God are creating in the same sense: "Then the One sitting on the throne spoke: 'Now I am making the whole of creation new' he said. 'Write this: that what I am saying is sure and will come true.' And then he said, 'It is already done. I am the Alpha and the Omega, the Beginning and the End'."[60]

Obviously, neither the God of the Covenant nor His works can be trapped in endless cycles and blind repetitions. With His essence only a cosmos and history are compatible which represent a once and for all process. As a result, the discrimination between those who shared this faith in God and those who opposed it is also final, irrevocable, and unrepeatable. The disastrous end of the latter is described as the ultimate consequence of idolatry: "But the legacy for cowards, for those who break their word, or worship obscenities, for murderers and fornicators, and for fortune-tellers, idolaters or any other sort of liars, is the second death in the burning lake of sulphur."[61] For those who cast their lot with the Maker and Restorer of all an eternal day is in store: "The throne of God and of the Lamb will be in its place in the city; his servants will worship him, they will see him face to face, and his name will be written on their foreheads. It will never be night again and they will not need lamplight or sunlight because the Lord God will be shining on them. They will reign for ever and ever."[62]

The solemn seriousness by which these declarations have been taken within Christianity cannot be pondered enough by anyone trying to fathom the real mainsprings without which modern culture, including its all-important scientific component, would be inconceivable. The Covenant was to become in a crucial sense the principal leaven in preparing that outburst of progress, the like of which history had not witnessed before. It will be the task of the next chapter to unfold a most telling aspect of the first phase of that process.

NOTES

1. A typical case is the discussion of the world view and culture of ancient Israel by P. Dupont-Sommer in R. Taton (ed.), *History of Science:* Vol. I, *Ancient and Medieval Science from the Beginnings to 1450*, translated by A. J. Pomerans (New York: Basic Books, 1963), pp. 122–132. More often than not, the contribution of some biblical beliefs to the emergence of science is simply ignored in books whose aim is to show the unique contributions of the people of the Old Testament to culture and thought. Among these books mention should be made of the following: D. B. MacDonald, *The Hebrew Philosophical Genius: A Vindication* (Princeton, N.J.: Princeton University Press, 1936), a true classic which was deservedly reprinted in 1965 (New York: Russel and Russel); *The Legacy of Israel*, a collection of essays edited by Edwyn R. Bevan and Charles Singer (Oxford: Clarendon Press, 1928), which is conspicuously short and superficial in its account of the great themes of biblical theology; "The Hebrews," a lengthy essay by W. A. Irvin in *The Intellectual Adventure of Ancient Man: An Essay on Speculative Thought in the Ancient Near East*, by H. and H. A. Frankfort *et al.* (Chicago: University of Chicago Press, 1946), pp. 221–360. The main point emphasized in Thorlief Boman's work is clearly conveyed in its title, *Hebrew Thought Compared with Greek*, translated from the German by J. L. Moreau (London: SCM Press, 1960).
2. Gn 2, 4–5. This and all subsequent quotations from the Bible are taken from *The Jerusalem Bible* (London and Garden City, N.Y. 1966) with the permission of the publishers, Darton, Longman & Todd, Ltd., and Doubleday and Co., Inc. From the many biblical references to Creator and creation, here only the most significant ones will be considered.
3. Is 2, 6–7.
4. Is 17, 7–8.
5. 2 Ch 33, 1–7.
6. Is 44, 19–20.
7. Is 40, 9–31.
8. Since, according to ancient observation, plants seem to sprout spontaneously from moist

ground, vegetation is considered in the biblical perspective as forming an organic unit with the soil.

9. The question whether the expression 'in the beginning' implies in that context the meaning of an absolute beginning is discussed in "Beginning," in *Encyclopedic Dictionary of the Bible* (A Translation and Adaptation of A. van den Born's *Bijbels Woordenboek*, Second Revised Edition 1954–57), edited by L. T. Hartmann (New York: McGraw-Hill Book Company, 1963), cols. 219–20.

10. For the semantic analysis of the verb *bara*, see J. van der Ploeg, "Le sens du verbe hébreu *bara*. Étude sémantiologique," *Le Muséon*, 59 (1946):143–57, and P. Humbert, "Emploi et portée du verbe *bara* (créer) dans l'Ancien Testament," *Theologische Zeitschrift*, 3 (1947): 401–22.

11. Gn 1, 28.

12. *Ibid.*

13. Gn 2, 1.

14. Ps 136, 4–9.

15. Ps 65, 8.

16. Ps 95, 4.

17. Ps 65, 9–13.

18. See Ps 8, 4; 19, 5–7; 104, 19; 148, 3.

19. See Ps 19, 3–5.

20. See Ps 104, 9; 148, 6.

21. See Jer 5, 24.

22. See Jer 31, 35.

23. The post-exilic origins of the Book of Job are now generally recognized. The most likely date of its composition is the early fifth century B.C.

24. Jb 9, 4; 9, 10. The same ideas are voiced later in the Book of Job by Bildad of Shuah, one of Job's visitors (Jb 25 and 26), though at a somewhat less elevated tone.

25. Jb 38, 1–7; 38, 31–35.

26. Jb 42, 1–6. In the quotation two lines between verses 3 and 5 have been omitted, as they are interpolation due to an early copyist's oversight.

27. Written most likely during the decades before the capture of Jerusalem by Pompey in 63 B.C.

28. The narration fills the whole seventh chapter of 2 M.

29. Like the two Books of Maccabees, this too, and the Book of Wisdom, are not in the Jewish canon of inspired books. Nevertheless, they are invaluable documents about Jewish thought during the last two centuries B.C.

30. Si 42, 15–26; 43, 1–37.

31. Written in Alexandria around the middle of the first century B.C.

32. Ws 7, 17–20.

33. Ws 7, 25–26.

34. Ws 11, 20–21.

35. Ws 11, 18.

36. Ws 13, 1–5.

37. Ws 13, 9.

38. Ws 14, 23–26.

39. Ws 14, 27.

40. Mt 10, 25.

41. Mk 12, 29.

42. 1 Co 8, 5–6.

43. Ac 17, 22–34.

44. Ac 17, 26–27.

45. Ac 17, 29.

46. Ac 17, 31.

47. Col 1, 15.

48. 2 Co 4, 6.

M

49. 1 Co 1, 31.
50. Ep 4, 13.
51. Rm 8, 38–39.
52. Rm 1, 18.
53. Rm 1, 20.
54. Rm 1, 21.
55. Rm 1, 22.
56. For the writings of Ignatius and Clement, see *The Apostolic Fathers* with an English translation by Kirsopp Lake (London: William Heinemann, 1912). Quotation is from Vol. I, p. 9. Clement, who first gives a magnificent description of the order and harmony of the universe created by God (sections xix and xx of his first letter), ties the ultimate salvation of the faithful through resurrection to the very fact of creation.
57. Although the authorship of John cannot be established with certainty, there can be no doubt about the book's genuine Johannine inspiration. Much of it was probably written around A.D. 95, though some of its parts may go back to Nero's time, or the late 60's. The book will be referred to as Rv.
58. Rv 18, 2–3.
59. Rv 21, 1–2.
60. Rv 21, 5–6.
61. Rv 21, 7–8.
62. Rv 22, 3–5.

CHAPTER EIGHT

The Leaven of Confidence

The rapid spread of Christianity throughout the Roman Empire has been the subject of many comments, much puzzlement, and heated debates. What cannot be controverted is the fact that at a remarkably early phase of its development Christian thought achieved a highly sophisticated awareness of its distinctive features with respect to some fundamental characteristics of classical culture and religiosity.[1] Not surprisingly, this ideological encounter first crystallized in Alexandria, the chief focus and haven of cultural activity and aspirations during late antiquity. It was there that history witnessed the emergence of the first school of Christian thought with Clement of Alexandria and Origen as its leading representatives. Clement and especially Origen were primarily interested in topics that had mainly to do with questions concerning Christian ethics, scriptural exegesis, and some fine points of theology. But a considerable part of their efforts aimed at vindicating the Christian message by showing its decisive superiority over a culture steeped in paganism.

In dedicating themselves to this double task Clement and Origen were as much innovators as followers. In a sense, their achievement consisted in giving full body to an already existing Christian intellectual tradition which showed a twofold aspect. On the one hand, it further articulated that part of the message of the Covenant which was primarily meant for those who already possessed the faith. On the other hand, the inevitable exposure to the pagan world necessitated reflections of more apologetical character with distinct bearing on cosmology. As to the former, it is important to note, in a cursory manner at least, the role of the belief in a Creator and Consummator of all in the formation of Christian religiosity and perseverance. It suffices to mention here *The Pastor of Hermas*,[2] an exhortational or devotional work, so popular in early Christian times as to create on occasion the impression that it belonged among the canonical books of the New Testament. The fundamental perspectives of the Covenant set the tone of the solemn admonition which is a high point of the work: "Lo, the God of powers, who by His invisible strong power and great wisdom has created the world, and by His glorious counsel has surrounded His creation with beauty, and by His strong word has fixed the heavens and laid the foundations of the earth upon the waters, and by His own wisdom and providence has created His holy Church, which He has blessed, lo! He removes the heavens and the mountains, the hills and the seas, and all things become plain to His elect, that He may bestow on them the blessing which He has promised them, with much glory and joy, if only they shall keep the commandments of God which they have received in great faith."[3]

The cosmological content of such a conviction was too explicit to remain unexploited in the encounters with those outside the fold professing a basically different explanation of the ultimate origin of the universe. In the *Address to the Greeks*[4] by Tatian, possibly the most impressive among the Christian apologetical writings of the second century, one finds not only the entertaining critique of the patent foibles of polytheism and idol worship. Tatian offers also some incisive reflections on the sole existence of God before He brought the world into existence "in the beginning": "For matter is not, like God, without beginning, nor, as having no beginning, is of equal power with God; it is begotten, and not produced by any other being, but brought into existence by the Framer of all things alone."[5]

Another apologist, Athenagoras, defined Christians as the ones "who distinguish God from matter, and teach that matter is one thing and God another, and that they are separated by a wide interval, for the Deity is increated and eternal, to be beheld by the understanding and reason alone, while matter is created and perishable . . ."[6] Not that Athenagoras meant to belittle the beauty of the universe. For him the world was "an instrument in tune, and moving in well-measured time," but he emphatically added that only that Being deserved worship "who gave [the world] its harmony, and strikes its notes, and sings the accordant strain."[7] The failure of philosophers to have a clear realization of this led them, according to Athenagoras, into patent inconsistencies about the origin and presumed permanence of the world. He saw a classic example of this in the doctrine of the Stoics who said "that all things will be burnt up and will again exist, the world receiving another beginning."[8] Such confusion could not be cleared up, argued Athenagoras, unless one recognized that "the efficient cause [God] must of necessity exist before the things are made."[9]

Apologetics is a complex art and permits more than one procedure. Exposure of the foibles of pagan practices represented one possible avenue, but there was also a more sophisticated approach. It consisted in showing that the best in pagan tradition came ultimately from the Covenant's wisdom codified by Moses. This procedure, practiced with great skill by Justin, the Martyr, led quite often to attributing to Greek philosophers positions they had hardly taken. Thus, in Justin's *Hortatory Address to the Greeks*[10] Plato was not, bafflingly enough, a representative of the eternity of the world. The author of the passage in the *Timaeus*, "time was created along with the heavens," learned, according to Justin, the beginning of the world in time from Moses.[11] Justin's reading of Plato is questionable, to say the least, but not his conviction about the fundamental importance of the opening phrase of Genesis. In his *First Apology*[12] Justin elaborated in the same vein the analogies of heathen mythologies with some Christian doctrines. The specific point he tried to make was that Christians should not be hated, as their beliefs were merely superior forms of ideas held also by pagans. But superior they were! To sense this one need only recall Justin's pointed comparison between the Christian belief in the end of the world and the Stoic doctrine of cosmic conflagrations: "And the Sibyl and Hystaspes said that there should be a dissolution by God of things corruptible. And the philosophers called Stoics teach that even God himself shall be resolved into fire, and they say that the world is to be formed anew by this revolution; but we understand that God, the Creator of all things, is superior to the things that are to be changed."[13]

Intimately connected in Justin's mind with the end of the world was, of course, the dogma of final judgment. The two formed one inseparable whole, and on it rested the reasonableness of cosmic and human history. He boldly voiced this tenet in his *Second Apology*,[14] addressed to the Roman Senate. He warned that august body of God's punishment because of their biased policy against Christians: "And that no one may say what is said by those who are deemed philosophers, that our assertions that the wicked are punished in eternal fire are big words and bugbears, and that we wish men to live virtuously through fear, and not because such a life is good and pleasant; I will briefly reply to this, that if this be not so, God does not exist; or, if He exists, He cares not for men, and neither virtue and nor vice is anything . . ."[15] The cogency and constructiveness of such reasoning could not make much impact in the Imperial Rome of Justin's time, dominated as it was by Stoic philosophy. Stoic moral doctrine contained, as Justin acknowledged, some admirable directives. Yet, it ultimately foundered because of the Stoic belief in an inexorable wheel of fate carrying along man and deity alike: "For if they say that human actions come to pass by fate, they will maintain either that God is nothing else than the things which are ever turning, and altering, and dissolving into the same things, and will appear to have had a comprehension only of things that are destructible, and to have looked on God himself as emerging both in part and in whole in every wickedness; or that neither vice nor virtue is anything; which is contrary to every sound idea, reason, and sense."[16]

Alongside paganism in its classical sense, another form of paganism, Christian in appearance, also posed a most serious challenge to Christianity during the latter half of the second century. The heresies recounted in Irenaeus' monumental work, *Against Heresies*,[17] represented in fact a complete parting with the fundamental propositions of Christianity. The Valentinians, Marcosians, Nicolaitans, Encratites, Borborians, Ophites, Sethians, and many others listed by Irenaeus, today are merely names familiar only to some specialists of church history. In Irenaeus' time they were at the peak of their influence. They were carriers of that religious syncretism and obscurantism which threatened with extinction whatever light the great philosophical schools of Greece succeeded in producing. Radical dualism and demonology, fatalism and reincarnation, emanationism and pantheism fused in those heresies with some Christian details, and were reinterpreted by an unbridled exercise in fantasy which barred any logic and consistency. The universe of those heretics was a maddening complex of demiurges and aeons that obeyed neither rhyme nor reason. One may only guess Irenaeus' mental nausea as he recalled the alleged production of all moist substance from the tears of Achamoth, the great female deity of some of his opponents. Her smile was the source of all lucid substance, her sadness gave rise to solid bodies, her terror produced mobile matter. Yet, the same ones who objected to a Creator who was not supplied with building material never tried to explain whence came to their Mother, a female from a female, "so great an amount of tears, or perspiration, or sadness, or that which produced the remainder of matter."[18]

A pastor of souls, Irenaeus must have known the futility of arguments addressed to wholly captive minds. But he had to defend the mental vision of his own flock. To achieve this, he thought it best to put emphasis on two points. One was the very basis of Christian belief, the faith in the Creator of all. The other was the fact, so clear to him already around A.D. 200, that only a firm adherence to the Church could safeguard that clear intellectual vision which derived from the very same faith. The first historical document which he recalled in this perspective was none other than that "most powerful letter to the Corinthians" dispatched by Clement, the bishop of Rome. As Irenaeus summed it up, the latter tried to resolve dissensions in Corinth, caused no doubt by an early wave of the same heretics, by impressing on the mind of the faithful in Corinth the notion of "the one God, omnipotent, the Maker of heaven and earth, the Creator of man."[19] A hundred years later the letter provided Irenaeus with a proof that genuine Christian doctrine was older than the views of "these men who are now propagating falsehood, and who conjure into existence another god beyond the Creator and Maker of all existing things."[20]

To mention after all this Irenaeus' insistence on a creation out of nothing, or his needling of the inner contradictions of the eternal recurrence through reincarnations, would be to belabour already familiar points. There are, however, some details worth recalling in the other great compendium of heretical doctrines written around A.D. 230, *The Refutation of All Heresies*[21] by Hippolytus, a disciple of Irenaeus. Unlike Irenaeus, Hippolytus emphasized that the conceptual vagaries of the heretics were an elaboration, or rather distortion, not of scriptural themes, but of the ideas and systems of philosophers and of sundry astrological doctrines and practices. Thus, in Hippolytus' work one finds an informative compendium of the doctrines of the Ionians, Pythagoreans, and the atomists with many interesting details about the later penetration of Babylonian, Persian, and Hindu thought into the Hellenistic mentality. Few developments appeared to Hippolytus as disconcerting as the practically complete harnessing of astronomy into the devious objectives of astrology. He grieved in particular that Ptolemy had not lived when the Tower of Babel was erected. By Ptolemy's calculations of celestial distances the maddening project to reach the heavens could have been nipped in the bud.[22]

Space does not allow us to summarize even briefly the many bizarre "cosmologies" surveyed and criticized by Hippolytus. Most of them were variations on the idea of the eternity of a divine world revolving through infinite aeons.[23] Their most immediate, or Greek, origins were traced by Hippolytus to the world conceptions of Empedocles, Parmenides, and Democritus. Hippolytus noted not only the adoption by the Stoics of Empedocles' idea that "all things consist of fire, and will be resolved into fire," and so on ad infinitum.[24] Hippolytus also recalled Empedocles' espousal of the tenet of the transmigration of souls even into fish, shrub, birds, and the oceans to boot. Through Hippolytus we also have the ancient story according to which Pythagoras, who instructed Empedocles about reincarnation, "asserted that he himself had been Euphorbus, who served in the expedition against Ilium, alleging that he recognized his shield."[25]

Hippolytus' account of Parmenides' philosophy also centred on the simultaneous assertion of the eternity of the world and of its endless conflagrations.[26] Hippolytus provided a valuable insight into his own picture about the historical roots of the idea of eternal returns as he began to discuss Democritus' cosmology, where the innumerable parts of an infinite world were in the perennial process of formation and dissolution. Democritus, he stated, propounded his system after "conferring with many gymnosophists among the Indians, and with priests in Egypt, and with astrologers and magi in Babylon."[27] No wonder that Hippolytus several times assigned the popularity of the Great Year to Chaldean influences.[28] The significance of such a clear awareness on Hippolytus' part about these matters cannot be emphasized enough. His case shows that the intellectually trained Christian could not help realizing that the major patterns of accepted world views contrasted sharply with the perspectives of the Covenant. A most telling aspect of this contrast was the meaningfulness which the belief in a Creator gave to the Christian world view, and the glaring lack of constructive meaning in the absence of such belief. Or as Hippolytus summed up the logical outcome to which Democritus' world view, inspired by Hindu gymnosophists and others, ultimately led: "This (philosopher) turned all things into ridicule, as if all the concerns of humanity were deserving of laughter."[29] This was precisely the type of laughter that reverberated with shudders.

Hippolytus obviously knew the difference between coarse and highly refined formulations of the very same fundamental error, the pantheistic eternity of the universe. He found both types seriously misleading in their own ways. Thus, he recalled in an incisive comment on Aristotle's contentions about deity and the universe: "The definition, however, which Aristotle furnishes of the Deity, is, I admit, not difficult to ascertain, but it is impossible to comprehend the meaning of it. For, he says, (the Deity) is a 'conception of conception'; but this is altogether a non-existent (entity). The world, however, is incorruptible (and) eternal, according to Aristotle."[30] Hippolytus had only contempt for Basilides, a heresiarch, who tried to transfer "the tenets of Aristotle into our evangelical and saving doctrine."[31] The touchstone of that saving doctrine was given with great solemnity by Hippolytus as he summarized in the concluding section both the pagan and Christian positions: "The first and only (one God), both Creator and Lord of all, had nothing coeval with Himself, not infinite chaos, nor measureless water, nor solid earth, nor dense air, nor warm fire, nor refined spirit, nor the azure canopy of the stupendous firmament. But He was One, alone in Himself. By an exercise of His will He created things that are, which antecedently had no existence, except that He willed to make them."[32]

When Hippolytus, often called the Origen of the West, achieved martyrdom about A.D. 236, the Alexandrian school of catechetics founded by Clement had already been for some time under Origen's direction. As to Clement, he was not so much interested in heresies as in the portrayal of the clarity of mental and moral perspectives secured by Christianity. Needless to say, he tried to make his point by constantly referring to the dark background represented by pagan antiquity. As emphasized in Clement's *Exhortation to the Greeks*,[33] the moral, social, and intellectual aspects of classical antiquity were inextricably interwoven with the debasing influences of the universal worship of idols. A chief result of idolatry consisted in a mental enslavement to the blind forces of nature. The classical concept of cosmos as a clearly structured harmony represented only one side of a coin the reverse of which reflected fear and despair in the face of inexorable and capricious forces in nature. In the same manner the mouthpieces of learning kept moving back and forth between rational statements and irrational myths. Christians, as Clement put it, refused to follow the pagan teachers who "though wise in their own conceit, have no more knowledge than infants."[34] This harsh stricture should be seen in the light of questions which Clement addressed to pagan educators: "Why, in the name of truth, do you show those who have put their trust in you that they are under the dominion of 'flux' and 'motion' and 'fortuitous vortices'? Why, pray, do you infect life with idols, imagining winds, air, fire, earth, stocks, stones, iron, this world itself to be gods? Why babble in high-flowing language about the divinity of the wandering stars to those men who become real wanderers through this much vaunted, – I will not call it astronomy, but – astrology?"[35]

Ordinary teachers merely followed in the footsteps of the leading investigators of the universe, the philosopher-scientists. "The host of philosophers turn aside," noted Clement, when faced with the question of the making of the universe. "They admit that man is beautifully made for the contemplation of heaven, and yet worship the things which appear in heaven."[36] Therefore, valuable insights into the harmony of the universe could not produce the desired purification of thought and attitudes. "Common opinion and custom," as Clement referred to superstitions and myths, had set the tone by making "slaves of those who follow them instead of searching after God."[37]

Slavery here meant a mental bondage that barred the emergence of a confident attitude toward nature. This is what Clement had in mind as he evoked the melodious tune of the harmony of a world coming forth from the hands of a rational as well as a benevolent Creator: "This pure song, the stay of the universe and the harmony of all things, stretching from the centre to the circumference and from the extremities to the centre, reduced this whole to harmony, not in accordance with Thracian music, which resembles that of Jubal, but in accordance with the fatherly purpose of God, which David earnestly sought. He who sprang from David and yet was before him, the Word of God, scorned those lifeless instruments of lyre and harp. By the power of the Holy Spirit He arranged in harmonious order this great world, yes, and the little world of man too, body and soul together; and on this many-voiced instrument of the universe He makes music to God, and sings to the human instrument."[38]

The passage is expressive not only because of its testimony to a robust confidence in human and cosmic existence, but also because of its witness to the crucially concrete role which the historical fact of Christ played in generating that confidence. The fact was gigantic and so was the encouragement it produced. On the one hand, it inspired the warning of a small David, the nascent Christianity, to the Goliath of antique culture: "let none of you worship the sun; rather let him yearn for the maker of the sun. Let no one deify the universe; rather let him seek after the creator of the universe."[39] On the other hand, it prompted a trust in the positive achievements of a pervasive though distinctly hostile culture, in its philosophy and science to be specific. After all, Christian belief in a personal, rational Creator pictured Him as being the weight, measure, and number of the universe, or rather "the only just measure, because He is always uniformly and unchangeably impartial, measures and weighs all things, encircling and sustaining in equilibrium the nature of the universe by His justice as by a balance."[40]

Such a firm conviction could only rejoice at the presence of genuine fragments of truth in antique learning. Among the many recounted by Clement, the one which he attributed to the Pythagoreans is worth quoting in full: the Pythagoreans say that "God is one; and He is not, as some suspect, outside the universal order, but within it, being wholly present in the whole circle, the supervisor of all creation, the blending of all the ages, the wielder of His own powers, the light of all His works in heaven and the Father of all things, mind and living principle of the whole circle, movement of all things."[41] This positive appraisal of the natural abilities of reason also implied that a faith corroborated by philosophical and scientific considerations was preferable to a simple faith. Clement voiced this conviction in the form of a stricture of those Christians who, satisfied with their spiritual riches, "do not wish to take a look at philosophy, or dialectic, and much less do they wish to learn the secrets of natural sciences. All they claim is the bare faith."[42]

The Christians Clement had in mind were not the simple, uneducated folks that especially in his time made up the bulk of the faithful. Not that the lowly status of most of his coreligionists had been an upsetting factor for him. Long before Clement, Saint Paul pointedly noted to the faithful of the bustling port city of Corinth that the great majority of Christ's followers could boast of no credentials with respect to descent, holdings, and learning. Humble origins and conditions of life were to remain for the most part a basic hallmark of Christian fellowship. This, however, did not suggest apathy about learning. Christians with education were expected to excel and face the just demands of an inquiring mind. Few of them have ever implemented this task with a daring and originality comparable to that of Origen, himself a pupil of Clement and his successor as the leader of the catechetical school in Alexandria.

As the title of one of Origen's masterworks, *On First Principles*,[43] indicates, Origen had a mind eager to explore basic issues. The book contains the first systematic discussions within a broadly theological framework of such fundamental philosophico-scientific problems as the freedom of the will and the origin and duration of the world. It was mainly these discussions that also cast on Origen the suspicion of unorthodoxy. He did his best to assimilate into a Christian synthesis everything that appeared to him grandiose in cosmological speculations. Coupled with this aspiration was an attitude exuding an unusual measure of cosmic compassion. Thus, Origen's faith in the benevolent Father and Creator of all led him to propose a process of gradual purification for all, even for the obstinately wicked. The mechanism serving this purpose consisted of a repeated existence throughout a great number of ages or worlds. Concerning the number of successive worlds Origen stated: "What may be the number or measure of these worlds I confess I do not know; but I would willingly learn, if any man can show me."[44]

The idea of transmigration of souls suggests Oriental influences that had been exerting considerable impact throughout the Hellenistic world. There is also some Oriental flavour in another favourite idea of Origen, the *katabolé*,[45] or the explanation of actual material existence as a result of gradual deterioration from higher levels of being. "There has been a descent from higher to lower conditions not only on the part of those souls who have by the variety of their own movements deserved it, but also on the part of those who have been brought down, even against their will, from those higher invisible conditions to these lower visible ones, in order to be of service to the whole world. For indeed 'the creation was subjected to vanity, not willingly, but by reason of him who subjected the same in hope'

[Rom. viii, 20], the hope being that both sun and moon, stars and the angels of God should fulfil an obedient service for the world; and it was for those souls which on account of their excessive spiritual defects required these grosser and more solid bodies and also for the sake of those others for whom this arrangement was necessary that the present visible world was instituted."[46] Such an approach to the laws underlying the realm of matter hardly seems conducive to a scientific exploitation, especially if one recalls that, according to Origen, the Artificer of all made the earth to be such as "to be able to hold all those souls which were destined to undergo discipline in it and also those powers which were appointed to be at hand to serve and assist them."[47]

This somewhat ghostly picture is not, however, without some saving features. Origen's *katabolé* is not an inexorable treadmill. First, the cause of punitive descent is a misdeed committed by the pure soul in a full exercise of its freedom of choice. Second, the process of repeated existences is not a blind, endless run. It is rather directed toward a shining goal, the restoration of all, which because of the goodness and justice of the Creator will certainly be achieved: "Souls are not driven," according to Origen, "on some revolving course which brings them into the same cycle again after many ages, with the result that they do or desire this or that, but they direct the course of their deeds towards whatever end the freedom of their individual minds may aim at."[48] Origen had only scorn for the idea of mesmerizing returns whether on the individual or on the cosmic level, and he justified his stance by displaying a firm grasp of the scientific side of the issue. Protagonists of identically returning worlds, he noted, demanded nothing less than sheer impossibility. It would be utter folly to expect with them that by pouring out twice the same bushel of corn on the ground the grains would fall in exactly the same position, or that the same grains would lie next to each other in both cases. Moreover, in view of the very large number of grains of corn, the exact repetition of the spreading of corn on the ground would not occur should the process be repeated through a fairly long time. Origen's conclusion readily followed: "It seems to me, then, impossible that the world could be restored again a second time with the same order and the same number of births, deaths and actions; but worlds may exist that are diverse, having variations by no means slight, so that for certain clear causes the condition of one may be better, while another for different causes may be worse, and another intermediate."[49]

But over and above such considerations there was an absolutely superior chain of events, the major steps of redemption culminating in Christ, which in their uniqueness barred any possible flirtation with the idea of eternal and purposeless cosmic cycles: "For if it is said that there is to be a world similar in all respects to the present world, then it will happen that Adam and Eve will again do what they did before, there will be another flood, the same Moses will once more lead a people numbering six hundred thousand out of Egypt, Judas also will twice betray his Lord, Saul will a second time keep the clothes of those who are stoning Stephen, and we shall say that every deed which has been done in this life must be done again."[50]

By recalling in such a context the whole history of salvation from the first parents to the emergence of the Church, the new family of God, Origen posed himself a problem which he tried to solve valiantly, though it clearly defied satisfactory explanation. The problem consisted in the correlation between cosmic and religious history. Origen assigned Christ's life, in line with the scriptural phraseology, to the age that represented the consummation of previous ages, but added in the same breath that other ages still would follow on a glorified level culminating in the complete restoration of all things. If this was already evasive from the viewpoint of chronology, no light was shed by Origen on the "age" in which the human race saw its beginning as the progeny of the first parents.

About one fundamental point Origen left no doubt. The chain of ages had an absolute beginning in the moment of creation, which also signalled the start of the flow of time.[51] Furthermore, Origen stated with similar firmness the idea of an absolute end for cosmic processes, and described it as a state exactly similar to the condition prevailing at the beginning. It is worth noting that Origen connected the finiteness of the universe along the parameter of time (its finite size had been the prevailing view) with the notion of intelligibility, anchoring it in God's ability to comprehend things. Here, Origen ended up in an uneasy position. By insisting that only those processes could be comprehended that have a beginning and an end, he seemed to imply that not even God could comprehend an endless process.[52] More specifically, he had to face the objection as to what God had been doing before the moment of creation. His reply, "God did not begin to work for the first time when he made this visible world, but that just as after the dissolution of this world there will be another one, so also we believe that there were others before this one existed,"[53] lacked what it claimed, logic, or at least consistency. For in the same context Origen excluded only the simultaneous existence of several worlds. He failed to recall what he had already previously stated, the absolute beginning and end of all worlds.

Such verbal exercise was perhaps all that those deserved who felt concern about God's "inactivity" before the moment of creation. More discriminating readers of Origen's work had ample reason to feel disturbed by his bold handling of such fundamental tenet of the faith as creation. Not that Origen had not toned down, as the years went by, the controversial character of some of his speculations. Two decades later, toward the end of his life, he faced once more the question of the eternal recurrence of all in his great masterpiece, *Contra Celsum*,[54] the refutation of a notable attack on Christianity by the Greek philosopher, Celsus, who had been active three generations earlier. Celsus' work, entitled "The True Word," is no longer extant, but its structure can be reconstructed with fair accuracy from Origen's reply.[55] The results show an impressive skill on Celsus' part in marshaling his arguments. His opening salvo aimed at discrediting the Christians whom he portrayed in the first part of his book as inconsistent, unsophisticated renegades of Judaism. Celsus obviously tried to capitalize on the anti-Jewish sentiments of Hellenistic times before he turned to the major issues of his task. They related to the claims that the basic Christian tenets were either contradicted by the fundamental propositions of Greek philosophy, or that they were at best inept borrowings from it. It is, indeed, a proof of Celsus' acumen that he articulated with utmost clarity the crucial point of conflict: according to him the doctrine of a personal Creator reinforced by the concept of an Incarnate God was wholly incompatible with the genuine idea of nature. Clearly, Celsus, in line with the best Greek philosophical tradition, could conceive of religion only as an ennobled form of naturalism. In his eyes nature was the supreme being outside of and above which nothing could exist. Nature was eternal and for ever recurring.

An outstanding trouble with all this was that matter had patently corruptible features which could not be considered divine. Thus, with the traditional inconsistency of his forebears in philosophy, Celsus fell back on doubling the number of eternal principles constituting nature. In addition to the divine aspect of nature, of which the superlunary regions were taken as the most palpable evidence, Celsus added another, equally eternal principle, rooted in the permanence of ordinary matter. From this followed the "principle of the conservation of evil" in nature, as the total quantity of matter was to remain the same throughout eternity. What was not the same in each moment was the amount of evil manifested through matter. But the total quantity of evil was the same for a so-called great cycle which included the identical return of all. Or to quote only one passage of Celsus from among the many preserved in Origen's reply: "It is not easy for one who has not read philosophy to know what is the origin of evils; however, it is enough for the masses to be told that evils are not caused by God, but inhere in matter and dwell among mortals; and the period of mortal life is similar from beginning to end, and it is inevitable that according to the determined cycles the same things always have happened, are now happening, and will happen."[56]

To be sure, Origen categorically opposed all of this, but his reply is worth seeing in detail as it is replete with considerations and convictions without which no sustained scientific investigation of nature is conceivable. They are remarkably free of the pitfalls of the outlook on nature advocated by Celsus, and should help bring into focus the indispensable role played by certain postulates in the full-scale emergence of science. Recalling the original setting of those postulates should seem extremely important in an age of feverish scientific activity where fascination with the intoxicating details of the superstructure can easily distract attention from the foundations. These are apt to be taken for granted to the point of slighting their permanent validity altogether under the impact of facile exercises in an allegedly scientific philosophy.

Origen's categorical rejection of the idea of matter as something evil implied far more than scoring a theological point. Underlying that rejection was an unshakable courage in the face of a nature that could appear inscrutable and fearsome to puny man. By assigning the source of all evil to the freedom of mind as distinct from the body, Origen asserted the inherent, ethical neutrality of matter, or rather its natural goodness as far as it existed and was governed by laws. About the "conservation of evil" in each great cycle, Origen emphatically noted that "if this is true, free will is destroyed."[57] How deeply he must have felt the seriousness of the case should be clear from the dramatic realism that pervades his illustration of the implications of Celsus' claim: "For if *it is inevitable that in the period of mortal life according to the determined cycles the same things always have happened, are now happening, and will happen*, it is obviously inevitable that Socrates will always be a philosopher and be accused of introducing new deities and of corrupting the youth; Anytus and Meletus will always be accusing him, and the council on the Aeropagus will vote for his condemnation to death by hemlock. Thus also it is inevitable that according to the predetermined cycles Phalaris will always be a tyrant, Alexander of Pherae will commit the same atrocities, and those condemned to the bull of Phalaris will always groan inside it."[58] A portrayal of this type with its gripping details can only provoke revulsion, and Origen was ready to exploit the effect in full. After repeating the foregoing phrase of Celsus he applied it to sacred history as well: "Then it is inevitable that according to the determined cycles Moses will always come out of Egypt with the people of the Jews; Jesus will again come to visit this life and will do the same things that he has done, not just once but an infinite number of times according to the cycles."[59]

These latter remarks naturally carried weight with Christians alone, but for them they also laid bare that layer of rock on which all flirtation with cosmic returns ultimately had to run aground. The once-and-for-all facts of the history of salvation, stretching from the Creation to the redemptive work of the only begotten Son of the Creator, formed for Christians a framework of thought which allowed for no compromise whatsoever. Origen knew only too well the psychological depths of Christian sensitivities on this point and it was, therefore, not necessary for him to verbalize the violent revulsion of Christians against the idea of cyclic returns. He displayed an equally sharp sense of psychology when he turned to the purely rational aspect of the question. Within the conceptual framework of cyclic returns, he noted, it was impossible to argue any cause in a convincingly consistent manner. Within that framework it was useless to make attempts at influencing history, and this held of Celsus' attempts as well. Or as Origen noted: "The same people will be Christians in the determined cycles, and again Celsus will write his book, though he has written it before an infinite number of times."[60] Writing a book against Christians, or against anything, could hardly appear more senseless, and no remark could better illustrate why the prospect of endless returns was utterly self-defeating.

It now remained for Origen to dispose of an ineffectual proviso of Celsus who desperately tried to take the sting out of the cosmic treadmill by exempting the gods and all "immortal aspects" of life from the blight of an inexorable process. What Celsus seemed to convey was that though there was an endless repetition of lives exactly similar, for instance, to that of Socrates, Socrates himself, in his innermost individuality at least, would not be involved in endless returns. But Celsus' proviso not only clashed with the Stoic doctrine of the universality of identical returns. It also represented an inept exercise in words and Origen did not hesitate to brand it as such by taking to task the Stoics as well: "In attempting to remedy the absurdities in some way the Stoics say that in every cycle all men will be in some unknown way indistinguishable from those of former cycles. To avoid supposing that Socrates will live again, they say that it will be some one indistinguishable from Socrates, who will marry some one indistinguishable from Xanthippe, and will be accused by men indistinguishable from Anytus and Meletus. But I do not know how the world can always be the same, and one world not merely indistinguishable from another, while the things in it are not the same but are indistinguishable."[61]

Origen was also at his debating best when he noted later in the work that it was absurd to argue against the Christian doctrine of resurrection while professing the doctrine of eternal returns. And once again he held up to mockery the endless return or "resurrection" of individuals by recalling that some Stoics "felt embarrassed" by the doctrine and suggested slight differences for each return. The majority view was, however, that

> in the succeeding period it will be the same again: Socrates will again be son of Sophroniscus and be an Athenian, and Phaenarete will again marry Sophroniscus and give birth to him. Therefore, although they do not use the word 'resurrection' at least they have the idea when they say that Socrates will rise again after originating from the seed of Sophroniscus and will be formed in the womb of Phaenarete, and after being educated at Athens will become a philosopher; and something like his previous philosophy will rise again and will similarly be indistinguishable from the one before. Moreover, Anytus and Meletus will again rise up as Socrates' accusers, and the council of the Aeropagus will condemn him. And, what is more ludicrous than this, Socrates will put on clothes which will be indistinguishable from those of the previous period, and will be in poverty and in a city called Athens which will be indistinguishable from that before. Phalaris will again be a tyrant, with a cruelty indistinguishable from that of the previous period, and will condemn men also indistinguishable from those before.[62]

It is in this connection that Origen lashed out at all supporters of the idea of eternal returns, among whom he named the Pythagoreans and the Platonists. "For when in certain fixed cycles the stars adopt the same configurations and relationships to each other, they say that everything on earth is in the same position as it was at the last time when the relationship of the stars in the universe to one another was the same. According, then, to this doctrine it is inevitable that when after a long period the stars come into the same relationship to one another which they had in the time of Socrates, Socrates will again be born of the same parents and suffer the same attacks, and will be accused by Anytus and Meletus, and be condemned by the council of the Aeropagus."[63] It was then right for Origen to turn the table on Celsus by asking him whether it would not be more logical to laugh at the proponents of eternal returns, including the "learned men among the Egyptians who have similar traditions respected," instead of jeering at Christians "for believing in paltry stories fit for old women?"[64] The blame for voicing "utterly absurd ideas" clearly lay at the door of those who, in Origen's words, "on grounds of their rational insight and dialectical speculations may not be lightly regarded."[65]

What, then, was ultimately wrong with a culture that could rightly be proud of so many "rational insights" and remarkable "dialectical speculations"? Origen gave the answer in a straightforward manner as he distinguished between two types of return, that proposed by the philosophers and that which formed the backbone of his theological speculations. The former amounted to a blind treadmill, the latter was goal-directed, and animated by purpose because everything that happened did so under the guidance of a rational and benevolent Creator. For this is the gist of Origen's question: "While as for us who say that the universe is cared for by God in accordance with the conditions of the free will of each man, and that as far as possible it is always being led on to be better, and who know that the nature of our free will is to admit various possibilities (for it cannot achieve the entirely unchangeable nature of God), do we appear to say nothing worthy of trial and study?"[66]

Origen could be even more explicit on the crucial role played by belief in the Creator in generating confidence for an understanding of the cosmos with a purposeful place for man in it. He did this by comparing his version of the succession of cosmic ages to the productive cycle of seasons:

> Even though everything had been arranged by Him at the creation of the universe to be very beautiful and very steadfast, yet nevertheless He has had to apply some medical treatment to people sick with sin and to all the world as it were defiled by it. In fact, nothing has been or will be neglected by God, who at each season makes what He should be making in a world of alteration and change. And just as at different seasons of the year a farmer does different agricultural jobs upon the earth and its crops, so God cares for whole ages as if they were years, so to speak. In each one of them He does what is in itself reasonable for the universe, which is most clearly understood and accomplished by God alone since the truth is known to Him.[67]

Concerning the ultimate source of that cosmic vision predicated on the fusion of truth and benevolence, Origen pointed to a central factor in Christian experience, the recognition of Jesus as the Incarnate Word of God. Once more, Origen did this with an eye on the doctrine of cosmic cycles. In its pagan version, he noted, there was no true place for resurrection, that is, for a transformation of life which preserves the continuity of the individual. His conclusion reverberates with the firmness of a conviction which set history on entirely new tracks: "For we know that even if heaven and earth and the things in them pass away, yet the words about each doctrine, being like parts in a whole or forms in a species, which were uttered by the Logos who was the divine Logos with God in the beginning, will in no wise pass away. For we would pay heed to him who says: 'Heaven and earth shall pass away, but my words shall not pass away'."[68]

Origen was only one of the countless who demonstrated the depth of that conviction by martyrdom. Their heroism promptly turned into a verification of the valour of the never-before-heard claims voiced by the Master from Nazareth. It was also the extraordinary nature of those claims that prompted zealous attention to keeping them in their utmost purity. What some latter-day historians simply classified as delusions leading to bloody fights around an iota were seen by the Fathers of Church as the wellspring of Christian endeavour which had to be protected at any price. Origen, the martyr, was one thing; Origen, the teacher, another. The latter did not always succeed in matching his good intentions with equally felicitous phrases. His well-meaning efforts to utilize the doctrine of cosmic cycles for Christian doctrine evoked the censure of Christian posterity which, consistently enough, could not permit a compromise, however slight, with the once-and-for-all nature of human life within the framework of the history of salvation.

All this comes through very clearly in a letter of Jerome, the sharp-penned critic of Origen in Christian antiquity. Jerome's lengthy letter to Avitus[69] is not a paragon of scholarly objectivity. Yet, it forcefully reveals the irreconcilable opposition between the idea of endless world-cycles and the Christian view of cosmic and human destiny. Jerome's righteous indignation may appear as far-fetched as Origen's flight of fancy. But Jerome spoke the voice of the mainstream of Christian consciousness when he denounced the succession of worlds, the transmigration of souls, and their changing from angels to devils and back, as an attempt to foist "upon the simple faith of Christians the ravings of philosophy."[70]

As to his disparaging remark on philosophy, it had ample justification when antique philosophy was taken in its concreteness, that is, as the actual milieu of thought pervaded by obscurantism of the grossest type. Concerning the faith of Christians he called it simple, not because it took refuge in the thinking of simpletons, but because it rested on an irrevocable commitment to a few simple statements about the universe. Those statements were originally recorded in the Bible which no one in antiquity interpreted with heavier reliance on the scholarly criteria of textual criticism than Jerome did. As an indefatigable inquirer into the manifold shades of meaning in biblical utterances, Jerome could not fail to realize that the uniqueness of the Bible largely rested in its account of the origin and destiny of the cosmos.

The details of that origin, as given in the six-day story of Genesis, were recognized by leading Church Fathers as being tailored to the times and circumstances in which the first chapters of Genesis were composed. Origen already remarked on the uneducated audience to which Moses had to speak,[71] and the same was emphasized in such classics of devotional exegesis as Saint Basil's homilies on the Hexaemeron,[72] or Saint Chrysostom's homilies on Genesis.[73] Of the two, Basil paid greater attention to the disparity of the biblical and the scientific (Aristotelian) world picture, but not to the point of losing sight of the unique cosmological message of the Bible on the radical contingency of the world. Basil, whose homilies remained very influential for the remainder of the patristic age,[74] emphatically singled out the first phrase of Genesis, "In the beginning God created the heaven and the earth," as the indispensable guideline for a basically correct interpretation of the physical world. It was against that phrase that every proposition which called this contingency in question was to be weighed and found wanting.

Educated in the best schools of Caesarea, Constantinople, and Athens, Basil urged young Christians to get the best education which at that time was still largely in the hands of pagan teachers. In his opinion, the clarity of the fundamental propositions of biblical cosmology and anthropology represented an effective protection against the "confusion" of philosophical opinions about man and world. In his homilies, Basil called upon the philosophers to agree first among themselves before making objections to the biblical doctrine.[75] Their most fateful error, Basil noted in the opening section of his first homily, was their infatuation with the circular motion of the heavens in which they found the supreme confirmation of their belief in the eternity and divinity of the world.[76] His remark, that every circle must have a starting or generating point, its centre, was to buttress his emphasis on the limitedness of the world. For Basil, the biblical doctrine of the world stood for a radical finiteness of the world in time. This implied for him not only an absolute beginning but also an end, the final transformation of all: " 'In the beginning God created.' It is absolutely necessary that things begun in time be also brought to an end in time. If they have a beginning in time, have no doubt about the end."[77]

In this perspective it was most natural for Basil to argue that changes (deformations and decays) in the world indicated its basic corruptibility. Unlike his pagan counterparts in philosophy, Basil, the Christian, was not overawed by the heavenly regions. He spoke with utter indifference about the supposedly divine fifth element, the ether, which he described as a mere hypothesis needed by faulty presuppositions.[78] This was not the only point where Basil's reasoning anticipated considerations that later helped science find its right track. The omnipotence of the Creator demanded, according to him, a parting with the Aristotelian dogma of the necessary finiteness and exclusiveness of the actual world.[79] Again, it was his belief in the Creator that prompted Basil's warning that no real explanation was provided by the method of a physical science which kept replacing one "basic" explanation with another without ever finding a really final one.[80] A similar motivation lay behind another of his remarks, namely, that even when an explanation was found for a particular phenomenon, the phenomenon itself remained as worthy of awe and admiration as it was before. The radical mysteriousness of existence and order was not to be pre-empted by any subsequent scientific attainment.

For this scientific development to get under way astrology had to cease dominating minds and emotions. Basil's incisive and relentless criticism of astrology in the light of the biblical doctrine of creation is a powerful illustration of a new outlook on the world, and forms a striking contrast with the murky mental atmosphere exuding from Ptolemy's *Tetrabiblos* and other astrological works which nipped science in the bud. That Basil dismissed out of hand the doctrine of "ages" or eternal returns[81] needs no detailed comments. For him the ultimate consideration about the world came from the ultimate destiny of man and in the biblical context this could only be a once-and-for-all process. The most fundamental of all questions concerned not the world but man's quest for happiness, which hardly made sense in that particular form of eternal cycles, the endless transmigration of souls: "Shun the idle talk of proud philosophers, who are not ashamed to regard their own soul and that of dogs as similar, who say that they were at some time women, or bushes, or fish of the sea."[82]

The audience to which these words were addressed had its large share of simple folks. With them a thorough airing of questions having philosophical and scientific ramifications was evidently not feasible. Systematic analysis of such questions could only be done in scholarly treatises of which, in patristic times, the most monumental in scope and breadth was *The City of God*[83] by Saint Augustine. It moulded more than any other book by a Christian author the spirit of the Middle Ages. Its pages were as many wellsprings of information and inspiration for the emerging new world of Europe about the meaning of mankind's journey through time. What the medievals learned in that book was, above all, the proposition that the physical universe and human history both had their origin in the sovereign creative act of God, which also established a most specific course and destiny for both. The intelligibility of human and cosmic existence portrayed in that framework represented a compactness and consistency the like of which had never before appeared

N

on the pages of any book. *The City of God* had, indeed, become the intellectual vehicle for a confidence which centuries later made possible the emergence for the first time of a culture with a built-in force of self-sustaining progress.

For the time being, it was not the rise of culture but its collapse that had to be given an explanation. The writing of *The City of God* was prompted by the sack of Rome in 410, an indisputable evidence of the throes of antique civilization. The next two decades brought more than enough proof that the days of classical antiquity were numbered, that Hellenistic learning and *pax Romana* were turning into smouldering embers waiting for final extinction. In such circumstances nothing was so natural as the desire to probe into the reasons of a wholesale calamity. The pagans pointed to the Christians, but it was easy to show, as did Augustine in the first part of *The City of God*, that pagans carried the sole responsibility for most of the ills and woes that presaged the demise of the Roman Empire.

Showing this represented only the negative part of Augustine's task. What he tried above all to prove was that the accelerating rate of cultural collapse palpably bespoke the onset of the final phase of cosmic and human history. No special learning is needed to see the primitiveness of Augustine's information about details. His resolute defence of a few thousand years as the life-span for cosmos and mankind alike makes the present-day reader uneasy. But Augustine had what most moderns armed with science (in more than one sense) do not possess: an unshaken confidence that regardless of a historic holocaust both cosmos and mankind were on a wholly meaningful course toward a supreme fulfilment.

Augustine's conviction rested on what has been aptly called the very core of all that his extraordinary mind produced: his awareness of the utter dependency of every creature on the creative act of a wholly transcendent Being. In a most logical way, the positive, or second part of *The City of God*, starts with extensive considerations about Creation that run through much of Book XI of Augustine's great classic. Among the points emphasized by him in this respect was the finiteness of the universe in time and space. Concerning the latter he quickly noted that an infinite space implied "Epicurus' dream of innumerable worlds,"[84] a situation hardly compatible with an unequivocal sense of purpose. As to the problem of infinity in time, Augustine's reply rested on his conviction of the basic goodness and beauty of the universe as issued from the hands of the Creator. Had the world existed from eternity, souls too would have existed from eternity and also their ability to sin. This implied the idea of a fall since eternity which, in Augustine's eyes, suggested the Manichean idea about an eternal principle of evil rooted in matter. To such a notion Augustine, the former Manichean, was not to give even unwitting support. For him the material universe was beautifully ordered and radically good, and he clung to that conviction with utmost resolve. It was in this connection that he took issue with Origen who, according to him, pictured the world as a place created for the punishment of souls guilty of evil.[85] For Augustine it was wholly baffling that "some even of those who, with ourselves, believe that there is only one source of all things, and that no nature which is not divine can exist unless originated by that Creator, have yet refused to accept with a good and simple faith this so good and simple a reason of the world's creation, that a good God made it good."[86]

Odd as it may seem, Augustine made no mention of Origen as he discussed in the next book of *The City of God* the creation of man in particular. Omitting any reference to Origen did not in any way indicate on Augustine's part leniency toward the doctrine of the cyclic return of man to life. Augustine's first strictures were directed at the least reprehensible form of cyclic returns. It pictured mankind, though not individuals, as having existed since times immemorial during which natural catastrophes had repeatedly reduced mankind to a few in number, so that cultural efforts had to start anew from the ashes, so to speak. Representatives of "cultural cycles," Augustine noted, "say what they think, not what they know."[87] But the matter touched deeper than the verification of the historical past when the contention was to be faced that the human race had come into existence on innumerable occasions because the number of past worlds was also beyond any count. Here Augustine's immediate answer was that the contention left the question of mankind's repeated emergence unanswered. Worlds burnt to ashes had no room for the survival of men who might restart the propagation of the race. Augustine also noted that those who found it difficult to accept the short past of mankind as indicated by the genealogies of the Bible, did not fare any better by stretching that past over many thousands of years. A finite past, however long, remained vanishingly small when compared with eternity.

Witty remarks turned into an animated and solemn protest when Augustine directed his attention to the classic form of the antique belief in the eternity of the universe, the doctrine of strictly identical, ever-recurring cycles. It was completely beyond him how anyone could find satisfaction in the prospect of going endlessly through cycles of "fantastic vicissitudes," that is, alternating forever between birth and decay, happiness and misery. Escape from that treadmill was in his view provided only by what he called "the straight path of sound doctrine"[88] which rested on the straightforward, once-and-for-all course prescribed by the Creator to His handiwork. Any suggestion that some expression of the biblical record of that "sound doctrine" hinted at cyclic returns stirred Augustine to an impassioned protest:

At all events, far be it from any true believer to suppose that by these words of Solomon those cycles are meant, in which, according to those philosophers, the same periods and events of time are repeated; as if, for example, the philosopher Plato, having taught in the school at Athens which is called the Academy, so, numberless ages before, at long but certain intervals, this same Plato and the same school, and the same disciples existed, and so also are to be repeated during the countless cycles that are yet to be, – far be it, I say, from us to believe this. For once Christ died for our sins; and, rising from the dead, He dieth no more. "Death hath no more dominion over him;" and we ourselves after the resurrection shall be "ever with the Lord," to whom we now say, as the sacred Psalmist dictates, "Thou shalt keep us, O Lord, Thou shalt preserve us from this generation." And that too which follows, is, I think, appropriate enough: "The wicked walk in a circle;" not because their life is to recur by means of these circles, which these philosophers imagine, but because the path in which their false doctrine now runs is circuitous.[89]

The passage should indicate the decisive strength by which the unique facts of salvation precluded for Augustine, the Christian, any departure from the firm conviction in the unique series of events of cosmic and human history. Flirtation with the idea of cyclic returns had to remain in the Christian perspective a vicious impiety to be avoided at any price. Or as Augustine exclaimed:

> What pious ears could bear to hear that after a life spent in so many and severe distresses . . . that after evils so disastrous, and miseries of all kinds have at length been expiated and finished by the help of true religion and wisdom, and when we have thus attained to the vision of God, and have entered into bliss by the contemplation of spiritual light and participation in His unchangeable immortality, which we burn to attain, – that we must at some time lose all this, and that they who do lose it are cast down from eternity, truth, and felicity to infernal mortality and shameful foolishness, and are involved in accursed woes, in which God is lost, truth held in detestation, and happiness sought in iniquitous impurities? and that this will happen endlessly again and again, recurring at fixed intervals, and in regularly returning periods? and that this everlasting and ceaseless revolution of definite cycles, which remove and restore true misery and deceitful bliss in turn, is contrived in order that God may be able to know His own works, since on the one hand He cannot rest from creating and on the other, cannot know the infinite number of His creatures, if He always makes creatures? Who, I say, can listen to such things? Who can accept or suffer them to be spoken?[90]

To thwart the dangers of such impiety Augustine did not hesitate to utilize the sharp edge of his debating ability. He faced head-on the claim that blessedness consisted in a most conscious acceptance of the prospect of eternal returns: "If they maintain that no one can attain to the blessedness of the world to come, unless in this life he had been indoctrinated in those cycles in which bliss and misery relieve one another, how do they avow that the more a man loves God, the more readily he attains to blessedness, – they who teach what paralyzes love itself?"[91] With his extraordinary penetration into the deepest recesses of human psyche he could promptly show that the course of cycles was not the road to happiness but to a wholesale paralysis of love on which all other human endeavour rested. The introspective power of Augustine showed itself at its best as he unmasked the psychological pitfalls of the belief in eternal returns:

> For what happiness can be more fallacious and false than that in whose blaze of truth we yet remain ignorant that we shall be miserable, or in whose most secure citadel we yet fear that we shall be so? For if, on the one hand, we are to be ignorant of coming calamity, then our present misery is not so shortsighted for it is assured of coming bliss. If, on the other hand, the disaster that threatens is not concealed from us in the world to come, then the time of misery which is to be at last exchanged for a state of blessedness, is spent by the soul more happily than its time of happiness, which is to end in a return to misery. And thus our expectation of unhappiness is happy, but of happiness unhappy. And therefore as we here suffer present ills, and hereafter fear ills that are imminent, it were truer to say that we shall always be miserable than that we can some time be happy.[92]

Man's unquenchable thirst for a final and irreversible state of happiness could not be portrayed better, nor could there be a more persuasive stepping stone to Augustine's bold claim that nothing exploded more forcefully the treadmill of endless cycles than did the fact "of the eternal life of the saints."[93] Within a rationalistic framework Augustine's inference amounted to mere wishful thinking, but within the context of the concreteness of the faith he espoused, "the true blessedness . . . which cannot be interrupted by any disaster"[94] served as the anchor and target of an entirely new orientation in history. Associating oneself with the new phase was the road to salvation which Augustine preached with unabashed zeal: "Let us therefore keep to the straight path, which is Christ, and, with Him as our Guide and Saviour, let us turn away in heart and in mind from the unreal and futile cycles of the godless."[95]

It was also with obvious satisfaction that Augustine looked at the task accomplished. He proudly registered that the doctrine of cycles had been exploded, disposed of, and escaped from,[96] and that this "was something of the greatest consequence, to wit, to secure entrance into eternal felicity."[97] Since the human soul was no longer threatened by the prospect of endless rebirths, there remained no reason, Augustine argued, why the physical universe should undergo cyclic destructions. This emphasis on the superiority of moral considerations over those concerned with the physical could make sense only within an outlook generated by the belief in that God whose creative activity was the outpouring of his infinite goodness. Countless are Augustine's references to the words, "And God saw that it was good," words that bring to a close each of the six steps of the biblical account of Creation. The same holds true of Augustine's recalling the phrase of the book of Wisdom: "Thou hast ordered all things in number, and measure and weight."[98]

The modern reader would hardly find congenial Augustine's insistent probings into the "deeper" significance of the sundry numbers occurring in the pages of the Bible. There is, however, a convincing quality in the tie forged by Augustine between the inherently quantitative aspects of the natural world and the existence of a Supreme Craftsman. The presence of quantitative correlations in any living body, lacking as it could be in obvious features of purposefulness, resolved at once the perplexity of his inquiring mind: "I must admit, I am unable to see why mice and frogs have been created, or flies and worms for that matter. I see, however, that all things, in their own way are beautiful . . . I cannot look at the body and members of any living creature without finding measure, number, and order displaying co-ordination in unity . . . Where all these features originate I do not understand, unless they are traced to that supreme measure, number, and order existing in the unchangeable and eternal sublimity of God. In everything where you find measures, numbers and order, look for the craftsman. You find none other than the One in whom there is supreme measure, supreme numericity, and supreme order, that is God, of whom it is most truly said that He arranged everything according to measure, and number, and weight."[99] In the true vein of a Christian Platonist, Augustine did not hesitate to affirm that man's uniqueness rested in part on his ability to discern geometrical shapes embedded everywhere in the material universe. Against the Manichean Faustus, who identified human intellect with visible light, he noted the absurdities of such a position and contrasted visible light with its Creator, the source of all existence: "It is from there that comes our principle of existence, our ability to know, and urge to love; from there comes to all irrational living beings the nature by which they live, the strength by which they feel, the mobility by which they search; from there also comes to all bodies the measure for subsistence, the number for decor, and the weight for proper disposition."[100]

Augustine's appreciation of quantitative relationships had, of course, no immediate consequences for the emergence of scientific method. His main concern went far beyond the acquisition of numerical data in particular and learning in general. What interested him most was the quest for happiness, and this implied far more than marshaling bookish details, a point well to remember in this age threatened by the tyranny of sheer learning and by the voracious storing of information. Possibly, he underestimated the role of man's mastery of nature by knowledge in the process of securing happiness. He took the view that the knowledge of natural sciences, astronomy in particular, could not help one much in understanding the biblical message, as it concerned not man's natural skill but his supernatural destiny.[101] On the other hand, he wanted no part of a study of the Bible which purposely ignored the well-established results of scientific studies. He put the matter bluntly: "It is often the case that a non-Christian happens to know something with absolute certainty and through experimental evidence about the earth, sky, and other elements of this world, about the motion, rotation, and even about the size and distances of stars, about certain defects [eclipses] of the sun and moon, about the cycles of years and epochs, about the nature of animals, fruits, stones, and the like. It is, therefore, very deplorable and harmful, and to be avoided at any cost that he should hear a Christian to give, so to speak, a 'Christian account' of these topics in such a way that he could hardly hold his laughter on seeing, as the saying goes, the error rise sky-high." Such a performance, Augustine remarked, would undercut the credibility of the Christian message by creating in the minds of infidels the impression that the Bible was wrong on points "which can be verified experimentally, or to be established by unquestionable proofs."[102]

While ignorance on the part of Christians was reprehensible, not every detail of knowledge about nature possessed, as Augustine was quick to note, the same measure of certainty. Beside incontrovertible facts there were probable hypotheses and simple conjectures. When some statements of the Bible collided with the latter, Augustine urged caution. A case in point was the question whether celestial bodies, stars in particular, were animated or not. As reason and observation provided no decisive evidence, nor did the Scriptures seem to be explicit, the matter was open to further inquiry.[103] When, however, a question appeared to be settled in a convincing manner by scientific reasoning, Scriptures had to be reinterpreted. Clearly, the biblical phrase about God stretching out the firmament as a tent (skin) clashed with the sphericity of the earth. This naturally demanded a spherical covering, which was also suggested by the motion of the planets and stars. Augustine was not reluctant to give reason its due: "The Bible contradicts those who affirm something which is false; for that is true which is asserted by divine authority and not that which is conjectured by human frailty. However if perchance, they [the heathen] should prove it [the sphericity of the heavens] with evidences that cannot be doubted, it remains to be shown that what is spoken of as a tent, does not contradict those true demonstrations."[104]

For Augustine the overriding issue invariably remained the dignity of the Bible as the outstanding embodiment of the supreme form of knowledge leading to eternal happiness. This is what reverberates in his analysis of another topic touching both on science and theology. The question concerned the reconciliation of the immobile firmament of the Bible with the rotating skies of astronomy: "For if the firmament stands firm, the stars, which are believed to be fixed therein, go around from east to west, those more to the north describing smaller circles around the pole, so that the skies would appear rotating like a sphere (assuming that there is another celestial pole), or like a discus (if there is no other celestial pole). My reply ... is that it would require much subtle and laborious reasoning to perceive which is the actual case; but to undertake and discuss these matters I have no time, nor is it needed by those whom I wish to instruct for their own salvation and for the benefit of the Church."[105]

Such were momentous remarks. When science came into its own Galileo found in them support against his antagonists and comfort for his own perplexities. He quoted such and similar patristic texts with obvious relish in his "Letter to the Grand Duchess Christina," his most sustained attempt to vindicate the simultaneous pursuit of knowledge and of an eternal happiness.[106] Not that temperamentally Galileo and Augustine had much in common. Galileo was overwhelmed by the clarity of the great book of Nature read with the eyes of reason, whereas Augustine saw clearly that the inquiring reason was always moved by some kind of faith, by a most general urge for happiness. Moreover, behind Galileo lay centuries of Christian culture, and he took for granted the Christian cultivation of happiness which came to Augustine only through a memorable struggle.

For Augustine the Christian pursuit of happiness represented a supreme asset that had to be protected from pitfalls of more than one kind. First, it had to be saved from a possible relapse into the monotony of cycles undermining man's sense of purpose; second, from an anti-intellectualism disdainful of the attainments of reason; third, from considering as something evil the material world, a creation of the all-good Creator whose fingerprints were evidenced by the disposition of everything according to weight, measure, and number. Most importantly, the pursuit of happiness rested on man's grasp of his own and of the material world's fundamental contingency, as everything rested on a sovereign, creative act of God.

A man with a restored sense of purpose, a man with an ability to discern intelligible patterns in the universe, a man aware of the vital difference between knowledge and happiness, a man confronting an external world not as an a priori product of his mind but accessible to the light of reason which itself was a participation in God's mind, such were some principal considerations which Augustine stressed throughout his literary career as a thinker and a Christian. For another thousand years, during the lengthy centuries of the Middle Ages, his writings remained a principal source of learning and reflection not without some decisive consequences for a new phase of human history steeped in the scientific enterprise.

To see the strength of optimism exuding from a firm belief in the Creator a brief look at Boethius will not be amiss, as his influence on the intellectual formation of the medievals was only second to the one exerted by the Bishop of Hippo. Boethius' tragic fate extinguished not only a great political vision, but also put an early end to his gigantic cultural plan, the translation into Latin of all the works of Plato and Aristotle. Nevertheless, his elementary treatises on geometry, arith- metic, astronomy, and music served, together with his translations of Aristotle's works on logic, as the main source of learning until the 12th century. His philoso- phical interests were matched by some fine theological treatises, but above all by his stature as a Christian. From his *The Consolation of Philosophy*[107] the medievals could learn how the Christian faith in the Creator gave meaningfulness to dark situations and courses of events as behind them the Source of all reason and bene- volence was believed to be weaving an intelligible pattern.

Supremely serene as the course of divine providence had to be, so was Boethius' discourse covering such agonizing questions as free will versus fate, and the apparent puniness of man's achievements compared to the enormous reaches of the universe. Three centuries earlier another great figure of Roman political life, Marcus Aurelius, had been at grips with similar problems, but could come up only with sombre if not icy resignation. His universe was at the mercy of the blind forces of a cosmic treadmill. Humans in that universe were as many momentary bubbles appearing and dissolving with a cyclic necessity over the dark expanse of an unfathomable sea of cosmic and individual destinies. Admirable as the calm of the Roman emperor was, it clearly lacked the confident aura of serenity which covers with a golden glow the pages of *The Consolation of Philosophy*. No wonder! For Boethius the correlation of elements forming the universe was the work of the Creator,[108] and the course of the world corresponded to a once-and-for-all process directed toward a meaningful destination.[109]

Living in the shadow of death meant for Boethius the anticipation of an eternal dawn, as was also the case with those who during centuries of cruel instability read and recited his great hymn to the Creator:

> O Thou, that dost the world in lasting order guide,
> Father of heaven and earth, Who makest time swiftly slide,
> And, standing still Thyself, yet fram'st all moving laws,
> Who to Thy work wert moved by no external cause:
> But by a sweet desire, where envy hath no place,
> Thy goodness moving Thee to give each thing his grace,
> Thou dost all creatures' forms from highest patterns take,
> From Thy fair mind the world fair like Thyself doth make.
> Thus Thou perfect the whole perfect each part dost frame.
> Thou temp'rest elements, making cold mixed with flame
> And dry things join with moist, lest fire away should fly,
> Or earth, opprest with weight, buried too low should lie.
> Thou in consenting parts fitly disposed hast
> Th' all-moving soul in midst of threefold nature placed,
> Which, cut in several parts that run a different race,
> Into itself returns, and circling doth embrace

The highest mind, and heaven with like proportion drives.
Thou with like cause dost make the souls and lesser lives,
Fix them in chariots swift, and widely scatterest
O'er heaven and earth; then at Thy fatherly behest
They stream, like fire returning, back to Thee, their God.
Dear Father, let my mind Thy hallowed seat ascend,
Let me behold the spring of grace and find Thy light,
That I on Thee may fix my soul's well cleared sight.
Cast off the earthly weight wherewith I am opprest,
Shine as Thou art most bright, Thou only calm and rest
To pious men whose end is to behold Thy ray,
Who their beginning art, their guide, their bound, and way.[110]

The impact of the belief in the Creator on the fate and fortunes of scientific quest went beyond the formation of a climate of existential confidence. When Boethius died, in 524, a contemporary of his was already proposing in the eastern part of the long divided Empire some startling considerations about the cosmos and its workings. The scholar in question was John Philoponus,[111] the most learned man of his time, whose adult life was spent in Alexandria. In the Museum there he first studied under Ammonius and served later as a teacher. The lecture style is certainly evident in Philoponus' famous commentaries on the major works of Aristotle. The internal evidence of those commentaries is less conclusive on their chronological position in the literary output of Philoponus about whose birth and death nothing is known with certainty. His conversion to Christianity probably took place shortly after he had written, in 517, his commentary on Aristotle's *Physics*.[112] Another major work of Philoponus, the *De aeternitate mundi contra Proclum*,[113] was composed in 529. His elaborate discussion of the biblical account of the creation, the *De opificio mundi*,[114] was perhaps written as late as c. 558, but the same evidence may also warrant advancing that date by some ten years.[115]

The closeness of his conversion and of the composition of his commentary on the *Physics* has been seized upon as a clue for the interpretation of the evolution of the thought of Philoponus, whose commentaries on Aristotle have no religious overtones. Typically Christian positions are not defended there, although distinctly pagan doctrines are not upheld either. It seemed, therefore, natural to assign those commentaries contain more than one crucial departure from Aristotle's funda- remained free of serious objections.[117] Most likely Philoponus' conversion to Christianity did not signal on his part a drastic turnabout concerning his attitude towards Aristotle. The pagan Philoponus was not an unqualified admirer of the Stagirite, nor was Philoponus, the Christian, a vehement antagonist of his. The commentaries contain more than one crucial departure from Aristotle's fundamental positions, yet, in the very last work of Philoponus, the strictly Christian *De opificio*, Aristotle is declared time and again the prince of physicists. Philoponus' now lost work, a general critique of Aristotle, could have hardly been what it is being claimed by some, a sweeping dismissal of almost everything that Aristotle held. Finally, his critique of the eternity of the world as championed by Proclus does not contain an elaborate statement of the Christian doctrine of creation in time. The absence of typically Christian utterances in some of Philoponus' commentaries may very well be due to a consistent policy on Philoponus' part to stay largely within the conceptual framework of the work which he commented upon or subjected to criticism.

While in earlier times Philoponus was best remembered as a zealous defender of monophysitism, in our times he is known as the one who could perhaps have freed physics from its Aristotelian bondage. There is something tantalizingly revolutionary in the resolve by which Philoponus stated in his commentary on the *Physics* that all bodies would move in vacuum with the same speed regardless of their weight (mass); that bodies with greatly different weights falling from the same height hit the ground practically at the same time; that projectiles move across the air not because the air keeps closing in behind them, but because they were imparted a certain "quantity of motion."[118] Today Philoponus' name is closely tied to a major recognition by twentieth-century history of science that Galileo had important and indispensable medieval forerunners, that the principles of momentum and inertia of moving bodies did not emerge during the seventeenth century as a Deus ex machina.

The foregoing statements of Philoponus are as many dismissals of the bases of Aristotle's physics. Aristotle's theory of motion did not lack critics in classical antiquity, but none of them was so incisive as Philoponus. The source of that incisiveness will probably never be determined with a certitude forestalling any dissent. It should, however, be worth recalling that conversions rarely are sudden, dramatic events. It is very likely that for years before his conversion Philoponus had been at grips with the major tenets of Christian faith. Thus, his departure from some specific tenets of Aristotle might have very well been due to a steady rapprochment of his mind to a final espousal of a creed in which the very first tenet is about the Creator of heaven and earth. That Philoponus did not disclose the deeper motivations of his radical innovations in physical theory could be ascribed to his adherence to the style of his commentaries. The lack of theological excursions in his commentary on Aristotle's *Meteorologica*, a work that he very likely wrote as a Christian, would seem to corroborate such a conjecture.

The principal support of such speculation comes from the subsequent evolution of Philoponus' thought. In his critique of Proclus, and especially in his explanation of the biblical account of creation, there are unmistakable evidences of the impact of the dogma of creation on Philoponus' reflections on some of the very same aspects of Aristotelian physics. This should be all the more significant as one of the *De opificio's* principal objectives was to counter the absurdly literal interpretation given to Genesis by Theodore of Mopsuestia. Against the latter's claim that all celestial bodies were moved by angels, Philoponus did more than to note the lack of scriptural support for Theodore. According to Philoponus it was still much better to ask in view of the omnipotence of the Creator, "Could the sun, moon, and the stars be not given by God, their Creator, a certain kinetic force (*kinetiké dunamis*), in the same way as heavy and light things were given their trend to move . . .?"[119] Such a question struck as much at the roots of Aristotelian cosmology as did Philoponus' insistence that the stars were not made of the ether but of ordinary matter (fire); that they differed in colour; and that the immensity of their number was beyond explanation.[120] About the ether as the material constituting the heavenly regions Philoponus limited himself to the remark: "Aristotle, who supposed the heavens to consist of some fifth corporeal substance, has already been refuted by us."[121]

This is not to suggest that Philoponus did not accept to the end the great majority of Aristotle's statements about the physical world. He went along with Aristotle even on such patently absurd contentions as the essential difference between stagnant and flowing waters,[122] or the identification of winds with exhalation and not with moving air.[123] On most questions of physics Philoponus was not a bold innovator as can be seen from his attitude toward the Aristotelian theory of light and vision.[124] But wherever Aristotelian physics and cosmology seemed to encroach on the Creator's prerogatives or on man's proper relation to his Creator, Philoponus instinctively parted ways with Aristotle. A good illustration of this is Philoponus' attitude toward the Aristotelian doctrine of the causal influence of heavenly motions on terrestrial phenomena. This he emphatically restated,[125] as he was also fond of reasserting the permanence and universality of physical laws in connection with eclipses.[126] But he rejected astrology as utterly incompatible with Christian faith, and as subversive to one's belief in the freedom of will and in morality.[127]

Philoponus' firm faith in the first tenet of the Christian creed made him thoroughly aware of the fundamental paganism of Aristotelian cosmology. The alleged divinity of celestial matter and the eternity of motion in a pantheistic world could not be accommodated in the Christian interpretation of the cosmos. The first crucial step toward a new theory of matter and toward a new theory of motion issued from a theological motivation which, however, could not serve as a guideline for specifics in physical theory. Philoponus himself served abundant evidence of this. It was purely accidental that Philoponus was led in the right direction as he argued with an eye on Genesis that the several heavens and numerous orbits postulated by Aristotelian and Ptolemaic astronomy had nothing to do with reality.[128] Zealous faith could easily mislead, as shown for instance by Philoponus' repeated insistence that the Mosaic cosmogony was in full accord with the scientific (Aristotelian) world picture. Thus, to mention only one example, Philoponus spared no effort to prove that for Moses the earth was not flat but spherical.

Clearly, the time was still far away when theologians were ready to do justice to the principle that Genesis was written for spiritual and not for scientific instruction. The first to state this programmatically was Philoponus himself, but he was also the first to ignore this wise principle. Still, as subsequent developments showed, sound theology could bring about its beneficial effects in spite of some theologians. A proof of this is the fact that long before Philoponus' writings became known to the medievals, the genuine doctrine of creation had not failed to generate anew the same crucial insights that once emerged in Philoponus' mind instructed by his Christian faith. To appreciate this in depth certain reasons for a long detour in the history of science should first be given a careful analysis.

NOTES

1. Of the very large literature on this question, the *Christianity and Classical Culture: A Study of Thought and Action from Augustus to Augustine* by Charles N. Cochrane (rev. ed.; Oxford University Press, 1944) still stands out as a classic. It gives, however, no account of the attitude of the Church Fathers towards Greek (and Roman) science. This attitude is usually dismissed as negative and obscurantist, but this can be maintained only on the basis of an eclectic and superficial reading of patristic texts. A careful and comprehensive study of all such texts relevant to the question is still wanting. Needless to say, such a study should be done by evaluating those texts in their historical and cultural context, and not in the light of seventeenth- and eighteenth-century developments and preconceptions.

2. Written around the middle of the second century. For a translation, see *The Ante-Nicene Fathers: Translations of the Writings of the Fathers down to A.D. 325*, edited by A. Roberts and J. Donaldson; American reprint of the Edinburgh edition, revised and annotated by A. Cleveland Coxe (New York: Charles Scribner's Sons, 1925), vol. II, pp. 9–55.

3. *Ibid.*, p. 10 (Book I, chap. iii).

4. Written in the early part of the second half of the second century. In *The Ante-Nicene Fathers*, vol. II, pp. 65–83.

5. *Ibid.*, p. 67 (chap. v).

6. In his *Plea for the Christians*, written around 177; in *The Ante-Nicene Fathers*, vol. II, pp. 129–48. Quotation is from p. 131 (chap. v). – At the same time he described the "multitude" as ones "who cannot distinguish between matter and God, or see how great is the interval which lies between them. . ." p. 135 (chap. xv).

7. *Ibid.*, p. 136 (chap. xvi).

8. *Ibid.*, p. 138 (chap. xix).

9. *Ibid.*

10. Composed, like the other works of Justin, during the two decades preceding his martyrdom in 165; in *The Ante-Nicene Fathers*, vol. I, pp. 273–89.

11. *Ibid.*, p. 287 (chap. xxxiii).

12. In *The Ante-Nicene Fathers*, vol. I, pp. 163–87.

13. *Ibid.*, p. 169 (chap. xx).

14. In *The Ante-Nicene Fathers*, vol. I, pp. 188–93.

15. *Ibid.*, p. 191 (chap. ix).

16. *Ibid.*, p. 191 (chap. vii).

17. Written between 182 and 188. In *The Ante-Nicene Fathers*, vol. I, pp. 315–567.

18. *Ibid.*, p. 370 (Book II, chap. x).

19. *Ibid.*, p. 416 (Book III, chap. iii).

20. *Ibid.*

21. In *The Ante-Nicene Fathers*, vol. V, pp. 9–153.

22. *Ibid.*, p. 30 (Book IV, chap. xii).

23. See *ibid.*, pp. 117–18 (Book VIII, chaps. i and ii).

24. *Ibid.*, p. 13 (Book I, chap. iii).

25. *Ibid.*

26. *Ibid.*, p. 16 (Book I, chap. ix).

27. *Ibid.*, p. 16 (Book I, chap. xi).

28. *Ibid.*

29. *Ibid.*

30. *Ibid.*, pp. 102–03 (Book VII, chap. vii).

31. *Ibid.*

32. *Ibid.*, p. 150 (Book X, chap. xxviii).

33. See especially chap. vi of the work in *Clement of Alexandria* with an English translation by G. W. Butterworth (London: W. Heinemann, 1919).

34. *Ibid.*, p. 153.

35. *Ibid.*

36. *Ibid.*, p. 143 (chap. v).

37. *Ibid.*, p. 163 (chap. vi).
38. *Ibid.*, p. 13 (chap. i).
39. *Ibid.*, p. 143 (chap. v).
40. *Ibid.*, p. 157 (chap. vi).
41. *Ibid.*, p. 163.
42. See the Greek original and French translation, *Les Stromates*, introduction by C. Mondésert, translation and notes by M. Caster (Paris: Éditions du Cerf, 1951), p. 79 (Stromate I. 43, 1).
43. Neither of the two extant texts, one Greek, one Latin, of the work, is original. The Latin version by Rufinus is much longer than the Greek, but shows also the traces of many arbitrary changes, as can be seen, for instance, from the lengthy passages of the work quoted in Saint Jerome's letter to Avitus. On these details see the Introduction to *Origen on First Principles*, translated into English, together with an introduction and notes by G. W. Butterworth (London: Society for Promoting Christian Knowledge, 1936).
44. *Ibid.*, p. 88 (Book II, chap. iii).
45. *Ibid.*, p. 239, note 5.
46. *Ibid.*, pp. 240–41 (Book III, chap. v).
47. *Ibid.*
48. *Ibid.*, p. 88 (Book II, chap. iii).
49. *Ibid.*
50. *Ibid.*, pp. 87–88.
51. *Ibid.*, p. 237 (Book III, chap. v).
52. *Ibid.*, p. 238.
53. *Ibid.*, pp. 238–39.
54. Subsequent references will be to the English translation with an introduction and notes by H. Chadwick, *Origen: Contra Celsum* (Cambridge: University Press, 1953).
55. For an informative discussion of the various attempts to reconstruct Celsus' work, see the essay by B. Pick, "The Attack of Celsus on Christianity," in *The Monist*, 21 (1911), pp. 223–66.
56. *Contra Celsum*, p. 236 (Book IV, § 65).
57. *Ibid.*, p. 237 (Book IV, § 67).
58. *Ibid.*, pp. 237–38. Words in Italics are quoted from Celsus' work.
59. *Ibid.*, p. 238.
60. *Ibid.*
61. *Ibid.*, p. 238 (Book IV, § 68).
62. *Ibid.*, p. 280 (Book V, § 20).
63. *Ibid.*, p. 280 (Book V, § 21).
64. *Ibid.*, p. 279 (Book V, § 20).
65. *Ibid.*
66. *Ibid.*, p. 280 (Book V, § 21).
67. *Ibid.*, p. 239 (Book IV, § 69).
68. *Ibid.*, p. 281 (Book V, § 22).
69. See pp. 238–44 in *A Select Library of Nicene and Post-Nicene Fathers of the Christian Church*. Second Series, Volume VI, *St. Jerome: Letters and Select Works* (Grand Rapids, Mich.: Wm. B. Eerdmans Publishing Company, n.d.).
70. *Ibid.*, p. 240.
71. See *Origen: Contra Celsum*, pp. 212–13 (Book IV, § 37).
72. *Saint Basil, Exegetic Homilies*, translated by Sr. A. C. Way, in *The Fathers of the Church: A New Translation*, Vol. XLVI (Washington, D.C.: The Catholic University of America Press, 1963), pp. 3–150.
73. *Homélies et discours sur la Genèse* in *Œuvres complètes*, edited by M. Jeannin (Bar-Le-Duc: L. Guérin, 1863-67), vol. 5.
74. Saint Ambrose's treatise on the Hexaemeron contains many passages from Basil's work for which Saint Jerome had only praises. It became one of the few works of the Greek Fathers that received a Latin translator well before the Middle Ages. Saint Gregory of

Nyssa, author of deeply philosophical and mystical writings on the Creation, was also much indebted to Basil, his older brother. See on this *Saint Basil: Exegetic Homilies*, p. viii.

75. *Saint Basil, Exegetic Homilies*, p. 40 (Hom. iii).
76. *Ibid.*, pp. 6–7 (Hom. i).
77. *Ibid.*, p. 7.
78. *Ibid.*, p. 18.
79. *Ibid.*, p. 40 (Hom. iii).
80. *Ibid.*, p. 15 (Hom. i).
81. *Ibid.*, pp. 34–35 (Hom. ii).
82. *Ibid.*, p. 119 (Hom. 8).
83. Subsequent references are to the translation by the Rev. Marcus Dods, *The City of God*, in *A Select Library of the Nicene and Post-Nicene Fathers of the Christian Church*, edited by Philip Schaff, Vol. II, *St. Augustine's City of God and Christian Doctrine* (Grand Rapids, Mich.: Wm. B. Eerdmans Publishing Company, 1956).
84. *Ibid.*, p. 207 (Book XI, chap. 5).
85. *Ibid.*, p. 217 (Book XI, chap. 23).
86. *Ibid.*
87. *Ibid.*, p. 232 (Book XII, chap. 10).
88. *Ibid.*, p. 234 (Book XII, chap. 13).
89. *Ibid.*
90. *Ibid.*, p. 239 (Book XII, chap. 20).
91. *Ibid.*
92. *Ibid.*, pp. 239–40.
93. *Ibid.*, p. 239 (Book XII, chap. 19).
94. *Ibid.*, p. 240 (Book XII, chap. 20).
95. *Ibid.*
96. *Ibid.*
97. *Ibid.*
98. Chapter xi, verse 20. See on this pp. 158–59 in the previous chapter.
99. *De Genesi contra Manicheos*, Book I, chap. 16, Latin text and French translation in *Œuvres complètes de Saint Augustin*, edited by Péronne *et al.* (Paris: Librairie de Louis Vives, 1873), vol. III, pp. 440–41. The passage concludes with the remark: "This way you will perhaps derive greater benefit when praising God in the littleness of the ant, than by crossing the river on a tall beast."
100. *Contra Faustum Manicheum*, in *Œuvres complètes*, vol. XXVI, p. 108 (Book XX, chap. 7).
101. *On Christian Doctrine*, in *The City of God and Christian Doctrine*, p. 550 (Book II, chap. 29).
102. *Sancti Aureli Augustini De Genesi ad litteram libri duodecim*, edited by J. Zycha, in *Corpus Scriptorum Ecclesiasticorum Latinorum*, vol. XXVIII, Sec. III, Pars 1 (Vienna: F. Tempsky, 1894), pp. 28–29 (Book I, chap. 19).
103. *Ibid.*, p. 62 (Book II, chap. 18).
104. *Ibid.*, p. 46 (Book II, chap. 9).
105. *Ibid.*, p. 47 (Book II, chap. 10).
106. For an English translation, see *Discoveries and Opinions of Galileo*, by Stillman Drake (Garden City, N.Y.: Doubleday, 1957), pp. 173–216.
107. Boethius, *The Theological Tractates*, with an English translation by H. F. Stewart and E. K. Rand, *The Consolation of Philosophy*, with an English translation of "I. T." (1609) revised by H. F. Stewart (Cambridge, Mass.: Harvard University Press, 1962).
108. *Ibid.*, p. 287 (Book III, chap. 12).
109. *Ibid.*, pp. 163–66 (Book I, chap. 6).
110. *Ibid.*, pp. 263–67 (Book III, chap. 9).
111. On Philoponus the most informative modern work is the translation into German of many passages from his writings by Walter Boehm, published with the latter's introduction and notes under the title, *Johannes Philoponus: Grammatikos von Alexandrien* (Munich: Verlag Ferdinand Schöningh, 1967). The concluding chapter in S. Sambursky, *The Physical World*

in Late Antiquity (New York: Basic Books, 1962) is devoted to Philoponus, but not without some tendenciousness.

112. Subsequent references will be to the second volume of the edition of the Greek text by H. Vitelli, *Joannis Philoponi in Aristotelis physicorum libros quinque posteriores commentaria* (Berlin: G. Reimer, 1888).

113. Edited by H. Rabe (Leipzig: B. G. Teubner, 1899).

114. Edited by G. Reichardt (Leipzig: B. G. Teubner, 1897).

115. On this and other related problems about Philoponus' life and literary output, see Gudeman's article "Johannes Philoponus," in *Paulys Realencyclopädie der classischen Altertumswissenschaft* (Stuttgart: J. B. Metzlersche Verlagsbuchhandlung, 1916), Vol. 18, cols. 1764–93.

116. A position defended by Gudeman.

117. As pointed out by E. Évrard, "Les convictions religieuses de Jean Philopon et la date de son Commentaire aux 'Météorologiques'," in *Académie Royale de Belgique: Bulletin de la Classe des Lettres et des Sciences morales et politiques*, 39 (1953): 299–357.

118. In *Aristotelis physicorum libros quinque posteriores commentaria*, pp. 681–82 and 682–84 (Corollarium de inani); and pp. 639–42 (ad Lib. IV, cap. viii).

119. *De opificio*, Book I, chap. 12; *ed. cit.*, (see note 114 above), pp. 28–29.

120. *Ibid.*, pp. 189, 79, and 117 (Book IV, chap. 15; Book II, chap. 13; Book III, chap. 4).

121. *Ibid.*, p. 118 (Book III, chap. 5). The same idea underlies his questions about the ethereal nature of celestial bodies in Book I, chap. 9.

122. *Ibid.*, p. 179 (Book IV, chap. 9).

123. *Ibid.*, pp. 61–66 (Book II, chaps. 2–3).

124. See on this S. Sambursky, "Philoponus' Interpretation of Aristotle's Theory of Light," *Osiris*, 13 (1958): 114–26.

125. *De opificio*, p. 15 (Book I, chap. 6).

126. *Ibid.*, p. 99 (Book II, chap. 21).

127. *Ibid.*, p. 121 (Book III, chap. 6). Philoponus paid no detailed attention to the idea of the Great Year. He mentioned it in a cursory manner in his commentary on Aristotle's *De generatione et corruptione* (see edition by H. Vitelli [Berlin: G. Reimer, 1887] p. 314), but not as an Aristotelian doctrine. In the same context there is also a short reference by Philoponus to the idea of reincarnation. Again, in his refutation of Proclus, Philoponus twice passed up the opportunity to comment on the Great Year and eternal recurrences (see Philoponus' rebuttal of Proclus' first and sixteenth argument, *ed. cit.*, pp. 4–5 and 578–79). This seems to corroborate the points already made about Philoponus' reverence for Aristotle whom he apparently did not wish to implicate in an absurd doctrine, and his determination to stay within the framework set by his opponent's argument. Proclus, as was already pointed out in a previous chapter, refrained from giving lengthy support to the doctrine of the Great Year, which some Neoplatonists of late antiquity, such as Porphyry, already recognized as absurdity.

128. *De opificio*, p. 115 (Book III, chap. 3).

CHAPTER NINE

Delay in Detour

Among the roundabout developments of intellectual history few can match in suspense the haphazard detours through which the Greek scientific corpus found its way to the new world of Europe. The fissures that fragmented the Roman Empire grew perhaps nowhere so deep and impassable as along the Adriatic. Athens and Rome had not only lost much of their cultural significance by the early seventh century. The contacts between the two had also diminished to a trickle and so did direct cultural communications between East and West. The remnants of Greek scholarship gravitated towards the East, in particular towards Persia. It was at Jundishapur in Southwest Persia that Syriac-speaking Nestorian Christians established in the sixth century their centre of learning where much of the Greek scientific, philosophical, and literary corpus had soon become translated into Syriac.

Jundishapur was hardly the hub of the world's great crossroads, but its influence reached even as far as Athens. When in 529 Justinian closed the Academy there, some of its best scholars continued their work in Persia. For all that, Greek learning remained in a kind of exile at Jundishapur where safety meant also a great deal of isolation. Geographic remoteness and political barriers impeded the speedy exchange of ideas on which the growth of knowledge heavily depends. All this suddenly changed when in 641 Persia became engulfed in the great wave of Muslim conquest. For the first time since Alexander the Great the Middle East was under one rule, as Syria, Mesopotamia, and Egypt had already fallen to the crusaders of the Prophet. The conquest of North Africa followed in rapid succession. In 711 the Arabs entered Spain and twenty-one years later they were storming the walls of Poitiers, the high-water mark of Arab conquest in France. Thus, within exactly 100 years after Muhammad's death much of the oikumené formed again a unity with incalculable possibilities for learning in general and science in particular.[1]

What made the new unity so special transcended the usual advantages coming with the political unification of a land area reaching into three continents. The Romans were undoubtedly better organizers, administrators, and road builders than the Arabs. Arabic, the new "lingua franca," was not more developed and effective than Greek and Latin. But the best the Romans could offer on the level of ideology was a vague political myth centred on the *pax Romana* and the high value placed on being a full-fledged citizen of the Empire. Apart from that the Roman Empire was a motley collection of cultures, traditions, races, religions, and last but not least of crass superstitions. In the newly emerging Muslim world diversity played a distinctly subordinate role. Differences of races remained largely what they were, but cultural distinction began to recede into the background. The Muslim conquest was primarily the spread and imposition of a new religious conviction codified in the Koran. For the first time in world history a giant and vigorous empire was steeped in a conviction that everything in life and in the cosmos depended on the sovereign will of a personal God, the Creator and Lord of all.

A most integral part of the Koran's monotheism was its ethics. From the Muslim creed there came a sense of existential purpose with special emphasis on the eternal destiny of the disciples of the Prophet. This in turn meant a new, history-oriented consciousness for the whole new Empire as most of its subjects became worshippers of Allah. Muslim ethics provided in turn a set of principles that formed the basis of a moral and social code eliciting a high degree of compliance. The Koran itself did not specifically encourage secular learning. As a composition its best feature did not lie in conceptual unity and compactness. In this respect it was no match to the books of the Bible, the true message of which it claimed to restore. Nor did the Koran effectively promote the study of the Bible among Muslims. But the Koran was a book with undeniable literary charm exuding the air of unflinching conviction and commitment. More importantly, it was a book read and recited day in and day out and as such it kept sparking in the Muslim mind a love for the written word in general.

It should, however, be pointed out that the continual study of the Koran was not the only factor in promoting intellectual curiosity among faithful Muslims. Well before the advent of Muhammad, Arabs took active part in serving the trade routes that connected Egypt with India, and Ethiopia with Constantinople. Muhammad himself was a merchant before he was seized with the consciousness of having a God-given mission. Arabs also acted as civil servants for states controlling parts of the Arabian peninsula or bordering on it. Thus, when Jundishapur became part of the Muslim world, there was no lack of Muslim Arabs who had some intellectual training and appreciation for learning. Jundishapur continued to flourish under the rule of the Umayyads, the first Muslim dynasty that ruled from Damascus.

It is a good indication of the cohesiveness of Muslim culture and of the love of learning among the followers of the Prophet that cultural endeavour suffered no setback as major political partitions developed in the Muslim world. First came the overthrow of the Umayyads, and the subsequent formation in 749 of the Persian caliphate with the Abbasids as rulers, who founded Baghdad in 762. Abd-al-Rahman, the sole survivor of the Umayyads, made his escape to Africa, and within a few years the foundations of the caliphate of Cordova were firmly laid. It was almost inevitable that a buffer state should emerge between Baghdad and Cordova, and this is what happened when the Fatimid caliphate was set up in Egypt during the first part of the tenth century.

In all three caliphates serious concern for the promotion of learning was in early evidence. The literally monumental proofs of this were the newly erected "Houses of Wisdom." During the reign of al-Mamun (813–33) an academy and observatory were set up in Baghdad. The Fatimid caliph, al-Hakim, who excelled also as an astronomer, established an institute of higher learning in the new city of Cairo in 966. In Cordova the caliph al-Hakim II (961–76) amassed more than 300,000 volumes for the library which almost immediately began to attract eager scholars from the Christian West. An impressive proof of Muslim interest in learning was

o

the hospitality extended to foreign scholars. Through the work of the Indian astronomer, Manka, visiting in Baghdad, the principal Hindu astronomical works were translated there into Arabic during the reign of the second Abbasid caliph, al-Mansur (754–75). About the same time Arabs became acquainted with the art of Chinese papermaking, and the manufacture of paper started in the Muslim world in a paper mill erected in Baghdad in 794. Systematic collection and translation of the Greek scientific and philosophical corpus got under way during the caliphate of Harun-al-Rashid. It was in large part the magnitude of this enterprise that made the foundation of the Academy of Baghdad inevitable a few decades later, in 828.

Arab efforts to translate the Greek scientific and philosophical works excelled both in scholarly carefulness and in the resolve not to miss any document. The latter point was well illustrated by the recollection of Hunayn ibn-Ishaq (c. 809–877), the famed translator of many of Galen's works. In his book, where he discussed the translation of some 130 works of Galen,[2] he recalled his travels all across Mesopotamia, Syria, Palestine, and Egypt to locate a Greek manuscript of Galen's book "On Proof." Only in Damascus could he find a text containing about half of the original. He also found that his copy did not fully match fragments of the work found by a fellow translator, Gibril. This eagerness to compare available copies casts a very favourable light on Hunayn's scholarship. His method was to make his translations from at least three different manuscripts of the same work to permit a fair reconstruction of the original.

Hunayn is also a most valuable witness of the fact that in his time Galen's works were studied in much the same manner as they were five–six centuries earlier. The translation by Hunayn, his son, and his nephew of Galen's enormous medical and philosophical output secured the complete domination by Galen's thought over medical practice and teaching in the medieval East and to a considerable degree also in the West well into the Renaissance. A hardly negligible factor in this was Hunayn's own prowess as a doctor and the author of medical treatises. He was the likely author of *The Book of the Ten Treatises on the Eye*, a pioneer systematization of ophthalmology.[3] Equally important was Hunayn's role as a teacher of medicine. It was the school of medicine founded by him that produced the greatest figure of Arab medicine, al-Razi (865–925). He is best remembered as the author of *A Treatise on the Small-Pox and Measles*, which has been reprinted more than forty times during the last four hundred years![4] The treatise contains the first clear description of the major symptoms of the two diseases and shows its author as a keen observer and respecter of facts.

About al-Razi's broad and perceptive approach to problems of medicine much is revealed by his treatises that carry titles such as, "Why ignorant physicians, laymen, and women have more success than medical men," "Why people prefer quacks and charlatans to skilled physicians," "Why frightened patients easily forsake even the skilled physician," and last but not least, "On the fact that even skilful physicians cannot heal all diseases."[5] Far more importantly, he wrote the medical encyclopedia, *Kitab al-hawi*, that is, "Comprehensive Book" which fittingly was called such because it included the whole of Greek, Syriac, and early Arab medical knowledge in addition to ample material from Persian and Indian medical sources. The importance of this work can be gauged from the fact that between 1486 and 1542 the Latin translation of such massive opus was printed five times. Without any exaggeration he could rightly be considered "the greatest clinician of Islam and of the Middle Ages."[6]

This leadership was true not only of one outstanding individual but of Islamic medicine as well. It was in that field that Islamic science displayed its most sustained efforts, revealed most of its realistic sense for the facts of observation, and served evidence of its practical genius. The finest monument of the last point is probably the great hospital founded in Cairo in 1284, long after Islamic science passed its zenith. Intensive interest in eye diseases went on unabated among Muslim physicians well into the fourteenth century. The *Kitab al-umda*, the last notable product of Arab ophthalmology, was written by al-Shadhili around 1375. Its novel details concerned the development of trachoma and the description of cancer of the eyelid. The work represented a fine capstone on an outstanding tradition, a chief figure of which was the caliph Ibn-abi-al-Mahasin. His treatise of ophthalmology, the *Kitab al-kafi fi al-Kuhl*, "The sufficient treatise on collyrium," written around 1265, is a systematic account of the anatomy of the eye, its diseases and treatments, including detailed discussion of some operations, especially the ones connected with cataract. As an eye surgeon he had an unrivalled reputation for removing cataracts even in cases when one eye was already lost.

But Islamic ophthalmology also provides a priceless insight into the problem of why Islamic science failed in the long run. It should seem paradoxical that the most incisive Arab ophthalmologist was none other than Ibn-Rushd (1126–98), known better under the name of Averroes, who was also the most resolute advocate that Aristotle's philosophy and world view had probably ever had. He broke new grounds for ophthalmology with his explanation of the role of the retina in the functioning of the eye. Obviously, his servile addiction to Aristotelianism left his mind unfettered when it came to practical problems of medicine, which Aristotle did not touch with his sweeping and usually wrong generalizations about nature. In turn, the practical problems of medicine, including eye diseases, raised no immediate questions about the foundations of scientific method. Up to a certain level the practice of medicine could flourish without the need for an entirely new outlook on the physical world and its regularity. A further illustration of this is the case of Ibn-Sina (980–1037), who is far better remembered as Avicenna, the famed philosopher. His million-word long *Qanun* ("Canon") had for centuries served as the standard textbook in Arab medical teaching. It is a storehouse of fine observations and pathological information. Its author was bold enough to turn against the authority of Euclid and Ptolemy, and claim that the rays of light passed from the luminous object to the eye. Yet, as it shall be shortly seen, no single Muslim thinker did more than Avicenna to nip in the bud the ultimate prospects of science in the Muslim world.

The cases of Averroes and Avicenna are principal illustrations of the symptom of which the whole history of Arab science is a classic paradigm. Muslim science made notable contributions to those parts of science which had, in the historical context at least, little or nothing to do with the laws of the physical world at large. Such was undoubtedly true of the Arab cultivators of mathematics.[7] The work of the best Arab mathematicians contradicts at the same time the often voiced view that the eventual decline of science among the Arabs was due to the practical bent of their minds. Ability to systematize and gist for abstract analysis are already evident in the work of al-Khwarizmi (died around 850), author of a treatise entitled *Hisab al-jabr w'al muzaqalah*.[8] It enjoyed such popularity in the medieval West that for centuries the author's name stood in a slightly changed form (algorism) for the subject matter. The work is divided into five parts. In the first two a list of quadratic equations is given together with the proofs of their solution through geometrical analogies. The third part is largely a discussion of the product of $(x \pm a)$ and $(x \pm b)$. In the fourth part there follow the rules for addition and subtraction of expressions which involve the unknown, its square, or its square root. This section also gives the rules for the calculation of square roots and the handling of squares in algebraic products. Selected problems bring the work to a close.

Ability to carry forward along original lines the material received from Greek mathematicians is also evident in the work of Tabit ibn-Korra (836–931), the renowned translator of the works of Euclid, Apollonius, Archimedes, and Ptolemy. The only fragment that survived of his original works is a fine chapter on the solution and properties of cubic equations. That the solution is given by geometrical methods indicates the keen interest of the Arabs in geometry. It also explains that the appearance of the rudiments of analytical geometry in a work of Omar Khayyam was the combined result of the maturation of mathematical tradition and a spark of genius. A world-renowned poet, Omar was also the author of an algebraic treatise in which he gave the first comprehensive classification of algebraic equations with respect to the number of terms. He did not consider negative, multiple, fractional, and imaginary roots, but made a most systematic effort to correlate algebra with geometry. Very valuable in this regard is his treatment of cubic equations for which he successfully sought solution by exploiting Apollonius' work on conic sections. He had every right to call attention to the originality of his accomplishments in the Introduction to his work: "Parts of this science deal with certain very difficult introductory theorems, the solution of which has eluded most of those who have attempted it. As for the ancients, none of their works on the subject have been preserved for us. Perhaps their studies and experiments did not lead to a solution of the problems, or their researches may not have compelled them even to examine them, or finally, their works on the subject, if there were any, may not have been translated into our language."[9]

Similar flashes of originality are also observable in the Arabic study of trigonometry inherited from the Greeks. Al-Battani (877–918) who was probably the most outstanding Arab scholar in the eyes of medieval and Renaissance men of science, derived the formulas $\sin \alpha = \tan \alpha / \sqrt{(1 + \tan^2 \alpha)}$ and $\cos \alpha = 1 / \sqrt{(1 + \tan^2 \alpha)}$. It was due to his influence that Hindu contributions to trigonometry, such as the replacement of chords by the sine, had become firmly established. The creativity of Abu-al-Wafa's (940–c. 997) trigonometrical work[10] should best be gauged from the formula he obtained: $\sin (\alpha + \beta) = \sin \alpha \cos \beta + \sin \beta \cos \alpha$. He also discussed the quadrature of the parabola and the volume of the paraboloid. The latent power of Arab mathematical genius is well attested also by al-Farabi's (c. 870–950) treatise on music where musical theory is reduced to a study of the various fractions of a chord of unit length and to the correlation among these fractions. Since the addition and subtraction of intervals corresponds then to the multiplication and division of fractions, one has at hand a correlation of notes which obeys a logarithmic law. Arab mathematical talent was also very instrumental in giving to the Hindu numerals and decimal notation a more explicit form, in which it could successfully challenge the cumbersome Roman numerals and arithmetic.

While mathematics and geometry have forbiddingly difficult branches, elementary algebra and basic Euclidean geometry are the embodiments of simplicity, especially when compared with the enormous complications of the physical world and its processes. The best attainments of the Greeks in physics were in statics and optics where simple forms of algebra and geometry found a natural and fruitful application. Partly for the same reason, the most valuable parts of Arab investigations in physics also deal with the question of balance and with the laws of the propagation of light rays. The books on balance by the three brothers, called Banu Musa (fl. c. 850), opened up interest in the works of Heron and Philo. The geometer, Tabit ibn-Korra, also wrote a book on balance. The continued interest in the subject is well evidenced by the fact that in the early twelfth century al-Khazini composed a remarkable treatise on mechanics. It also dealt with questions of hydrostatics, including various aspects of capillarity, and contained interesting speculations about the weight of air.

In optics the first notable Arab treatise is al-Kindi's (fl. 850) summary of the main results obtained by Euclid, Heron, and Ptolemy, together with an interesting discussion of the rainbow. The first Arab physical treatise was, not surprisingly, the *Kitab al-manazir* ("Book about Optics") of Ibn-al-Haitham (965–1038), or Alhazen, as the medievals called him.[11] It was he who extended the investigation of the laws of reflection from plane to concave and parabolic surfaces. He also constructed such metal reflectors which in turn led him to the discovery of spherical aberration. His other original feats were the location of the focus of a paraboloid and the use of the camera obscura. Most importantly, he made a fresh start in the study of refraction. Although he did not discover Snell's law, he correctly connected the change of path to a change in the velocity of light as it passed into a new medium. In his speculative analysis of refraction he proposed in a rudimentary form the law of inertia, the law of the composition of forces, and the law of least time. A remarkable feature of his work was the emphasis on experimentation. It was careful and persistent observation that led him to the finding that Ptolemy's law, which made the angle of refraction proportional to the angle of incidence, was true only for small angles. He also experimented with magnifying glasses and came close to the modern theory of convex lenses.

That such a pioneering and creative work failed to inspire sustained and vigorous interest among Arab investigators of optical phenomena should help bring into focus some crucial aspects of the scientific enterprise among the Arabs. With respect to questions about the real world it not only displayed a baffling inertia but also got enmeshed in the quagmire of crass superstitions. A concise illustration of this is that *al-khemi*, the Arabic name of the study of materials and compounds, came to stand for alchemy, the very embodiment of obscurantism. The classic summary of Arab alchemy by the Brethren of Purity,[12] of whom more will be said later, illustrates what happens when mystical and astrological proclivities gain easy dominance over a whole branch of science. The cultivation of "chemistry" earned, indeed, a bad name even among some Arabs, and was denounced time and again as a largely worthless enterprise by some of the leading Arab scholars.

Such a defence of the "light of reason" was not, however, without some irony of its own. Al-Kindi, a vehement critic of alchemy, was also a staunch defender of astrology.[13] He lent special attention and credence to the doctrine of the world-year (Great Year) and of shorter cycles determining the succession of rulers, religions, reigns, and physical catastrophes. This seems to indicate that he owed much of his expertise to astrologers among vanquished Persians.[14] For the latter, astrology patently served the cause of political revenge. The Syrian Theophilos, member of another subjugated minority in the Muslim empire, who served as astrologer under the first Sasanid caliphs, reminded the court in Baghdad, as Ibn-Khaldun recalled the daring feat,[15] that the Muslim era would be limited to 960 years. Ibn-Khaldun, whose speculations about logical calculus earned some praise in the *Ars magna* of Raymundus Lullus, gave a detailed account of how the conjunctions of Saturn and Jupiter have specific recurrences along the zodiac in 20, $12 \times 20 = 240$, and $4 \times 240 = 960$ years.

As a zealous Muslim, Ibn-Khaldun refused to admit the demise of the Muslim world by 1530, or 960 years after the Hegira. Nor did al-Kindi take kindly to such a prospect. But both he and Ibn-Khaldun did their best to fit major historic events that occurred during the sixth, seventh, and eighth centuries into the minor cycles of 20 and 240 years. Such speculations were seized upon with undisguised pleasure by not a few Arab scholars. Al-Kindi, who was the great encyclopedist of the ninth-century world, was the mentor of Abu-Mashar (fl. 870), the most famed of Arab astrologers. He carried the art of casting horoscopes for individuals and nations to its logical extreme. His "Book of the Revolution of Birth Years"[16] is a painfully repetitious series of statements on how the fate of nations, rulers, religions, and individuals is determined by the position of the various planets along the twelve signs of the zodiac.

It was typical that the manifold absurdities of the work did not receive an immediate and well deserved rebuttal among Muslim scholars. Well over a century later al-Biruni denounced in his famous *The Chronology of Ancient Nations* "the follies committed by abu-Mashar,"[17] but only to claim that it was perfectly reasonable to compute cycles and to list the horoscopic significance of each day from ancient calendars. The third chapter of his book dealt rather significantly with the sundry "eras" reckoned by ancient civilizations. Al-Biruni carefully recalled mainly Persian and some Hindu speculations that tried to trace the date of the deluge to the conjunction of Jupiter and Saturn. What he wrote in this connection about Abu-Mashar's efforts deserves to be quoted in full:

> Now, when he thought that he had well established the computation of this sum according to the method, which he has explained, and when he had arrived at the result that the duration of those periods, which astronomers call "*star-cycles*," was 360,000 years, the beginning of which was to precede the time of the Deluge by 180,000 years, he drew the inconsiderate conclusion, that the Deluge had occurred once in every 180,000 years, and that it would again occur in the future at similar intervals.[18]

As a devout Muslim, al-Biruni could accept neither the idea of perennial recurrences of great cosmic and historical events nor the idea of individual palingenesis. In the same context he noted that the claims of Abu-Mashar and others "revile those who warn them that the hour of judgment is coming, and who tell them, that on the day of resurrection there will be reward and punishment in yonder world."[19] Obviously, the belief in an absolute and cosmic end was as much a basic part of Muslim orthodoxy as was the belief in an absolute beginning for creation. Al-Biruni took pains to emphasize that only divine revelation could provide definite information about the origins of the world.[20] Insistence on eternal recurrences would, according to al-Biruni, bring discredit to the science of astronomy, and he hastened to point out that the idea of Great Year had no real support in astronomy. He called attention to the "discrepancy of the [planetary] cycles" by which he meant their incommensurability.[21] Again, in al-Biruni's opinion, the idea of eternal cycles rested on the notion of the eternity of motion which he resolutely rejected: "It is well known among philosophers and others, that there is no such thing as an *infinite* evolution of power ($\delta\acute{v}\nu\alpha\mu\iota s$) into action ($\pi\rho\hat{\alpha}\xi\iota s$), until the latter comes into real existence. The motions, the cycles, and the periods of the past were computed whilst they in reality existed; they have decreased, whilst at the same time increasing in number; therefore, they are not *infinite*."[22]

Evidently, for al-Biruni, his Muslim faith firmly imposed the finiteness of the world in time and prevented him also from attributing a divine eternity to the heavens. He recalled with intense dislike the opinion of those for whom time consisted of cycles "at the end of which all created beings perish, whilst they grow at their beginning: that each such cycle has a special Adam and Eve of its own, and that the chronology of this cycle depends upon them."[23] He also censured as "foolish persuasion" the view of those for whom "time has no *terminus quo* at all."[24] But the very same faith did not make him affirm the absolute beginning of the world as a creation out of nothing, nor did the same faith make him perceive something specific about the Creator as the source of law and consistency in the physical world. The former point can be seen in al-Biruni's strange concession to the "possible" eternity of those cycles. Concerning Abu-Mashar's claim that the creation was marked by a very specific conjunction of all planets in the first part of Aries, al-Biruni noted: "It is quite possible that these (celestial) bodies were scattered, not united at the time when the Creator designed and created them, they having these motions, by which – as calculation shows – they must meet each other in *one* point in such a time (as above mentioned)."[25]

Furthermore, al-Biruni was eager to illustrate the point with the graphic analogy of living beings placed on the rim of a huge wheel and moving with respect to each other at various velocities. "If you then ask the mathematician as to the length of time *after* which they would meet each other in a certain point, or *before* which they had met each other in that identical point, no blame attaches to him, if he speaks of billions of years."[26] Such computations did not imply, al-Biruni insisted, that those situations actually occurred time and again, that those beings actually lived through the same correlations again and again, and that the same would occur to them in the future at regularly repeated intervals. But then, if all this had nothing to do with reality, why was it necessary to indulge in calculations of this type? Did al-Biruni himself not remark about the great cycles of the Persians and the Hindus: "If anybody would construct such cycles on the basis of the observations of Ptolemy, or of the modern astronomers, he might do so by the help of the well known methods of such a calculation, as in fact many people have done, e.g. Muhammad ben Ishak ben Ustadh Bundadh Alsarakhsi, Abu-al-wafa Muhammad ben Muhammad Albuzajani, and I myself in many of my books, particularly in the *Kitab-al-istishhad bikhtilaf al'arsad.*"[27]

This remark should help in understanding the second point mentioned above, namely, the failure of Muslim thinking to go beyond the Aristotelian and Neo-platonist positions as they reflected on the Creator as the source of the laws of the physical world. Infatuation with cosmic and celestial cycles also meant a tacit or at least subconscious acceptance of a deterministic role of the celestial world on terrestrial processes and human affairs. This in turn could not fail to undercut man's urge to explore the world, ferret out some of the secrets of its working, and try to dominate it. In al-Biruni's voluminous treatise on astrology[28] there is compressed between the opening and closing praise of Allah a systematic presentation of all the relevant information needed for the successful practice of a craft that had nothing to do with the veracity of the true God. If an al-Biruni, the paradigm of learnedness among Muslims, could not see the irony of his invoking Allah's approval on his astrological discourse, how could the same be perceived by Muslim scholars of lesser stature? Astrology and astronomy remained in the closest unity as far as most Arab scholars were concerned, and inside that unity speculation about cycles retained a place of honour. An illustration of this is the specific connotation which grew around the words *tasyrat*, *intiha'at* and *fadarat* in Arab astronomical literature. All three came to stand in one way or another for the Great Year, or world year, or great cycle.[29]

A chief characteristic of astrology is its thorough inconsistency. Astrology is not so much a system as an ever burgeoning set of capricious aperçus grafted on disconnected observations. Astrology is a revelling in the momentary and in the concrete with no real concern for the causal and unequivocal interconnectedness of things, events, and processes. Astrology aims at satisfying the will and the emotions, not the intellect. Clearly then, astrology could easily remain part of a cosmology which did not aim at a rational interpretation of the world at large, but at the satisfaction of a culture emphasizing the concrete and the volitional. The main features of Islamic cosmologies bring this out in a forceful manner. The most notable cosmological syntheses achieved by Islamic thinkers represent a luxuriant fusion of rational and mystical, of observational and cabbalistic elements.[30] The world there is always a huge living organism, animated by the world-soul, and vibrating both in its whole and in its uncounted parts with unfathomable volitions.

The heavily voluntaristic and organismic world picture that dominated the cosmology of medieval Muslim scholars was in one respect a heritage from the Hellenistic past. It also agreed with the world view of the Persians and the Hindus, from whom the Arabs eagerly learned. Although several of its features stood in opposition to fundamental tenets of the Koran, Muslim scholars made that world picture their own. The Koran claimed to contain the sound guidelines for a proper understanding of the biblical revelation, and faithfully echoed several major biblical themes about the universe. These were the notions about the world as a providential dwelling place prepared by the Creator for man, as a harmonious entity beneficial to man, and as a paramount sign of the existence and goodness of the Creator and of the Creator's absolute sovereignty over all. In the Koran, such themes are also intimately tied, in close analogy to the Bible, to the moral dimension of man's existence principally epitomized in his duty to prepare for the final judgment. The Koran continually voices the biblical view that the most reprehensible offence against the Creator is idolatry. Warning against idolatry (which in the Koran includes alongside the fetish worship the trinitarian dogma of Christianity) forms the general backdrop of most of the references in the Koran to Creator and creation.

A reminder against idolatry and about resurrection introduces the charming description of the nomadic (Arab) man's immediate surroundings as a handiwork of the Creator in Sura XVI, entitled "The Bee."[31]

> And he hath subjected the night and the day to your service; and the sun, and the moon, and the stars, which are compelled to serve by his command. Verily herein are signs unto people of understanding. And *he hath also given you dominion over* whatever he hath created for you in the earth, distinguished by its different color. Surely herein is a sign unto people who reflect. It is he who hath subjected the sea *unto you*, that ye might eat fish thereout, and take from thence ornaments for you to wear; and thou seest the ships ploughing *the waves* thereof, that ye may seek *to enrich yourselves* of his abundance, *by commerce*; and that ye might give thanks. And he hath thrown upon the earth *mountains* firmly rooted, lest it should move with you, and *also* rivers, and paths, that ye might be directed: and *he hath likewise ordained* marks *whereby men may know their way*; and they are directed by the stars. *Shall God* therefore, who createth, *be* as he who createth not? Do ye not therefore consider? If ye *attempt to* reckon up the favours of GOD, ye shall not *be able to* compute their number; GOD *is* surely gracious *and* merciful; and GOD knoweth that which ye conceal, and that which ye publish. But the *idols* which ye invoke, besides GOD, create nothing, but are themselves created. *They are* dead, *and* not living; neither do they understand when they shall be raised. Your GOD *is* one GOD. As to those who believe not in the life to come, their hearts deny *the plainest evidence*, and they proudly reject *the truth*.[32]

The description of the world at large contains the same lesson:

> Preach that there is no GOD, except myself; therefore fear me. He hath created the heavens and the earth, to *manifest his* justice: far be that from him which they associate *with him!* He hath created man of seed; and *yet* behold he is a professed disputer *against the resurrection*. He hath likewise created the cattle for you; from them ye have wherewith to keep yourselves warm, and *other* advantages; and of them do ye *also* eat. And they are likewise a credit unto you, when ye drive them home *in the evening*, and when ye lead *them* forth to feed *in the morning*: and they carry your burdens to a *distant* country, at which ye could not *otherwise* arrive, unless with *great* difficulty to yourselves; for your LORD *is* compassionate and merciful. And *he hath also created* horses, and mules, and asses,

that ye may ride thereon, and for an ornament *unto you*; and *he likewise* created *other things* which ye know not. *It appertaineth* unto GOD to instruct *men* in the *right* way; and *there is* who turneth aside from the same: but if he had pleased, he would certainly have directed you all. It is he who sendeth down from heaven *rain* water, whereof ye have to drink, and from which plants, whereon ye feed *your cattle, receive their nourishment. And* by means thereof he causeth corn, and olives, and palm-trees, and grapes, and all *kinds* of fruits, to spring forth for you. Surely herein is a sign *of the divine power and wisdom* unto people who consider.[33]

Muhammad is warned time and again about the reluctance of many to take heed of the message he was given to preach, namely, that the world *is* a workmanship of God:

Do not the unbelievers therefore know, that the heavens and the earth were solid, and we clave the same in sunder; and made every living thing of water? Will they not therefore believe? And we placed stable *mountains* on the earth, lest it should move with them; and we made broad passages between them for paths, that they might be directed *in their journeys*: and we made the heaven a roof well supported. Yet they turn aside from the signs thereof, *not considering that they are the workmanship of God*. It is he who hath created the night, and the day, and the sun, and the moon; all *the celestial bodies* move swiftly *each in its* respective orb.[34]

The reference to celestial orbs is obviously a trace of astronomical lore. Astrological parlance might have provided a line in that section of Sura III where the inevitable punishment of the infidels is stated:

Think not, O *prophet*, that they shall escape from punishment, for they shall suffer a painful punishment; and unto GOD *belongeth* the kingdom of heaven and earth: GOD is almighty. Now in the creation of heaven and earth, and the vicissitude of night and day, are signs unto those who are endued with understanding; who remember GOD standing, and sitting, and *lying* on their sides; and meditate on the creation of heaven and earth, *saying*, O LORD thou hast not created this in vain; far it be from thee: therefore deliver us from the torment of *hell* fire.[35]

The Koran did not, of course, advocate astrology which represented the very opposite to the Koran's impassioned insistence on the sovereignty of God. That astrology remained nevertheless an eagerly practiced art throughout the Muslim world should primarily be blamed on the proverbial inconsistency of humans. It should, however, be a cause of some reflection that the Koran failed to inspire a single, extensive, rationally argued treatise against astrology. This is a striking illustration of the more general fact that the Koran did not provide the necessary mental encouragement and guidelines for a rational approach to the universe. The reason for this lies in the overly voluntaristic and moralistic tone of the Koran, and more particularly, in its emphasis on the will of the Creator. In the Koran no conspicuous effort is made to tie the sovereign decisions of God to His nature, that is, to His rationality. In other words, the will of God seems to be above any norm, however sound and intrinsically just may a norm appear to human reasoning.

As the Koran has it, Allah sovereignly guides some along the right way, and the faithful had to remind themselves: "Say, unto GOD *belongeth* the east and the west: he directeth whom he pleaseth into the right way."[36] The phrase, "he guideth whom he pleaseth," is in fact one of the most repeated phrases by which the Koran emphasizes the absolute sovereignty of God. This sovereignty sets the tone of all aspects of God's relationship with man and the whole of creation. The dictum, "he forgiveth whom he pleaseth and punisheth whom he pleaseth,"[37] describes the Lord of the moral order in the same manner which admits no qualification, as does the phrase, "he createth what he will," describe the Lord of the physical world. Undeniably, something akin to capricious despotism may seem to come through in the phrase, "GOD causeth to err whom he pleaseth, and directeth whom he pleaseth."[38] True, the God of the Koran is the "Merciful One," but the Koran fails to elaborate on the crucial claim: "Verily GOD will not wrong *any one even* [by] the weight of an ant."[39]

The prevailing tone in the Koran is set by omnipotent voluntarism if not wilfulness. The nature of God and the relevance of that nature to the characteristics of the world is hardly touched upon in the Koran. A classic case is Sura XXXV, entitled "The Creator."[40] It is a warning against falling into idolatry. It contains a comparison between the utter impotency of idols ("they do not have the power even over the skin of a date-stone") and the power of God ("He sendeth the winds and raiseth the clouds"); between the prosperity of the devout and the ultimate discomfiture of the evildoer. As an evidence of the reliability of God far greater emphasis is put on the often witnessed failure of idolators than on the reliable performance of God's great handiwork, nature. The latter point is disposed of in the single verse: "He causeth the night to succeed the day, and he causeth the day to succeed the night; and he obligeth the sun and the moon to perform their services: each *of them* runneth an appointed course."[41] The very same chapter, devoted to proving that God is most reliable in his ordinances, especially in his rewards and punishments, reminds the faithful once more of the point so characteristic of the God of the Koran: "Verily, GOD will cause to err whom he pleaseth, and will direct whom he pleaseth."[42]

Phrases like these had an ominous ring which reverberated far beyond the confines of theology. Grappling with the problem of the relation between the God of the Koran and the world of human experience became ultimately a struggle about the scientific quest. On the face of it the opening salvo of the conflict could not have been more purely theological. In 827, the caliph al-Mamun issued the proclamation that the Koran, contrary to the traditional orthodox view, was not in its actual Arabic form an identical replica of the heavenly original, but was rather created in time. The condemned position, that the Arabic text of the Koran was co-eternal with God, was clearly an absurdity and rested on a blind acceptance of the parlance of the Koran which in a patently figurative sense described Allah sitting on a throne, and made hyperbolic statements about his seeing, hearing, and speech. The broader point of the caliph's proclamation consisted, therefore, in a recognition that the anthropomorphic style of the Koran about God was subject to rational criteria.

The proclamation was actually made under the influence of the Mutazalites, a group of learned Muslims eager to satisfy religion as well as reason.[43] Their very name, originating from the word *itazala* or withdraw, indicated that they wanted to avoid the extreme of anthropomorphic theology as well as that of pure rationalism exemplified by Aristotle's metaphysics and by the various Neoplatonist systems. The need for a middle course made itself felt through a number of theological issues which the Koran's forceful and absolutist language kept steadily generating. Thus, if it was true that Allah's creating function was exactly as overpowering as described in the Koran, then man's failures and evil actions too were a creation of Allah. This, however, meant that Allah was not only the producer of evil, but also that justice broke down when man was punished here or hereafter.

To solve this question, it was not possible to fall back on the Koran's rather disconnected, rhapsodic utterances. Nor was there much help either in the Koran's emphasis on Allah's exclusive power if a balanced view was to be achieved on what underlay the question of just punishment, namely, the general relationship between the Creator as a necessary and ultimate Being, and man, a being obviously endowed with the ability to reason and to make free decisions. The Koran's inadequacy as a source of balance in judgment became evident through the comportment of the Mutazalites as well as through the stance eventually taken by their antagonists, the Mutakallimun. The former were adamant against any compromise with traditional orthodoxy which treated rather lightly the demands of reason. The latter, although well versed in the art of syllogistic discourse (their name was a derivative from the word *kalam*, or to speak in a formal manner), had use for philosophy only as long as it helped undermine their opponents' position. The Mutakallimun saved Muslim orthodoxy but merely in a negative sense. They succeeded in discrediting their opponents as rationalists, skeptics, and agnostics, but failed to construct a system in which the relation of faith and reason, revelation and experimental evidence, Creator and creature, would have received a balanced account.

The Mutakallimun can, therefore, hardly be considered counterparts of medieval Christian scholastic theologians. Unlike the latter, they failed to form clearly identifiable schools, and they were rather unwilling to give reason its due in a positive and constructive sense. Orthodox Muslim theology turned into a discursive defence of mysticism, and mysticism is certainly a field where preferences of the individual thinker, of his subjectivism in short, can readily take the leading role. Typically enough, the first great representative of Muslim orthodoxy, following its temporary eclipse after al-Mamun's edict, was none other than al-Ashari (873–935), one of the greatest Muslim mystics. His firm adherence to the spirit and parlance of the Koran produced what is the most genuinely Muslim world view. For him and his disciples, the Asharites, the world consisted of a set of points in space and of atomistic moments of time.[44] What held these units, according to him, together was not some law of nature but the will of God. Consequently, a particular configuration of time and space units did not define the configuration to follow. Configurations, which ultimately represented the world, followed one another in al-Ashari's system in virtue of the will of God, who could freely change the course of events according to His sole pleasure. The theological roots of this world picture are forcefully expressed in the Creed of al-Ashari: "We believe that God created everything by bidding it 'Be' . . .; that nothing on earth, whether a fortune or a misfortune, comes to be, save through God's will; that things exist through God's fiat; that no one can perform an act prior to its performance, or be independent of God or elude His knowledge . . .; that there is no creator save God; and that the deeds of the creatures are created by Him and predestined by Him . . .; that the creatures can create nothing but are rather created themselves."[45]

The Asharite world view was as completely un-Greek as it was also wholly unbiblical, but it reflected all too well the Koran's concept of God and world. The time atoms and space atoms of the Asharites had the veneer of science though only in retrospect. If scientific endeavour were ever to have a more hostile soil it was a discourse about a world made of disjointed units that obeyed no law other than the wholly unfathomable will of Allah. A hundred and fifty years later the prospects for science improved not a whit as traditional Muslim orthodoxy received its greatest defender, al-Ghazzali (1058–1111), aptly surnamed Hujjatu-l-Islam or "Islam's convincing proof." His own intellectual career mirrored the agonizing vacillation and final choice made by Muslim orthodoxy between faith and reason. His first passion was secular erudition, but it was in mysticism where he found his peace. Mysticism in his case meant a total surrender to the God of the Koran and to the absolute supremacy of His will in any matter however paradoxical it might appear to human reasoning.

The dictates of reason resting on the recognition of a causal connection between events seemed to al-Ghazzali to infringe on several points on the absolute prerogatives of God as portrayed in the Koran. He, therefore, made a frontal attack on the Mutazalites in a book which became a milestone of Muslim though not only by its contents but also by its evocative title: *Tahafut al-falasifah* ("Incoherence of the Philosophers").[46] It is a matter of speculation whether the Mutazalites went indeed as far as to deny flatly the ability of the Creator to override the causal flow of events by miracles. Al-Ghazzali said, however, just that and took a position which was tantamount to the abandonment of a causal connection between any and all events. He declared that fire and burning, light and sunrise, medicine and healing, "or any other set of events observed to be connected together in Medicine, or Astronomy, or Arts, or Crafts . . . are connected as the result of the Decree of God (holy be His name), which preceded their existence."[47] By this he meant that human reason had to stop at the observation of simultaneity, or of immediate sequence, and forego the obvious inference to causality. Listing such cases as the transformation of something into ashes by fire, or the emergence of an embryo through fertilization by sperm, he insisted that "all these things are observed to exist *with* some other conditions. But we cannot say that they exist *by* them . . . On the contrary, they derive their existence from God . . . So it is clear that existence *with* a thing does not prove being *by* it."[48]

Problem XVII in al-Ghazzali's book, from which the foregoing quotations are taken, has the title, "Refutation of their belief in the impossibility of a departure from the natural course of events." Other chapters of the book do not have the same immediate bearing on scientific reasoning, but they are very revealing in two respects. First, they illustrate the failure of the Mutazalites to steer a middle course between faith and reason. Second, they also reveal al-Ghazzali's resolve to deny to reason the ability to conclude to the existence of the Creator, to some of his attributes, and to the created character of the world. The Mutazalites might have been properly taken to task for their contention that reason could establish the eternity of the world. This was clearly a yielding on their part to a patently pantheistic feature of the Aristotelian and Neoplatonist world view. But al-Ghazzali failed to realize that his mysticism, too, was courting with age-old pantheistic notions when he chided his opponents for their inability to prove that the heavens were alive.[49]

Traditional Muslim orthodoxy could show its thoroughly unscientific posture not only in the manner in which science was ignored by al-Ashari and al-Ghazzali. The Brethren of Purity (Ikhwan al Safa), a quasi-monastic fraternity, reached out for scientific lore, but only as part of their program of mysticism. The result was equally stifling for a prospective emergence of genuinely scientific mentality and endeavour.[50] Their encyclopedic summary of knowledge (Rasa'il) consists of fifty-two treatises grouped into four books, of which the first deals with mathematical and educational topics, the second with natural sciences, the third with the theory of knowledge, and the fourth with theology. It would be mistaken to conclude from this that the Brethren of Purity were too keen on keeping separate the rational from the mystical, the observational from the cabbalistic, the logical from the poetical. Their preference was exactly in the opposite direction. The main pillars of their world view were, in addition to the Koran, the Pythagorean number mysticism, the Greek conception of the world as a huge organism, the doctrine of world-soul, and the hierarchical ordering of the universe. Thus, they explained creation as a process analogous to the self-generation of integers. The numbers 1 to 9 signified for them the Creator (1), the World-Intellect (2), the World-Soul (3), the Original Matter (4), the Absolute Body (5), and the four elements (6–9).[51]

It should be easy to see that such a system had much in common with an emanationist, pantheistic conception of the world. Not that the Brethren of Purity did not insist on the transcendency of the Creator, and on the temporal inception and finite duration of the world.[52] Yet, these assertions failed to grow into a much needed purifying and corrective set of principles, either with respect to the world view of the Brethren of Purity or to the spirit of Muslim science in general. This is perhaps best seen in the manner in which astrology is condemned in the Rasa'il. It is only the art of charlatans that is repudiated, because they try to predict the future without any careful "scientific" study of the planets. There is, indeed, a supreme irony in the disclaimer of the Brethren of Purity that without previous study the unknown future is an open book only for God.[53]

Consequently, those with expertise in "reliable" astrology could confidently turn to the interpretation of past, present, and future on the basis of the position of the planets. In fact, there is hardly a chapter in the Rasa'il without astrological information. There the processes characteristic of the macrocosmos and microcosmos are not ruled by laws of nature, but by the cyclic motion of the sun, moon, and the planets. The six planets were identified with the six phases of history represented by the six major prophets, Adam, Noah, Abraham, Moses, Jesus, and Muhammad. The Last Judgment was to occur, according to the Brethren of Purity, at the moment when the six planets came together again. This moment also was pictured by them as the Creation by God of a new "day."[54] What this really meant was not explained in sufficient detail, but no effort was made either to disconnect the six phases of history with the doctrine of the Great Year which the Brethren of Purity closely tied to the course of historical and cosmic events. They insisted that as the apogees and nodes of the planets passed from one sign of the zodiac to another every 3000 years major changes also occurred in civilizations. These changes implied huge geophysical transformations as well, such as the replacement of continents by seas, of flatlands by mountains, when in every 12,000 years the planetary shift was complete from one quadrant of the zodiac into another.

Even if the Brethren of Purity did not go as far as to assign a rebirth to everything in every 48,000 years, their acceptance of a cyclic, inexorably occurring course of cosmic and human events was unreserved. In their eyes all short terms were cyclic: "Everything which in this world is produced quickly lasts a short time, disappearing rapidly [to be reborn anew]; and that depends upon a motion of the universal sky which is rapid, of short duration, and returns quickly to its beginning."[55] No different was the case with the large-scale patterns: "It is by these primary and intermediate causes [the revolution of stars and planets] that the domination of the world passes from one people to another, that cultures as well as desolation may be transported from one quarter of the earth to another. All these events happen by virtue of the power determining the conjunctions which occur in regulated times and circumstances."[56]

It should be easy to see that in such an outlook no sufficient room was left for a rational investigation of the laws of the universe. The connection of all sorts of events with the position of planets demanded imagination, wishful thinking, and phantasy far more than cool reasoning. The Muslim notion of the Creator was not adequately rational to inspire an effective distaste for various types of pantheistic, cyclic, animistic, and magical world pictures which freely made their way into the Rasa'il. There astronomy, for instance, was considered a science not so much derived from observation and calculation as from the "revelation" given to Hermes Trismegistus who allegedly travelled as far as Saturn to gather the true science of the heavens. The prominence given to Hermes Trismegistus should be highly revealing to anyone familiar with the mystical obscurantism dominating those anonymous writings compiled in late Hellenistic times. No wonder that the Rasa'il could not represent a forward step in science. It rather illustrates the conceptual morass that can also be found in the writings of Paracelsus and Bruno, chief advocates during the Renaissance of Hermes Trismegistus and cabbalism.

The failure of the Mutazalites to cultivate science in a hopeful harmony of faith and reason, the thoroughly antiscientific reaction of the Mutakallimun, the disappearance of truly scientific perspectives in the mysticism of the Brethren of Purity, are major evidences of the sad prospects for science within Muslim religious ambiance. The same logic ran its full course in the position taken by Averroes, one of the most incisive and influential Muslim thinkers of all times. With an uncommon resolve he took the only remaining option: the cultivation of science in total independence from any tenet of the Muslim creed. To what extent he advocated the principle of double truth has been the source of many conflicting interpretations. What is beyond any doubt is that he became the chief inspiration of Averroists, a school of philosophers best known for their intransigent adulation of Aristotle and for their obstinate insensitivity to contrary evidence. Averroes could, however, hardly be surpassed for his unconditional surrender to the Aristotelian way of discoursing about the world. According to him Aristotle invented physics and gave it a form to which nothing further could be added. He viewed it as "miraculous" and well-nigh divine that after fifteen hundred years no error of any consequence could be found in Aristotle's physics.[57] He attributed a sort of infallibility to Aristotle as he claimed that no anatomical observation of Galen could contradict Aristotle's conclusions because these "are universal demonstrations, taken from propositions that are natural, primordial, universal, and essential. Now it is the property of such demonstrations that they cannot be contradicted by any sense perception."[58]

It was also Averroes' contention that religion and philosophy should be kept absolutely separate, or else doubts might arise both about faith and Aristotle's doctrine.[59] In stating this, Averroes unwittingly revealed the deepest reason of the bondage into which his thinking became entangled. Although a deeply devout Muslim, he was unable to take some of the Koran's major tenets about the Creator and the world as guidelines for reasoning. They might have perhaps led him out of the labyrinths of the Aristotelian interpretation of the world in which the obvious was at times talked away with hardly a second thought in order to prevent a thorough reconsideration of the imposing but hastily erected edifice. The edifice was the Aristotelian system of the world resting on some eagerly exploited observations and on the systematic slighting of many others.

The real attractiveness of that world picture was not its superficial agreement with observations, but that it could also be deduced from a few basic postulates. It was one of the greatest lures ever presented to human reason which can all too readily fall prey to the mirage of self-explaining and final solutions. Apriorism means freedom from endless trial and error in the search for truth; it means a facile insertion of man and of his mind into a pantheistic whole and into its eternal rhythm going through forever the same patterns.[60] Aristotelian physics and especially its doctrine of motion were integral parts of that pantheistic outlook of which the eternity of the world was a cardinal tenet. A new era in science could only come by a reopening of the case of Aristotle. Opting for Aristotle with the resolve and influence of Averroes and at a time when Muslim culture began to pass beyond its zenith, stole therefore from the Arabs a major if not last chance to make the crucial breakthrough toward modern science. With Averroes the horizons became rigidly restricted. The Koran and Muslim faith displayed once more its pathetic inability to be a guide not only to emotion and moral conduct but also to reason. The "rehabilitation" of reason, as advocated by Averroes, was equivalent to a dismissal of the Koran as a repository of rationally investigable tenets. The major evidence of this is the *Tahafut al-tahafut* ("Incoherence of the Incoherence"),[61] or Averroes' attack on al-Ghazzali's critique of the use of reason. There, in discussing miracles reported in the Koran, Averroes wrote:

The learned among the philosophers do not permit discussion or disputation about the principles of religion, and he who does such a thing needs, according to them, a severe lesson. For whereas every science has its principles, and every student of this science must concede its principles and may not interfere with them by denying them, this is still more obligatory in the practical science of religion, for to walk on the path of the religious virtues is necessary for man's existence, according to them, not in so far as he is a man, but in so far as he has knowledge; and therefore it is necessary for every man to concede the principles of religion and invest with authority the man who lays them down. The denial and discussion of these principles denies human existence, and therefore heretics must be killed. Of religious principles it must be said that they are divine things which surpass human understanding, but must be acknowledged although their causes are unknown.[62]

A philosophy of nature, systematically ignoring the question of miracles and all they imply, such was Averroes' ideal. If the science of Aristotle or the science of ancients, to use his expression, was so eminent, it was because they did not discuss miracles "although they were known and had appeared all over the world."[63] The implication was that Aristotle and ancient philosophers in general were familiar with Old Testament miracles (a view often repeated in late antiquity and early Middle Ages by Christians and Muslims), but they wisely ignored them as far as philosophy was concerned. However that may be, it was certainly not true that ancient philosophers did not "discuss any of the things which are said to happen after death."[64] But life after death could easily be part of a syndrome going on with absolute necessity. Miracles meant precisely that no law of nature was intrinsically and absolutely necessary and eternal. Miracles that were true, or held to be such, could be powerful reminders against attributing to given laws of nature and to the parts and whole of nature an eternal necessity which they demonstrably did not possess. Of this crucial aspect of the problem Averroes remained conspicuously unaware.

Averroes also failed to see that the real issue was not al-Ghazzali's occasionalism, but the elucidation of the contingency of the world. The need for it presented itself in Muslim consciousness in the measure Arab scholars assimilated the Greek scientific explanation of the world, or more particularly, the Aristotelian world view. Al-Farabi, "the second teacher," whose achievements in musical theory were already mentioned, earned his fame not only by his paraphrase of the ideas of Aristotle, the "first teacher." Al-Farabi also made utterances that became a milestone in the history of philosophy.[65] His great ambition was to calm the waters of a deeply troubled Muslim intellectual ambience, for which, as al-Farabi noted, the most vexing problem had to do with the concept of creation.[66] The problem arose from the encounter of the Muslim faith in creation, or more exactly in a creation out of nothing, with the Aristotelian tenet of a necessary, eternal world.

Al-Farabi was fully aware of the fact that the Muslim creed imposed a radical difference from the viewpoint of existence between God and any other being. The most often quoted passage of the Koran in this respect was verse 88 of Sura XXVIII, which states that "Everything shall perish except Him self."[67] Al-Farabi obviously had this passage in mind as he wrote: "He rules supreme, that is, He has the power to make a being non-existent, namely, to deprive of existence those

P

beings which as such deserve annihilation; for everything vanishes except He."[68] In diametrical opposition to such a tenet stood the Aristotelian doctrine in which the universe was a necessary being in every respect except perhaps for the place it occupied. Al-Farabi correctly saw that in order to solve the dilemma, distinction should be made between beings that are necessary and beings that are only possible with respect to existence.[69] But al-Farabi failed to reserve for God exclusively the category of necessary beings. For him the heavenly regions and immaterial beings were also eternally and necessarily produced by God, or rather emanated from Him. In his classification only sublunary or temporal beings were considered as "possible," that is, contingent with respect to existence.[70]

Al-Farabi was not the first Muslim scholar who found nothing strange in identifying creation with necessary emanation. Before him, the same was done in the so-called *Theology of Aristotle*,[71] and also in the *Rasa'il* of the Brethren of Purity. Clearly, there must have been something specifically indecisive in the Muslim theological doctrine of creation, or else it becomes well-nigh impossible to find explanation for the early emergence and endurance of such a syncretism in the minds of leading Muslim scholars. The same indecisiveness is intimated by the fact that Muslim scholars failed to perceive in the doctrine of contingency the very basis of a satisfactory formulation of scientific method. If the superlunary world, or the dominating part of the universe, was a necessary entity, then its laws too were to be thought of as eternally necessary. Apriorism then had its theological sanctioning with the simultaneous loss of interest in an experimental investigation of the universe. If the latter procedure through trial and error needed theological encouragement, it could only come from a resolute recognition of the radical contingency of the whole and the parts of the universe. Such a conception of the universe could easily generate a searching stance with respect to external nature, and even motivate an enthusiastic exploration of it. The recognition of this had not, however, taken roots in the Muslim world. Or as one of the leading Islamologists of our times felt impelled to state in his analysis of the Muslim attitude toward the world: "Research per se, as an effort to widen man's insight into the mysterious ways of the Creator, is not experienced as a means of glorifying God."[72]

Behind this failure lies the sharp dichotomy between reason and will that characterizes the Koran's doctrine of the Creator. From that dichotomy came the split mentality which kept producing heavily lopsided solutions for the coordination of knowledge and faith, science and religion, physics and metaphysics. The same dichotomy is also evident in the cultivation of an esoteric and exoteric teaching on the part of many Muslim scholars grappling with the problem of the relation between uncreated and created beings. A classic case is Avicenna, who had few rivals either as a physicist or as a metaphysician among the Arabs. As a philosopher he echoed with great appreciation of what was a halting step on al-Farabi's part toward the recognition of the contingency of created beings. Yet, in Avicenna's case the deepest layer of the mental soil was as unreceptive for the idea of contingency as was the case with al-Farabi or with most Muslim scientist-philosophers of any distinction. In spite of their professed faith in the Koran's God, they all came under the full sway of the Aristotelian doctrine of the eternity and necessity of the heavens. While Arab scientists could give a half-hearted or equivocal support to the creation out of nothing, the idea of a creation in time found no real echo among them.

It should be easy to see that appreciative reflection on an absolute beginning in time for everything might have readily led to a sympathetic consideration of the possibility of much smaller beginnings. At any rate, a trust in the possibility of beginning was needed in more than one sense to achieve emancipation from the straitjacket of Aristotelian physics and from its idea of motion. The truth of this is dramatically evidenced in what is perhaps the most puzzling single paradox of Islamic science: the elaboration of a critique by Avicenna of a special aspect of the Aristotelian doctrine of motion, and the effectiveness with which Avicenna, the metaphysician, helped to lock Muslim thought in the tracks of strict Aristotelianism. Avicenna's brilliance did in the Arab East what Averroes did in the West. If not for metaphysics, at least for the fundamental questions of science, Avicenna's impact "discouraged original investigations and sterilized intellectual life." But G. Sarton, the author of the foregoing statement,[73] failed to draw attention to a most important reason for Avicenna's negative influence. The reason is rather arcane, namely, Avicenna's esoteric doctrine about Creator and creation. As Averroes reported, Avicenna really believed in the identity of the heavenly bodies with God.[74]

Avicenna's case should seem the more revealing and frustrating because his notion of impetus was in one respect even more advanced than that of John Philoponus, the great sixth-century Christian scholar. While Philoponus claimed that in a void "violent" motion would spend itself sooner or later, that is, come to a halt, Avicenna argued that such motion would go on for eternity. There was, however, one crucial difference between Avicenna and Philoponus. For the latter the discussion of violent motion in the void served to prove its possibility. Moreover, Philoponus' belief in creation prompted him to attack major tenets of Aristotelian physics. With Avicenna the surmise of the idea of inertial motion represented an insight wholly severed from reality. And so it did with other Muslim scholars who picked up his conjecture on the point. As an expert investigator of the question wrote: "What was still needed to be done was that Avicenna and his successors should cast off the habits of Peripatetic thought, and build their physics on a limiting case . . . The Arab philosophers did not tackle this task, they did not even perceive it."[75]

It should be clear that the possibility of that new perception, or the sighting of new horizons, could hardly arise in a thinking in which Plato and Aristotle proved to be a greater enticement than a Bible overshadowed by the Koran. In Avicenna's world, which is eternal and necessary, everything is neatly placed along a chain of emanation from the First Cause. It is a grandiose and seductive system which became the best known part of Avicenna's philosophy. In it God is said to generate the First Intelligence, out of which proceeded the Second. It is with the Second Intelligence that the purest materialization, the formation of the outermost sphere, took place. The Third Intelligence was embodied in the sphere of fixed stars. Intelligences of lower order generated in turn the spheres of the five planets, of the sun and of the moon. While this last Intelligence was too weak to generate another Intelligence, it produced both the world of terrestrial matter and acted as an "intellectus agens" for all men.[76]

This last point, which logically cannot but minimize the independence and originality of thought of the individual, should forcefully evoke the blind alley in which some of the keenest thinkers among the children of the Koran had been trapped. The reasons for this and for the ultimate decline of science in the Arab world are undoubtedly perplexing. Even more so is, however, the attitude of those who prefer to ignore the problem or answer it with platitudes. Clearly, not much is gained by noting that the question of the early decline of scientific interest among the Arabs is one "which raises very obscure problems of general psychology about which no one has yet put forward any very definite theory,"[77] and leaves it at that. The platitudinous also comes through in the comment of the noted historian of Arab medicine, Max Meyerhof, who blamed the "religious persecutions" of Muslim scientists from the twelfth century on for the subsequent standstill of science in the Arab world.[78] The alleged reason of that persecution, according to him, was the feat that scientific studies "lead to loss of belief in the origin of the world and in the creator."[79] Whatever the historical truth of general and continual persecution against science in the Muslim world, such a persecution has conceptual aspects that demand a closer analysis if it is explanation that is really looked for.

This deeper look was not attempted by W. Hartner[80] in his thematic exploration of the problem of the stillbirth of science in the Muslim world. Of the four factors singled out by him, two, the influence of Arab courts and the inferiority complex of Muslims with respect to the rapidly growing medieval Europe, do not seem to have played a major role. The other two, the acceptance by Muslim scholars of scientific (Aristotelian) dogmas and the hostility of Muslim religionists toward secular scholarship, are in fact begging the question as they are more statements of fact than explanations. To hint, as Hartner did, that in late medieval Europe the gradual relaxation of clericalism meant the liberation of science, still does not dispose of the question within the Muslim context. Any adequate reading of history would show that religious, nay dogmatic, convictions did not peter out in the West as clericalism declined. As far as the Islam is concerned, there was no priesthood there. Muslim clericalism, if there was such a thing, could only be the resolve of devout Muslim scholars to reassert the supremacy of the religious outlook, attitude, and world view characteristic of the Koran. Consequently, the reasons for the negative and decisive role of religion in the decline of Arab science should be sought in some tenets specific to that religiosity. There may also lie the real reasons for the acceptance of certain "scientific dogmas" as well.

It should not be surprising that historians of science or culture who look askance at religion would neglect this strictly theological side of the question. Perhaps they shy away from the task because of their misgivings about the idea of religious truth. The investigation of the problem does not, however, even remotely imply a choice among religious truths. The problem can conveniently be handled within the historical framework, and in order not to anticipate the themes of subsequent chapters, within the framework of Muslim culture itself. It was there that the whole life of Moses ben Maimon, or Maimonides, was passed. Between Cordova where he was born in 1135 and Cairo where he died as the caliph's physician in 1204, stretched half of the Muslim world. Its days were numbered though, as its western and eastern flanks were soon to be lost. The Christian *reconquista* took Cordova in

1236, Seville in 1248, and ten years later Mongolian hordes devastated Baghdad. Cairo was to remain for the next two centuries the major centre where studious Arabs could safely pursue their work.

There and elsewhere Muslim intellectuals turned inward. Their self-imposed isolation from the outer, and especially from the Western world could hardly be more complete, either in intensity or in endurance. Not until the late nineteenth century did modern technology and medicine begin their penetration into the Muslim world, but Muslim scholars showed no willingness for an open encounter with modern science. Thus, until very recently the heliocentric system was taught as a hypothesis in the school of philosophy at the Great Mosque in Cairo. In view of such persistent mental inertia, if not stubbornness, Maimonides' great philosophical work, the *Guide for the Perplexed*[81] written in Arabic, was not to change the hopeless course which Muslim thought was following. Not that devout Arabs could not have read with tremendous intellectual profit Maimonides' penetrating critique of the abuse of reason by official Muslim theology, which followed closely the tenets of the Mutakallimun and leaned heavily toward occasionalism. What Maimonides offered was a careful balance between faith and reason, between Bible and science, a balance which never became a reality within the medieval Muslim world where scientist-philosophers came under the sway of necessitarianism. As a result, science among the Arabs largely remained, apart from some minor though valuable modifications, in the stage in which they found it in the old Greek and Syrian manuscripts.

The historian of science and particularly the student of Arab science has, then, the obvious task to explain how a book like *The Guide for the Perplexed*, especially its second, or philosophical and scientific part, could be written in the Muslim milieu and yet evince a mentality far superior to it. This is all the more remarkable as Maimonides showed no condescendence toward Muslim and Greek scholars. He did his best to give them the benefit of doubt as can be seen, for instance, in his famous "Letter on Astrology."[82] There he went as far as to charge with abuses in the matter only Chasdeans, Chaldeans, Egyptians, and Canaanites. According to him the "wise men of Persia" always recognized the utter falsehood of astrology. About the "wise men of Greece, Persia and India" he mentioned only their composition of works about the "exceedingly glorious science" of astronomy.[83] Such was an extremely generous appraisal in view of Maimonides' claim that he read every work on "judicial astrology" translated into Arabic.[84] Those translations certainly included Ptolemy's *Tetrabiblos*, to mention only one example, which, like its Hindu and Persian counterparts, exuded that very idolatry from which Maimonides tried to save Jews scattered all over the Arab world. The idolatry implied in astrology turned around the proper relation between Creator and man. In astrology there was no room either for God, creating freely, or for man acting freely and with responsibility.

But precisely because of the biblical doctrine of creation, Maimonides could not take the seemingly obvious way out of the dilemma and call into doubt the universality and permanence of physical laws. Maimonides kept insisting on the all-encompassing influence of starry heavens on everything in the sublunary world, save, of course, man's freedom to act. Such a stance corresponded to keeping a most delicate dynamical balance, and Maimonides proudly pointed out in his "Letter on Astrology"[85] that his *Guide for the Perplexed* contained all the mental exertions needed to justify the paradoxically skilful balancing. By twelfth-century standards, Maimonides' work was indeed the model of equilibrium in which both reason and faith were given their due. That this could be done cannot be explained without recognizing the overriding role that was played in Maimonides' thinking by the concept of the Creator and of the *creatio ex nihilo*, as contained in the Bible and vigorously upheld in Jewish tradition. The Mutakallimun, too, believed in both, but faith in the God of the Koran was not to inspire the picture of the universe in which laws and causal connections dominated. The most, noted Maimonides, the Mutakallimun were willing to admit about lawfulness in the universe was that it resembled human habits, such as the customary riding of the king of a city through its streets. Still, a king could readily break his habits, and so could any or all parts of the universe shift to a different "habit." As Maimonides pointedly remarked about the principal contention of the Mutakallimun, "the thing which exists with certain constant and permanent forms, dimensions, and properties, only follows the direction of habit . . . On this foundation their whole fabric is constructed."[86]

The defence of the permanency of the laws of nature was seen by Maimonides as a most natural corollary of the main purpose of his book: the full exposition of the Scriptural account of the creation and its defence against philosophical objections.[87] He took pains to call attention to the disastrous pitfall which his stance could entail. The permanency and indestructibility of nature was also a basic tenet with Aristotle, who argued that the universe was, therefore, "the necessary result of causal relation." The Aristotelian view, as Maimonides hastened to note, included a "certain amount of blasphemy."[88] For devout Jews it was clearly an anathema that "the Universe came into existence, like all things in Nature, as the result of the laws of Nature."[89] The biblical account demanded a middle course between the more or less overt occasionalism of the Mutakallimun and the pantheistic necessitarianism of Aristotle endorsed by the Mutazalite tradition. Or as Maimonides concisely put it, with respect to the fixity of the laws of nature one had to agree with Aristotle "in one half of his theory."[90] His explanation of this summarized an age-old faith, but its wisdom was also the voice of the future: "For we believe that this Universe remains perpetually with the same properties with which the Creator has endowed it, and that none of these will ever be changed except by way of a miracle in some individual instances, although the Creator has the power to change the whole Universe, to annihilate it, or to remove any of its properties. The Universe had, however, a beginning and commencement, for when nothing was as yet in existence except God, His wisdom decreed that the Universe be brought into existence at a certain time, that it should not be annihilated or changed as regards any of its properties, except in some instances; some of these are known to us, whilst others belong to the future, and are therefore unknown to us. This is our opinion and the basis of our religion."[91] This was also the basis on which the future of science was finally secured, when for the first time in history a whole culture espoused the genuinely biblical doctrine of Creation as its very spiritual and intellectual foundation.

NOTES

1. On Arab history and culture in general a classic work of reference is P. K. Hitti, *History of the Arabs from the earliest Times to the Present* (1937; 8th ed.; London: Macmillan, 1963). The most scholarly monograph on Arab science is Aldo Mieli, *La science arabe et son rôle dans l'évolution scientifique mondiale*, originally published in 1938 and recently reprinted with a new bibliography and analytical index prepared by A. Mazahéri (Leiden: E. J. Brill, 1966). The section "Arabic Science" by R. Arnaldez and L. Massignon in R. Taton (ed.), *History of Science*, Vol. I, *Ancient and Medieval Science*, translated by A. J. Pomerans (New York: Basic Books, 1963) does not seem to substantiate its authors' claim about the crucial contribution of Arabs to science. Much useful material can be found in *The Legacy of Islam*, edited by Sir Thomas Arnold and Alfred Guillaume (London: Oxford University Press, 1931).

2. See G. Bergsträsser, *Hunain ibn Ishaq über die syrischen und arabischen Galen-Übersetzungen* (Leipzig: F. A. Brockhaus, 1925), § 115 (p. 39 of translation). Ibn-Ishaq also noted (see §§ 13, 14 [pp. 8–9]) that he revised his translations as years went by and better manuscripts came into his possession.

3. The Arabic text was edited with an English translation and glossary by Max Meyerhof (Cairo: Government Press, 1928).

4. Translated from the Arabic by William A. Greenhill (London: Printed for the Sydenham Society, 1848). In the introduction by the translator, there are listed 35 printings of the work between 1498 and 1787 in Greek, Latin, English, French, and German.

5. See "Science and Medicine," by M. Meyerhof in *The Legacy of Islam*, p. 323.

6. As stated by G. Sarton in his *Introduction to the History of Science*, Vol. I, *From Homer to Omar Khayyam* (Baltimore: The Williams and Wilkins Co., 1927), p. 609.

7. The most comprehensive account of Arab mathematicians and their works is the massive monograph by H. Suter, *Die Mathematiker und Astronomen der Araber und ihre Werke*, in *Abhandlungen zur Geschichte der mathematischen Wissenschaften* (Leipzig: B. G. Teubner, 1900). For a much shorter and accessible discussion, see the essay by Baron Carra de Vaux, "Astronomy and Mathematics," in *The Legacy of Islam*, pp. 376–98.

8. Available also in English translation under the title, *The Algebra of Mohammed ben Musa*, edited and translated by F. Rosen (London: Printed for the Oriental Translation Fund, 1831).

9. *The Algebra of Omar Khayyam*, translated with commentaries by Daoud S. Kasir (New York: Columbia University, 1931), p. 43. The last three types of equation which Omar solved by two hyperbolas were of the type $x^3 + cx^2 = bx + a$; $x^3 + bx = cx^2 + a$; $x^3 + a = cx^2 + bx$.

10. See A. von Braunmühl, *Vorlesungen über Geschichte der Trigonometrie* (Leipzig: B. G. Teubner, 1900–03), vol. I, pp. 54–61; and Carra de Vaux, "Astronomy and Mathematics," in *The Legacy of Islam*, pp. 389–90.

11. The Arabic text with a German translation was edited by J. Baarmann, "Abhandlung über das Licht von Ibn al-Haitam," *Zeitschrift der deutschen morgenländischen Gesellschaft*, 38 (1882):195–237.

12. Lengthy sections of it are available in German translation by F. Dieterici, *Die Naturanschauung und Naturphilosophie der Araber im zehnten Jahrhundert: Aus den Schriften der lautern Brüder* (Berlin: Verlag der Nicolai'schen Sort.-Buchhandlung, 1861), pp. 95–140.

13. See O. Loth, "Al-Kindi als Astrolog," in H. Derenbourg *et al.*, *Morgenländische Forschungen: Festschrift Herrn Professor Dr. H. L. Fleischer zu seinem fünfzigjährigen Doctorjubiläum am 4. März 1874* (Leipzig: F. A. Brockhaus, 1875), pp. 261–309.

14. As meticulously illustrated in the essay of E. S. Kennedy and B. L. van der Waerden, "The World-Year of the Persians," *Journal of the American Oriental Society*, 83 (1963): 315–27. This essay based on a careful study of the original sources contrasts sharply with the tendencious manner in which the Pan-Babylonian School tried to exploit Zoroastrianism to its aims around the turn of the century. See, for instance, the articles, "Ages of the

World (Babylonian)," "Ages of the World (Zoroastrian)," by. A Jeremias and N. Söderblom respectively in *Encyclopaedia of Religion and Ethics*, edited by J. Hastings (New York: Charles Scribner's Sons, 1908), vol. I, pp. 183–87 and 205–10. While Zoroastrianism assimilated much of the Hindu doctrine of cycles, Zoroaster himself seems to have been a believer in the final victory of good over evil. (See E. Herzfeld, *Zoroaster and his World* [Princeton, N.J.: Princeton University Press, 1947], vol. I, pp. 20–21.) There is no reason, however, to conclude that the biblical doctrine of a once-and-for-all creation and historical process is a post-exilic borrowing from Zoroaster's thought and tradition.

15. For details, see C. A. Nallino, "Astrologia e astronomia presso i Musulmani" (1908), in *Raccolta di scritti editi e inediti*, edited by Maria Nallino, vol. V. *Astrologia, Astronomia, Geografia* (Rome: Istituto per l'Oriente, 1944), p. 15.

16. The Latin translation made by Joannes Hispalensis and usually referred to under the title, *Albumasar de magnis conjunctionibus, annorum revolutionibus ac eorum profectionibus, octo continens tractatus*, was first printed by E. Ratdolt (Augsburg, 1489). Another printing followed in Venice in 1515. References will be to the former printing, whose colophon reads: Opus albumazaris de magnis conjunctionibus explicit feliciter: magistri johannis angeli viri peritissimi diligenti correctione: Erhardias Ratdolt viri solertis eximia industria, et mira imprimendi arte: qua nup venetiis: nunc auguste vindelicorum excellit noiatissim pridie kal. Aprilis. 1489. Albumasar is the Latinized form of Abu-Mashar.

17. The subtitle of the English translation by C. E. Sachau reads: *An English Version of the Arabic Text of the Athar-ul-Bakiya of Albiruni, or "Vestiges of the Past"* (London: W. H. Allen, 1879), p. 31. The book was written around 1000. Reference was already made in the first chapter to al-Biruni's opposition to the Hindu doctrine of yugas.

18. *Ibid.*, p. 29.

19. *Ibid.*, p. 31.

20. *Ibid.*, pp. 29–30.

21. *Ibid.*, p. 31.

22. *Ibid.*, p. 30.

23. *Ibid.*, p. 115. He also recalled that "other people, again maintain that in each cycle a special Adam and Eve exist for each country in particular, and that hence the differences of human structures, nature, and language [are] to be derived." *Ibid.*

24. *Ibid.*

25. *Ibid.*, p. 30.

26. *Ibid.*

27. *Ibid.*, p. 29.

28. *The Book of Instruction in the Elements of the Art of Astrology*, by Abu'l-Rayhan Muhammad ibn Ahmad al-Biruni. Translation facing the Arabic original by R. Ramsay Wright (London: Luzac & Co., 1934).

29. For further details, see the scholarly discussion by Edward S. Kennedy, "Ramifications of the World-Year Concept in Islamic Astrology," in *Actes du dixième Congrès International d'Histoire des Sciences, Ithaca 26 VIII 1962 – 2 IX 1962* (Paris: Hermann, 1964), pp. 23–45.

30. As rather proudly brought out by Seyyed Hossein Nasr, a Persian historian of science, in his *An Introduction to Islamic Cosmological Doctrines: Conceptions of Nature and Methods Used for Its Study by the Ikhwan al Safa, al-Biruni and Ibn Sina* (Cambridge, Mass.: The Belknap Press of Harvard University Press, 1964). Bafflingly enough, he presented that unsavoury mixture as the only antidote against what he called the materialism and inhumanity of modern Western science.

31. Subsequent quotations from the Koran are from the translation by George Sale, *The Koran; commonly called the Alkoran of Mohammed* with explanatory notes selected by F. M. Cooper (New York: A. L. Burt, n.d.).

32. Verses 12–23, pp. 224–25.

33. Verses 2–11, p. 224.

34. Sura XXI, verses 31–34, p. 265.

35. Sura III, verses 185–89; p. 101.

6. Sura II, verse 137, p. 69.

37. Sura V, verse 21, p. 122.
38. Sura XIV, verse 4, p. 216.
39. Sura, IV, verse 44, p. 107.
40. An alternate title of that Sura is "Angels,"; see pp. 339–43.
41. Sura XXXV, verse 14, p. 340.
42. Sura XXXV verse 9, p. 339.
43. On the theological background and ramifications of this crucial conflict in Muslim intellectual history, see A. J. Wensinck, *The Muslim Creed: Its Genesis and Historical Development* (Cambridge: The University Press, 1932), chaps. iii, iv and v.
44. The philosophical and scientific aspects of this trend were aptly treated in Majid Fakhry, *Islamic Occasionalism and Its Critique by Averroes and Aquinas* (London: George Allen & Unwin, 1958), and Salomon Pines, *Beiträge zur islamischen Atomlehre* (Gräfenhainichen: A. Heine, 1936).
45. Quoted in M. Fakhry, *Islamic Occasionalism*, pp. 56–57.
46. Translated into English by Sabih Ahmad Kamali (Lahore: The Pakistan Philosophical Congress, 1958).
47. *Ibid.*, p. 185.
48. *Ibid.*
49. See chap. xiv, pp. 163–67.
50. This is, of course, not admitted by S. H. Nasr, whose *Introduction to Islamic Cosmological Doctrines* contains the most detailed account available in English on the world view of the Brethren of Purity.
51. *Ibid.*, pp. 51–52. The actual list from 6–9 is more complicated than given here in the text.
52. *Ibid.*, p. 55.
53. *Ibid.*, p. 82.
54. *Ibid.*, p. 83.
55. *Ibid.*, p. 80.
56. *Ibid.*
57. See Averroes' preface to his commentary on Aristotle's *Physics* in *Aristotelis opera cum Averrois commentariis* (1562; reprinted Frankfurt-am-Main: Minerva G.m.b.H., 1962), vol. IV, p. 5r.
58. In his commentary on Aristotle's *De partibus animalium; ibid.*, vol. VI, p. 139r.
59. In the commentary on the *Physics; ibid.*, vol. IV, p. 36v.
60. Needless to say, Averroes fully upheld Aristotle's views on the eternal recurrence of everything generic. See his *Paraphrasis super librum de Generatione et corruptione; ibid.*, vol. V, p. 396r.
61. Translated from the Arabic with introduction and notes by Simon Van den Bergh (London: Luzac, 1954).
62. From "The First Discussion about the Natural Sciences," *ibid.*, p. 322. Averroes further elaborated on this theme in his "The Decisive Treatise, Determining What the Connection Is between Religion and Philosophy," readily available in English translation in R. Lerner and M. Mahdi (eds.), *Medieval Political Philosophy: A Sourcebook* (New York: The Free Press of Glencoe, 1963), pp. 163–86.
63. *Tahafut al-tahafut*, p. 322.
64. *Ibid.*
65. See on this E. Gilson, *History of Christian Philosophy in the Middle Ages* (New York: Random House, 1955), pp. 185 and 638.
66. See the German translation by F. H. Dieterici of his "The Bezels [Foundations] of Philosophy," in al-Farabi's *Philosophische Abhandlungen* (Leiden, E. J. Brill, 1890), p. 1.
67. *Ed. cit.*, p. 313.
68. See *Philosophische Abhandlungen*, p. 83; see also pp. 87 and 67.
69. *Ibid.*, pp. 57 and 66.
70. This whole question was carefully treated in the essay by E. L. Fackenheim, "The Possibility of the Universe in al-Farabi, Ibn Sina and Maimonides," *Proceedings of the American Academy for Jewish Research*, 16 (1946–47): 39–70.

71. See translation with notes by F. H. Dieterici, *Die sogennante Theologie des Aristoteles* (Leipzig: J. C. Hinrichs, 1883).

72. G. E. von Grunebaum, *Islam: Essays in the Nature and Growth of a Cultural Tradition* (2d ed.; London: Routledge & Kegan Paul, 1961), p. 112.

73. *Introduction to the History of Science*, vol. I, p. 711.

74. *Tahafut al-Tahafut*, p. 254.

75. S. Pines, "Les précurseurs musulmans de la théorie de l'impetus," *Archeion*, 21 (1938): 303. See also his "Quelques tendances antipéripatéticiennes de la pensée scientifique islamique," *Thales*, 4 (1940): 210–19.

76. See the Fifth Tract of the First Part in *Algazel's Metaphysics: A Medieval Translation*, edited by J. T. Muckle (Toronto: St. Michaels' College, 1933), pp. 119–29.

77. As did Baron Carra de Vaux in his essay "Astronomy and Mathematics," in *The Legacy of Islam*, p. 397.

78. In his essay, "Science and Medicine," in *The Legacy of Islam*, p. 337.

79. *Ibid.*

80. "Quand et comment s'est arrêté l'essor de la culture scientifique dans l'Islam?", in R. Brunschvig and G. E. von Grunebaum (eds.), *Classicisme et déclin culturel dans l'histoire de l'Islam: Actes du Symposium International d'Histoire de la Civilisation Musulmane* (*Bordeaux 25–29 Juin 1956*) (Paris: Éditions Besson-Chantemerle, 1957), pp. 319–37.

81. Subsequent references are to the translation from the original Arabic text by M. Friedländer (2d rev. ed., 1904; New York: Dover, 1956).

82. See translation by R. Lerner in *Medieval Political Philosophy*, pp. 227–36.

83. *Ibid.*, p. 230.

84. *Ibid.*, p. 229.

85. *Ibid.*, pp. 229, 231.

86. *Guide for the Perplexed*, Part I, chap. 73, p. 128.

87. *Ibid.*, p. 211 (Part II, chap. 28).

88. *Ibid.*

89. *Ibid.*, p. 201 (Part II, chap. 27).

90. *Ibid.*, p. 210 (Part II, chap. 28).

91. *Ibid.*, pp. 210–11 (Part II, chap. 28).

CHAPTER TEN

The Sighting of New Horizons

Beginnings, if they are truly such, can reveal at one stroke the whole story to follow. Adelard's *Quaestiones naturales*,[1] which easily marks the true dawn of medieval science, starts with phrases that are pregnant with the future. This is all the more remarkable as nothing could have been more natural for Adelard than to acquiesce in his nephew's insistence that the spontaneous appearance of plant life in a dishful of dried soil was strictly miraculous. Genuine devotion for miracles was, during the Middle Ages, second only to a rampant craving for the miraculous, and nature had enough mysteries of its own to feed appetites eschewing the exigencies of reason. Against this background there stands out with monumental incisiveness in Adelard's rejoinder the firm vindication of the prerogatives of both the Creator and of His creation: "It is the will of the Creator that herbs should sprout from the earth. But the same is not without a reason either."[2] He stuck to the same point when pressed by his nephew for whom the natural explanation based on the doctrine of the four elements seemed to leave something to be desired and to be remedied only by a recourse to God's universal effectiveness: "I do not detract from God," went Adelard's reply. "Whatever there is, is from Him and through Him. But the realm of being is not a confused one, nor is it lacking in disposition which, so far as human knowledge can go, should be consulted. Only when reason totally fails, should the explanation of the matter be referred to God."[3]

The autonomous character of nature and the possibility of its extensive rational investigation were points emphasized by Adelard on more than one occasion. He noted that the "spontaneous generation" of plants was possible because nothing was ever lost in nature notwithstanding some processes (decay, evaporation, etc.) indicating the contrary.[4] Equally significant was his emphasis on one's reliance on the eyes of reason to infer, for instance, the extremely fast propagation of light.[5] His belief in the world as a product of an infinitely intelligent Creator made Adelard also aware of the subtlety of nature that would not readily yield to the probings of human intellect.[6] In the same vein his admiration for Arab and Greek learning found a healthy counterbalance in his faith in the Creator. It prevented him from sharing the worshipful attitude of Greek philosopher-scientists and of many Arab scholars toward the starry firmament. Not that Adelard disputed a very exact, causal influence of stars on terrestrial phenomena. He even spoke of the extraordinary intelligence of stars (planets) evident in their progress and retrogress. Such intelligence could be called divine, but he firmly distinguished it from God: "If the question is about the true God," he argued, "who is the universal cause of all, unimaginable, immutable, infinite, it would be abominable to call the outermost sphere God in that sense."[7]

Adelard of Bath (fl. 1125) was one of those eager medievals who went on long and arduous journeys in the quest of learning. He travelled across the Muslim world as far as Aleppo and through his efforts as translator medieval Europe got its first access to Arab trigonometry, to the description of astrolabs, and to Euclid's geometry. His contacts with Muslim scholars certainly made him familiar with their frustrating struggle to reconcile reason and faith. Struggles of this type were not a speciality of the Muslim and Jewish world. A priceless glimpse into medieval Latin Christianity is allowed by a remark of Adelard's nephew who noted that many of his contemporaries denied God, or identified Him with Nature.[8] The presence of pantheistic, atheistic, and agnostic proclivities at a time when "the centuries of faith" were approaching their zenith should not be surprising. The high Middle Ages[9] were intellectually a turbulent era. While orthodoxy could be imposed through patently unevangelical methods, man's inner assent proved itself to be doggedly elusive to enforcement. Individual faith in its turn represented a commodity to be fought for continually. Opposing man's acceptance of a transcendental belief with all its dynamics and tensions, was the innate inclination in man to seek repose in an identification with nature as the ultimate entity. Side by side with the startling experience of man's freedom there was also the wishful desire to be carried effortlessly on the waves of nature's events. As one would expect, neither of these experiences could provide an exhaustive answer to all aspects of man's position in nature. Man's freedom obviously has its constraints, and so is the soaring of his mind hampered by his bodily condition. A carefree yielding to the course of nature is in for rude jolts that force man to fight for survival and for a deeper understanding of his surroundings.

Medieval man, strong as many of them could be in faith, formed no exception to all this. As to the scientific understanding of the inexorable processes of nature, they too were tempted by the mirage of fatalism. Its chief vehicle was astrology, the branch of speculation in which observation and vagueness found an alliance never matched in any other field. The vagueness of the dicta of astrology should suggest that it depends on and also fosters a state of mind for which the indefinable and inevitable are synonymous. Astrology claims a complete insight into the recondite intricacies of nature and human destinies, while at the same time it disdains a partial though reliable grasp of the workings of nature. Medieval investigators of nature fell short of their goals in the measure in which they fell under the sway of the fundamental tenets of astrology.

An early case in point is that of Bernard Silvester (fl. c. 1150) from the School of Chartres, whose *De mundi universitate*,[10] a treatise on the macrocosmos and the microcosmos, appears, when compared with Adelard's work, a morass replete with astrological pantheism. His predicament shows all too well the lure which astrology could exert even as the Middle Ages began to evolve as a distinctly Christian phase of culture. The eleventh and twelfth centuries witnessed an onrush of translations into Latin of Arabic astrological treatises and even of pieces of Hermetic literature.[11] Viewed against this background, it is impossible not to see the purifying role of the biblical account of creation as it appears, for instance, in the exposition by Thierry of Chartres (d. c. 1155).[12] His aim was to show that underlying the phrases of Moses there was an unobjectionable physics corresponding to the best in Plato's *Timaeus*. Admiration for that work and its spirit could have easily forced a tacit relapse into Greek animism and pantheism. But this possibility was patently foreclosed when the biblical faith in creation was stated at the outset with no vacillation: "The intention of Moses was to show that the creation of things and the formation of men was by the only one God, to whom alone worship is due. The utility of the work [by Moses] is the acquisition of knowledge about God through His handiwork."[13] Such a firmness in faith could then pour new wine into old skins, that is, new perspectives into old frameworks of explanation. The four causes of Aristotle were old, but they now served a new insight: "Of the substances composing the world there are four causes: efficient, or God; formal, or God's wisdom; final, or His kindness; material, the four elements." But consciousness formed by that faith felt it necessary to point out in the same breath, lest some misunderstanding should arise, that all elements, that is, all things composing this world, "should have Him as their Creator, because they all are subject to change and can perish."[14]

By elements Thierry of Chartres meant all things composing the universe. Since his faith did not permit the divinity of the heavens, he explained the stars and the firmament as being made of water and air.[15] The naturalness with which he discussed the watery composition of stars is worth noting in connection with an avowed admirer of Plato's *Timaeus*. A parting with the hallowed dichotomy between heaven and earth represented no longer an insuperable difficulty, precisely because of one's faith in the Creator. Strengthened with such a faith, Thierry would indulge in listing statements of classical authors about the "spirit of the world" as something analogous to the "spirit of God" hovering above the chaos at the outset of creation. Still, Thierry had no illusion about the difference between chaff and wheat. His belief in an absolute beginning of the universe also prompted in him a provocative explanation of the circular motion of upper air (firmament) and stars. Such rotation of a very light material was, according to him, the recoil from a very solid axle or central point, the earth. The great speed of the heavens was for him much like the flight of a projectile: "when a stone is thrown, its impetus is ultimately due to the hold of the thrower against something solid [the ground]; the more firm is the stand of the one who throws, the more impetuous is the throw."[16]

In this one can see at best a vague anticipation of the impetus theory out of which classical physics was to be born. The conceptual context, the recall of an absolute beginning at the moment of Creation, was, however, a feature present no less in the embryonic than in the fully developed form of that theory. The same faith in creation was also at play as Thierry emphasized the utility of a mathematical approach to the investigation of the universe. "There are four kinds of reasons that lead man to the recognition of the Creator: the proofs are taken from arithmetic, music [harmony], geometry, and astronomy."[17] Clearly, if the Creator arranged everything according to number, measure, and weight, man's understanding of the world had to reflect a mathematical character or, in other words, science was to be based on mathematics.

The weight of such considerations, Platonic in some sense but biblical and Augustinian in their actual effectiveness, can best be seen in the writings of Grosseteste (c. 1168–1253), Bishop of Lincoln, and possibly the first chancellor of the University of Oxford. Grosseteste, a most influential figure of medieval scientific thought,[18] could not have been more explicit about anchoring his methodology of science in the notion of God as a Creator. This also served as a factor of safeguard in his acceptance of light as the basic material substratum. In the absence of faith in the Creator such a distinctly Neoplatonist step could have very well trapped him in pantheistic, a priori speculations about the universe. His awareness of the limitations of created intellects made him recognize, however, that it was one thing to enjoy the certitude of mathematical (geometrical) notions and another to discover them unerringly in nature, mathematical as its structure could be. The mathematical nature of the universe followed most directly from its being made of light, which Grosseteste held to be pre-eminently mathematical in character. The spread of light from one point along straight lines, its refractions and reflections, its obvious affinity with heat, were as many indications for him that geometrical patterns ruled all processes in nature: Or as he put it in his *De lineis*:

> The usefulness of considering lines, angles and figures is the greatest, because it is impossible to understand natural philosophy without these. They are efficacious throughout the universe as a whole and its parts, and in related properties, as in rectilinear and circular motion. They are efficacious also in cause and effect (*in actione et passione*), and this whether in matter or in the senses, and in the latter whether in the sense of sight, where their action properly takes place, or in other senses, in the operation of which something else must be added on top of those which produce vision . . . But the effects are diversified according to the diversity of the recipient. For when received by the senses this power produces an operation in some way more spiritual and more noble; on the other hand when received by matter, it produces a material operation, as the sun by the same power produces diverse effects in different subjects, for it cakes mud and melts ice.[19]

All this could be best put to use in optics. There, especially in the course of his investigations of the rainbow, Grosseteste developed the application of his methodology which included such seminal programmes as induction, falsification, and verification. His critique of Aristotle's and Seneca's explanation of the rainbow by reflection and his original account of it by refraction represented an indispensable step toward the adequate solution offered less than a century later by Theodoric of Freiberg (d. 1311). While Grosseteste considered the role of clouds as a whole, he imagined them to approximate one sphere and emphasized that the process forming a rainbow could be studied by observing the passage of light through a spherical glass bulb filled with water. Obviously, these were powerful suggestions toward the recognition of the role of individual raindrops in the process.

Details like these should, however, be remembered together with their intimate connection in Grosseteste's thinking with the notion of the Creator. A telling evidence of that intimate connection is the very able summary of Grosseteste's theory of scientific measurement by William of Alnwick, regent-master of the Oxford Franciscan House in the 1310's. According to him, Grosseteste considered all measurements made by man as intrinsically imperfect, as they were based on conventional units and not on counting the infinitely small, indivisible units (points) contained in every extension:

For how are we to know the number or quantity of that line which the first Measurer has measured? That quantity he reveals to no man, nor can we measure the line by means of infinite points, because they are neither known nor determined (*finita*) to us, as they are to God by whom they are comprehended. Whence this method of measuring is for us as uncertain as the first . . . Therefore there is no perfect measure of continuous quantity except by means of indivisible continuous quantity, for example by means of a point, and no quantity can be perfectly measured unless it is known how many indivisible points it contains. And since these are infinite, therefore their number cannot be known by a creature but by God alone, who disposes everything in number, weight and measure.[20]

The ring of the last phrase should be sufficiently evocative of the role of the biblical faith in the Creator in Grosseteste's thinking. His still unedited treatises, the "Hexaemeron" and the "De universitatis machina,"[21] are further documentary indications of the measure in which his scientific methodology depended on the idea of the Creator as a wholly rational, personal Planner, Builder, and Maintainer of the universe. The rationally clear atmosphere of Grosseteste's writings should not, however, be taken as typical of thirteenth-century discourses on the parts and whole of the universe. The real situation rather resembled an almost unmanageable mixture of insights, and rumours, of sound principles and fantastic tales, of critical sense and baffling credulity, of reason and magic. The best single illustration of this is probably the *De universo* of William of Auvergne,[22] who died in 1249 as bishop of Paris. The massive opus bespeaks both the proverbial darkness of an age groping for a way out of the jumble of phenomena, and also the miracle of an age, which, unlike all previous cultures, did not wholly succumb to the magical and irrational in its quest for understanding. The incessant references to the magical in the *De universo* have been the subject of an informative study,[23] but the same cannot be said about the equally pervasive presence there of the Christian doctrine of Creator and creation.

That presence was truly a saving one. It served as a firm barrier against the conceptual chaos that began to make itself felt in medieval thought from the middle part of the twelfth century. The chief carriers of the morass were the translations into Latin of such works as the *Centiloquium*, or a hundred astrological aphorisms, falsely ascribed to Ptolemy, his *Tetrabiblos*, Albumasar's *Isagoge Minor*, *Liber conjunctionum*, and *Flores astrologiae*, to mention only some outstanding titles.[24] The art of *interrogationes* and *electiones*, or the astrological method of discovering hidden factors and of determining propitious times for various actions, had before long become an obsession in medieval courts. In addition, there came the far more disreputable preoccupation with judicial astrology, or the forecasting of the future, individual and social, to say nothing of necromancy. All this implied in various degrees an outlook on the world where rationality could not prevail, or where a rational investigation of nature was doomed to ultimate failure. It is in this light alone that a proper appraisal is possible about the continual polemics in the *De universo* against Manicheism, fatalism, pantheism, star worship, and similar betrayals of the proper nature of man's rationality and ultimate destiny. The chief representatives of these aberrations were identified in the *De universo* as the Saracens (Arabs), the Greeks of old, and the Hermetic philosophers.

The notion of Great Year received a particularly lengthy discussion in the *De universo*.[25] This indicated not only the strength of the lure for the medievals of an ever recurring order of things, but also William's awareness of the fact that the idea of Great Year represented the farthest reaching embodiment of the non-Christian world view. The importance of the topic in William's eyes was made overriding by the circumstance that, as he put it, "great is the reputation of those who held this doctrine," and he mentioned among others Plato and Aristotle. "It is therefore proper," he reminded his reader, "that you should acquire the truth and certainty in the matter, lest it should appear that you dissent on the basis of superficial and weak reasoning from the opinion of so many and great authorities."[26] The six arguments in support of a Great Year of 36,000 years, and their refutation, were presented by William in great detail, but in themselves they are not particularly instructive. They are repetitious and concern the various recurrences in nature. William rebutted all these "proofs" by noting that either the recurrence in question was not identical, or that it was not a true recurrence. As one may suspect, proofs and rebuttal had one thing in common, namely, a very defective information about nature and its workings.

Uncertainty and vagueness about the facts of nature was a typical feature of the times of William of Auvergne. Thus, it was all too easy to supplement meager information about facts with speculations that could readily go over into sheer phantasy. Yet, even in such cases the ramifications of the Christian belief in the Creator could provide some precious guidelines for reasoning about phenomena. A fine piece of evidence in this regard is William's discussion of various "irradiations"[27] by which he largely meant cases of telepathy both in men and animals. A debated point concerned the alleged ability of dogs to recognize a thief from a distance. William seemed to accept the ability in question although he argued that it could not be attributed to the olfactory prowess of dogs. Thieves, William noted, do not smell in a way exclusive to them. Then he advised his reader: "In such and similar questions you should not trust the procedure of the inexperienced who in all cases, whose causes they do not know and are unable to investigate, take a facile recourse to the omnipotence of the Creator and call all such things miracles."[28]

The second part of the advice aimed at saving the rights of reason not from a commodious reliance on the Creator, but from that equally disastrous pitfall into which the devout among learned Muslims fell, noted William with an eye on the Mutakallimun: "Similarly, you must also part with those who in such matters take refuge in almighty God's most imperious will, and wholly abandon these questions as insoluble, and feel themselves at ease when they say that the Creator willed it that way, or that His will is the sole cause of all such things." The fatal error of such a stance had to do, as William remarked, with the failure of distinguishing between primary and secondary causality. "They err intolerably, first, because they assign only one solution to all such phenomena; second, because when asked about the cause, they refer to the remotest cause, although such questions are so varied that they cannot be settled by one single solution."[29]

Obviously, what was needed was an unremitting search coupled with an unconditional respect for some guidelines marking the major pitfalls in reasoning about nature, a contingent entity. These guidelines could be formulated in several ways. One of them was exemplified in the gigantic effort of Aquinas to bring reason and faith into a stable synthesis.[30] The moderate realism which characterizes Aquinas' theory of knowledge, and the analogy of being which forms the essence of his metaphysics, proved to be the classic stance of balance in reasoning. His resolve to give reason its due in the highest possible measure meant, however, in the actual context, an overly generous acceptance of the Aristotelian system, the epitome of rational explanation of the world at that time. In this search for conceptual stability, Aquinas was also motivated by the sad predicament of Muslim theologians and philosophers and by their highly unsettling impact on a Christian Europe going through its birthpangs.

Accordingly, the first major work of synthesis by Aquinas, the *Summa contra gentiles* (completed in 1257), aimed at countering the occasionalism and fatalism contending with one another within Muslim theology and philosophy. The task, as can be guessed, centred on questions about the Creator and the nature of human intellect. The stratagem demanded that Aquinas should not be found wanting in his admiration for Aristotle, the Philosopher. In fact, Aquinas departed from Aristotle only in cases where the Christian creed allowed under no circumstance for a compromise. This attitude of Aquinas was carried over in full into his *Summa theologica* (completed in 1273), a work in which synthesis, not polemics, dominated. The surprising extent to which Aquinas went in accepting Aristotle's cosmology and physics can be seen by taking a look at only one chapter in his massive opus, the 91st Question in its Third Part, where he discussed "The Quality of the World after the Judgment."[31] The topic, imposed by the concluding tenet in the Christian creed, meant a most acute confrontation with the very heart of Aristotle's cosmology and theory of motion. The contents of the five articles of Quaestio 91 show that the presence of cyclic features in the world was un unassailable truth for Aquinas, who firmly reasserted the efficient causality of a rotating sky on everything in the sublunary world. He found no fault with the generic return of physical patterns, including plants and animal species. He also went along with Aristotle on the point that the cosmos would of itself go on forever through endless begettings of individuals.

That Aquinas still had not become a hapless prisoner of the Aristotelian world view was due to his awareness of the guidelines set by the Christian creed about the cosmos. Against Empedocles' claim about a cyclic rejuvenation of the cosmos he noted that the new heaven and earth were supernatural, "just as grace and glory are above the nature of the soul."[32] Against the coupling of the precession of the equinoxes with the cyclic theory of the world, his principal argument was that this would allow the exact calculation of the moment of the world's end, in patent contradiction to the Gospel. He opposed the idea of an infinite endurance for the world through endless cycles on the ground that this would also mean that the number of the elect would become infinitely large: "But this is not in keeping with our faith, which holds that the elect are in a certain number preordained by God, so that the begetting of men will not last for ever, and for the same reason, neither will other things that are directed to the begetting of men, such as the

Q

movement of the heaven and the variations of the elements."[33] For Aquinas, a Christian and a Saint, the ultimate *raison d'être* of the cosmos consisted in its subordination to man's eternal, unique and supernatural destiny.

This last point should also reveal a distinctly negative impact which can be exercised by tenets of the Christian creed about research concerning the destiny and duration of the world. They not only can save physical theory from an imprisonment into Aristotelian or other a priori postulates, but they can also create the illusion that some all-encompassing "final solutions" have been acquired about the physical world in the scientific sense. Aquinas is, indeed, notable for his lack of appreciation of experimental investigation. His case is, however, more that of individual temper and preference than of methodological dictates. His master, Albertus Magnus, was a most enthusiastic advocate of experimental investigation, and he found in the contingency of the world the justification to his prolific collection of data concerning natural history.[34] There was no difference between disciple and master as far as the ominous cloud of the doctrine of eternal recurrences was concerned. Albertus' dissertation on *De fato*[35] shows not only his awareness of the issues at stake for humanity, but also his familiarity with the history of the question. He referred to Plato, Aristotle, the Stoics, Ptolemy, the Arab astronomers, especially to Albumasar, and, of course, to the Church Fathers. Christian consciousness had already achieved a firm tradition in the matter.

Emphasis on experimentation was a new wine which easily could prove heady. It could produce firebrands, and Roger Bacon was one of them.[36] Evaluations of his place in the history of science oscillate between lopsided encomiums and studied neglect. The first extreme was usually adopted by those ready to take great visions for actual accomplishments. They were joined, ironically enough, by those for whom the beginnings of science coincided with the apocryphal story of Galileo and the tower of Pisa. For these the unusual friar is the classic example of a great mind struggling in the fetters of institutional obscurantism. The other extreme, the stance of silent treatment, is usually taken by those who grudgingly have come to recognize that Galileo never dropped balls to test the law of free fall and that, what is perhaps more reprehensible, he did not refer to his medieval predecessors to whom he owed so much. This is not to suggest that Roger Bacon was a forerunner of Galileo as far as the laws of motion are concerned. But Bacon's impetuous crusading to secure the service of science on behalf of the Christian faith has much of the boldness and drama that became the hallmark of Galileo's career. Within ten years of the composition of the *Opus majus*[37] in 1267 he was imprisoned on suspicion of holding novel views.

Friar Roger was certainly not censured for his emphasis on the basic unity, interconnectedness, and interdependence of all branches of learning. There could be nothing wrong about his reasoning that since the Creator was one and there was only one creation, its understanding too had to form one single body of truth.[38] Again, he merely echoed the Church Fathers' somewhat naive interpretation of cultural history according to which all the science of the heathen had come from Moses, or if not, it had to be considered a form of "natural" revelation. Nor was anything shocking, in an age of great ferment, in his insistence that the Church should make the most of the Greek scientific corpus which was being rapidly recovered and translated in Bacon's lifetime. Theologians of his time could only nod in agreement on reading his warnings about the difference between final and efficient (secondary) causes, a distinction that intended to render its due to supernatural destiny as well as to temporal endeavour. They should have felt gratified by his assertions of the forever partial character of man's knowledge about the world and by his stricture of Aristotle's claims about a priori, definitive verities concerning the processes of nature.

Bacon might have stunned his contemporaries by his visionary references to contraptions by means of which men would fly, speed across dry land, and see faraway objects as if they were at arm's length, – but dreaming was not necessarily harmful. His concoction of a magic powder with never-before-experienced explosive property was a different matter, yet many an alchemist enjoyed the good will of both political and ecclesiastical potentates. At any rate, the gunpowder seems to have been the only real experimental success of the one whom some called the Father of experimental science. His continual reference to the need of experimenting had much to commend itself, but others, like Albertus Magnus, deserved no less credit on that score. There was nothing revolutionary in his, at times inordinate, praise of mathematics. To speak of mathematics as the most certain of all forms of human knowledge was a fashion of the time, and everybody saw proof of this in the superior exactness of astronomy over all other branches of science.

Bacon was not the first, nor the last, to be trapped by the glitter of perfection, but his case has a particular moral. Admiration of an outstanding perfection can easily turn into sweeping generalizations and this is precisely what happened to him. The vista of the unfailing retracement of their courses by celestial bodies imposed on his mind the idea of an inexorable determinism of events. True, he did his best to safeguard man's freedom and moral responsibility. His prolific analysis of the influence of stars and planets could, however, easily undercut his otherwise sincere persuasion about man's uniqueness in the inexorable turnings of nature's great machinery. His case shows also the difference between the hapless capitulation of most Arab commentators of Aristotle to the idea of cyclic determinism and the unwavering refusal of their Christian counterparts to consider serious compromise on that crucial issue.

In Bacon one can see the enormous lure of a world view in which the cyclic course of celestial bodies ruled and eventually brought back whatever happened. Still, his boundless admiration for Greek and Arabic science prompted him to claim that all philosophers of Greece and Rome, and Muslim scholars as well, had rejected the deterministic influence of stars. He even presented the author of Tetrabiblos, Ptolemy, as a staunch defender of human freedom.[39] Bacon's ground for that startling performance consisted in the distinction between the general trend of each individual life and the single actions of man. According to him, the temperament of the individual was completely determined by the influence of stars while single decisions were not.[40] Collective actions, such as battles, represented a general trend that could be strictly predicted together with its outcome. And so were the relative fortunes of religions. Not surprisingly, the Opus majus contains a section on the application of astrology to Church government. By this Bacon not only meant the forestalling of disastrous projects such as the Children's Crusade through the casting of accurate horoscopes,[41] but also the recognition of great cycles in the history of religions. According to him the association of Jupiter with any of the other planets had a special significance. The closeness of Jupiter to Saturn, Mars, Sun, and Venus respectively, was responsible for the flourishing phase of Judaism, Chaldeism, and of Egyptian and Saracen worship. As he believed that it was Mercury's turn to be associated with Jupiter, he predicted a convincing victory of Christianity over Mohammedanism.[42]

Bacon could not have been more emphatic about the existence of cycles in human history closely paralleling various lunar and planetary cycles. He suggested that history showed cyclic patterns of 20, 240, and 960 years,[43] and he even believed that the time of the coming of the Antichrist might be ascertained from the study of such and similar cycles. The same study was, in his eyes, the key to the determination of the moment of creation, namely, that it had taken place on the vernal or autumnal equinox.[44] But the moment of Creation and the arrival of the Antichrist represented absolute limits for him. He rejected the possibility of several "ages" and of several worlds simply because the belief in the Creator and His work was patently incompatible with an endless recourse along the chain of causality.[45] As a result, he did not take the precession of the celestial axis, the period of which he put at thirty-six thousand years,[46] as evidence of a Great Year. Had he done so, he merely would have followed in the footsteps of Siger of Brabant whose reasoning is a classic illustration of the ultimate consequences of an infatuation with the eternity of the world.

Siger of Brabant's professed concern lay in securing the autonomy of philosophy from theology.[47] His irrepressible advocacy of the cause even secured his lasting glory in Dante's *Paradiso*[48] where he shares the honours accorded by the poet to twelve outstanding individuals who displayed an extra measure of loyalty to their sense of mission. Autonomy of philosophy meant for Siger an unreserved loyalty to Aristotle's thought and the results spoke for themselves. As a consistent or rather radical Aristotelian, Siger felt it his duty to ignore the obvious bearing of the dogma of Creation on some basic philosophical issues. Consequently, the world was for him, as it was for Aristotle, a necessary emanation from the First Cause. With the first and fundamental line of difference between Creator and creature gone, other differences were also rapidly vanishing. Individuals became wholly absorbed in their species, so that only one rational soul was shared by all, and in fact by the whole realm of being, the First Cause not excluded.

In such a framework of thought nothing could be more natural than the abolition of some differences with respect to time. In that great, pantheistic, organismic universe of things, equivalent in Siger's autonomous philosophy to a supreme animal, every little detail and event was believed to come back with inexorable circularity. Significantly enough, it is in Siger's opuscle, *De aeternitate mundi*, that the claim occurs that "whatever existed, it will come back in the same kind, including opinions, laws, religions, and everything else, because the inferior [sublunary] things cyclically recur due to the superior [superlunary] cycles, although no memory is left of some of the recurrences because of their great antiquity."[49] What Siger added at that point could hardly sound convincing to those familiar with the tenor of this reasoning: "We are, however, making these statements, not as true ones, but as ones expressing the opinions of the Philosopher." Siger certainly grasped the gist of Aristotle's thinking, but the autonomy of philosophy swiftly slipped through his fingers. What he gained was the smug confines of a dead-end out of which only a firm attachment to the propositions of faith about the creation of the world could rescue philosophy.

In the question whether the world was eternal or not, the concern about the autonomy of reason came to a head no less than did the loyalty of the Christian to his faith. About the message of the Christian dogma on creation there could be no misgiving: the universe had a beginning and would have an end. For the believer this could have presented an almost insuperable temptation to espouse certain considerations, suggesting the finiteness of the world in time, as being conclusive. But the most learned among the faithful took consistently the position that not reason but revelation alone could settle a matter that truly needed settling. Even an Aquinas, hardly a polemicist by nature, stepped into the breach to discredit in a public lecture on the eternity of the world[50] those whom he called "murmurantes." And murmer they did. Their virulent agitation could hardly be stilled by abstract discourses, sound as they were. Effective reorientation could not come about except through recourse to the deepest resources of faith. This had to imply something dramatic if the guideposts of faith for reason were to be planted with a historic impact.

The dramatic event took place on March 7, 1277, when a list of 219 propositions was condemned by Étienne Tempier, bishop of Paris.[51] The wide spectrum of questions over which a firm judgment was passed clearly indicated that the claim about the perennial recurrence of everything in every thirty-six thousand years (Prop. 92), and about the various aspects of the alleged eternity of the world (Prop. 83–91), represented only a most revealing part in a far broader and deeper issue. What ultimately was at stake was man's rather newly acquired awareness of the contingency of the world with respect to a transcendental Creator, source of all rationality and lawfulness in the macrocosmos as well as in the microcosmos. Even a cursory review of the condemned propositions can show that what philosophy needed was not an "autonomy" based on the exclusion of some very real experiences, religious as they could be in character. The true need of philosophy consisted in vistas pointing far beyond the confines of pantheistic monism, however ancient its pedigree could be in the history of man's speculations about the universe and his destiny in it.

The vindication of the Creator's attributes opened up far-reaching possibilities for the interpretation of the cosmos. The recognition of the possibility of several worlds (Prop. 27); the rejection of the superlunary material as animated, incorruptible, and eternal (Prop. 31–32); the admission of the possibility of a rectilinear motion for celestial bodies (Prop. 66); the rejection of their actual motion as if sparked by animal desire (Prop. 73); the rejection of the celestial orbs as organs equivalent to the eyes and ears of the human body though not as parts of a celestial machinery (Prop. 75); the rejection of the deterministic influence of stars on individuals from the moment of birth (Prop. 105); the rejection of the necessary production of the "first matter" from the celestial one (Prop. 107); all these decisions followed intimately from the effort to safeguard the abilities and exclusive rights of the Creator against any compromise dictated by a narrow rationalism. Those decisions also shaped the state of mind and conceptual groundwork for a revolutionary new approach toward the understanding of the workings of celestial and terrestrial bodies alike.

In a sense the decree of 1277 stated nothing basically new. Nor was the action of a bishop of Paris binding on the universal Church. The matter was primarily a university affair as Siger taught for over a decade at the University of Paris where Étienne Tempier, as bishop of Paris, was charged with the duty of maintaining orthodoxy. The prominence of the University of Paris, the large number of points covered in the decree, and its timing, endowed it nevertheless with special significance. The decree is a classic manifestation of the firmness of medieval Christians already in possession of the Greek philosophical and scientific corpus. They made their stand in the conviction that their belief in the "Maker of Heaven and Earth" imposed on them radical departure from some basic assumptions of Greek learning and world view.

One may, therefore, look with Duhem at the decree as the starting point of a new era in scientific thinking,[52] provided it is kept in mind that the decree expressed rather than produced that climate of thought which Whitehead once rightly presented as the most crucial ingredient for the eventual creation of modern science. His views were expressed in a long passage which, because of its classic Whiteheadian stamp, deserves to be quoted in full:

I do not think, however, that I have even yet brought out the greatest contribution of medievalism to the formation of the scientific movement. I mean the inexpugnable belief that every detailed occurrence can be correlated with its antecedents in a perfectly definite manner, exemplifying general principles. Without this belief the incredible labours of scientists would be without hope. It is this instinctive conviction, vividly poised before the imagination, which is the motive power of research: – that there is a secret, a secret which can be unveiled. How has this conviction been so vividly implanted on the European mind?

When we compare this tone of thought in Europe with the attitude of other civilisations when left to themselves, there seems but one source for its origin. It must come from the medieval insistence on the rationality of God, conceived as with the personal energy of Jehovah and with the rationality of a Greek philosopher. Every detail was supervised and ordered: the search into nature could only result in the vindication of the faith in rationality. Remember that I am not talking of the explicit beliefs of a few individuals. What I mean is the impress on the European mind arising from the unquestioned faith of centuries. By this I mean the instinctive tone of thought and not a mere creed of words.

In Asia, the conceptions of God were of a being who was either too arbitrary or too impersonal for such ideas to have much effect on instinctive habits of mind. Any definite occurrence might be due to the fiat of an irrational despot, or might issue from some impersonal, inscrutable origin of things. There was not the same confidence as in the intelligible rationality of a personal being. I am not arguing that the European trust in the inscrutability of nature was logically justified even by its own theology. My only point is to understand how it arose. My explanation is that the faith in the possibility of science, generated antecedently to the development of modern scientific theory, is an unconscious derivative from medieval theology.[53]

Half a century has passed since these words startled a distinguished audience at Harvard University and indeed the whole intellectual world. The magnitude of the shock merely corresponded to the impenetrable density of a climate of opinion for which the alleged darkness of the Dark Ages represented one of the forever established pivotal truths of the "truly scientific" interpretation of Western intellectual history. Not being a professional historian of science Whitehead could not be blamed for not having perused the first five volumes of Duhem's *Système du monde*. Its extraordinary wealth of documentation might have very well raised in Whitehead's mind some doubts about the validity of the concluding phrase of his long statement. At any rate, the spectacular flow of studies on medieval science touched off by Duhem's monumental work provided among other things ample evidence that the medieval faith in the scrutability of nature had its logical justification in the medieval theology about Creator and creation, and that the faith in the possibility of science *is a most conscious derivative* from the tenets of medieval theology on the "Maker of Heaven and Earth."

Obviously, such a crucially productive climate of opinion could not be the work of a single decree. It was the fruit of a commonly shared belief nurtured in part by a uniform educational system, consisting of universities, cathedral schools, and monasteries, the like of which neither Greece, nor Rome, nor any ancient great culture was ever able to produce. In the medieval educational system, especially in the medieval universities, the readiness to innovate was a far more widespread phenomenon than a historiography steeped in the intellectual militancy of the Enlightenment would have ever wanted to consider. The flux of that creativity was not, of course, steady with respect to both time and locality. In the medieval University of Paris, too, periods of creativity alternated with decades of stagnation, a symptom which still holds true of any institute of learning or research. But probably the flourishing period of no other university had influenced as decisively the course of intellectual history as was the case with the golden period of the University of Paris spanning much of the fourteenth century.

The importance of that period had been portrayed in great detail in Duhem's pioneering investigations on the medieval origins of classical physics.[54] It was he who almost single-handedly inspired a vigorous interest in medieval science with the result that the main lines of development now seem to be firmly established. Among these the most pivotal are the medieval anticipations of the concepts of inertia and momentum, the medieval assertion of the possibility of a three-dimensional infinite vacuum, the groping for quantitative in addition to qualitative accounts of physical processes, and finally the realization of a basic need for experimentation if progress was to be made in understanding and conquering nature. Duhem's own conclusions concerning the specifics needed revisions following a more detailed and more complete look at the manuscripts, some of which had come to light only after his untimely death. It seems that more justice is done to the record by assuming a middle position between Duhem's maximalism and a mini-malism harking back to the virulent antimedievalism of the Enlightenment for which everything before Galileo was dark, as far as science was concerned.

The various conceptual, cultural, and technical details of the medieval roots of classical physics are today available in a vast literature.[55] Still, there seems to be wanting an adequate appraisal of the intimate connection which existed in the thinking of the medieval "physicists" between their portentous conjectures and their firm belief in a personal Creator. How this belief provided them with fruit-ful directives can be seen from their comments on *De caelo* (*On the Heavens*), the programmatic exposition by Aristotle of his own world view. The classic and most influential case is John Buridan's *Quaestiones super quattuor libris de caelo et mundo*[56] which in its slightly modified version due to Albert of Saxony, a disciple of Buri-dan, had an unmistakable influence on Galileo.[57] Buridan's work shows that the radical departures from the main proposition of Aristotelian cosmology and physics were made with most direct references to the Christian belief about the fundamental relations between Creator and creatures. Thus, the road from the closed world of Aristotle to the infinite space of classical physics as a framework of unimpeded inertial motion was staked out by Buridan's emphasis on God's unrestricted ability to impart rectilinear motion to the whole world, a proposition striking at the very roots of Aristotelian cosmology.[58] Again, one gets an early glimpse of the infinitely numerous worlds of classical physics and astronomy on finding Buridan reaffirm his faith in the power of the Creator in the face of Aristotle's insistence on the exclusive uniqueness of one single world.[59] The Aristotelian dichotomy between superlunary and sublunary matter was dealt a decisive blow, and the unitary approach of classical physics to earthly and heavenly bodies was foreshadowed, when Buridan, inspired by his faith, discussed the sub-stance of stars in a manner which patently deprived them of the divine and imper-ishable characteristics which Aristotle attributed to them.[60] While Aristotle denied that the heavens could decay, Buridan was quick to remind his reader that the Creator even has the ability to annihilate the world.[61]

Buridan could not have been more aware of the fact that his highly critical approach toward Aristotle derived from his Christian belief. In discussing the question of the immobility of the uppermost heavens, or the so-called empyreal heavens, Buridan flatly stated that "in many an instance one should not believe Aristotle who made many propositions contrary to the Catholic faith because he wanted to state nothing except what could be derived from considerations based on what is seen and experienced."[62] Not that Buridan wanted to downgrade the role of reason and experience. He warned lucidly against confusing the respective competence of faith and reason. In discussing the question of whether the plurality of heavenly motions can be directly traced back to God, Buridan offered an explanation that did justice to both theology and natural science. As to the latter he noted that "in natural philosophy one should consider processes and causal relationships as if they always came about in some natural fashion; therefore, God is no less the cause of this world and of its order, than if this world were eternal."[63]

Belief in a Creator whose powers were not limited to the features of the actually observed world contributed, therefore, most effectively to the liberation of critical thinking from the shackles of Aristotelian science. Buridan's work illustrated in many telling details the vigour of this new critical spirit. To see this, one need only read his discussions of projectile motion and of the fall of bodies. In both cases he abandoned fundamental postulates of Aristotelian physics. In the former case it was the idea of the continuous contact of the mover with the moved body which was rejected; in the latter it was the idea of attraction on the body by its "natural" place which suffered the same fate. As his own explanation Buridan offered the notion of impetus, or a quality implanted in the moving body by the source of motion, and for the case of attraction he outlined the notion of gravity as innate in all bodies. If Buridan rejected the daily rotation of the earth it was for experimental reasons and not because of Aristotle's arguments which he showed to be inconclusive. What seemed to clinch the argument for him was the case of an arrow shot vertically upward, which, as he remarked, could not fall back to the same spot if the earth actually rotated.[64] Against Aristotle Buridan maintained the continuous though relatively small changes in the position of the earth's centre of gravity, which implied that the earth was actually moving with respect to the centre of the universe. Distant as these advances remained from the mature form of classical physics, they represented an enormous step away from Aristotle's physics and in the right direction at that.

In making full use of the liberating effect of his belief in the Creator, Buridan produced some insights with a truly modern ring. The classic case was his clear hint at what came to be called the inertial motion of heavenly bodies in classical physics. After reviewing the merits of his impetus theory with respect to various terrestrial motions, he outlined its usefulness for celestial mechanics. Thus, in the same breath, Buridan spoke of broad jump, of planetary motion, and of the Creator who is the ultimate factor imparting given quantities of motion to various parts of the universe:

[the impetus then also explains why] one who wishes to jump a long distance drops back a way in order to run faster, so that by running he might acquire an impetus which would carry him a longer distance in the jump. Whence the person so running and jumping does not feel the air moving him, but [rather] feels the air in front strongly resisting him.

Also, since the Bible does not state that appropriate intelligences move the celestial bodies, it could be said that it does not appear necessary to posit intelligences of this kind, because it would be answered that God, when He created the world, moved each of the celestial orbs as He pleased, and in moving them He impressed in them impetuses which moved them without His having to move them any more except by the method of general influence whereby He concurs as a co-agent in all things which take place; "for thus on the seventh day He rested from all work which He had executed by committing to others the actions and the passions in turn." And these impetuses which He impressed in the celestial bodies were not decreased nor corrupted afterwards, because there was no inclination of the celestial bodies for other movements. Nor was there resistance which would be corruptive or repressive of that impetus. But this I do not say assertively, but [rather tentatively] so that I might seek from the theological masters what they might teach me in these matters as to how these things take place.[65]

Buridan made his last remark with tongue in cheek. He evidently wished to reaffirm the line of demarcation which he carefully observed between a metaphysical and a natural analysis of the phenomena. It was with undeniable originality that he staked out his own field of investigation and he held to it with exemplary consistency throughout a long and most influential teaching career. Contemporaries of Buridan pointed at his novel approach on more than one occasion. They must have sensed that Buridan's way of thinking signalled a distinct departure from traditional patterns of thought. Later developments showed that Buridan was indeed a pioneering forerunner of Galileo and the earliest major representative of the mechanistic spirit of classical science.

As was already mentioned, subsequent generations read Buridan's judicious commentary on Aristotle's *On the Heavens* in the version of Albert of Saxony. But Buridan's originality made itself felt mainly through the writings of his most outstanding disciple, Nicole Oresme (1323(?)–1382), who completed his illustrious career as Bishop of Lisieux.[66] This last honour came to Oresme as a reward for the many years of service on behalf of Charles V both before and after he ascended the throne in 1364. An enthusiastic patron of learning, Charles did his best to keep Oresme's extraordinary abilities on the highest level of productivity. It was at Charles' behest that Oresme had undertaken the translation and interpretation of three major works of Aristotle, the *Nicomachean Ethics*, the *Politics*, and *On the Heavens*, during the 1370's. Together with the *Politics*, Oresme also translated and interpreted a spurious work of Aristotle, the *Economics*, a fact which throws light not only on the interests of his royal patron but also on the versatility of Oresme's mind. It seems that the attention of the royal family was drawn to Oresme in 1359 at the latest by the remarkable works which he produced while at the Collège de Navarre, of which he became the grand master in 1356. Among those works are not only the routine commentaries on standard theological and philosophical texts required for academic promotions, but discussions replete with original insights covering such a range of topics as monetary theory, astronomy, astrology, geometry, and algebra. Concerning these discussions the originality of Oresme's mind is best expressed in his prolific effort to apply the quantitative method even at phenomena and experiences which come largely under the realm of psychology. What he looked for in his *De configurationibus qualitatum et motuum*[67] was a universal mathematical method equally applicable to physical changes as well as to changes occurring in man's inner experiences. What all this amounted to was in a sense a call for a quantitative psychology and esthetics.

A scholar of such calibre could confidently be expected to give a worthy account of his abilities when Charles V gave him the task of translating and explaining three major works of Aristotle. Of these, the *On the Heavens*, was scientific and it served for Oresme as the vehicle of expressing his long-nurtured reflections on some fundamental issues facing the scientific investigator of nature. The work today is recognized as a great classic of scientific literature. Opinions are, however, far from being unanimous about the measure of Oresme's role in the rise of modern science. The encomiums heaped on Oresme by Duhem[68] and H. Dingler[69] had recently yielded to a pronounced reserve on the part of D. B. Durand,[70] A. Koyré,[71] M. Clagett,[72] and E. Grant,[73] of whom the last two earned their scholarly reputation in part by publishing critical editions of Oresme's major scientific works.[74]

Obviously, in most evaluations of broad historical impact one can never aim at an exactness which would of its own precision command unanimity. Precise determination of the starting points of new ideas, trends, and outlooks is well-nigh impossible, even when a generous allowance is given to the individual historian's own preferences to look at the historical process, the pace of which hardly ever keeps a steady rate. Nevertheless, when the individual historian's preferences are carefully weighed, the discrepancies of opinions often turn out to be differences of perspectives rather than of facts. The gropings of Oresme and others would of necessity seem clumsy by comparison with some statements of Galileo. In a sense it may even appear prohibitive to make an attempt at a comparison. The situation may resemble the enormous difference which is between the primordial application of wheels at the dawn of civilization as a means of transportation, and the engineering sophistication embodied in the wheels of a modern jet plane. The finished product of a long development may appear as a new class of its own. Thus, there is ample justification in emphasizing the newness of Galilean science. At the same time the first primitive cartwheel deserves to be considered as a product of genius which made all subsequent wheeled carriages possible.

The dichotomy between these two approaches of appraisal seems to be irreducible because of a fundamental inexactness inherent in the process of conceptualization. Thus, in a sense the problems of dating the start of the new or Galilean science will defy an exact solution. Moreover, in a positivistic climate of thinking such as ours, the tangible products will always have greater persuasiveness than their often hazy adumbrations. In addition, there seems to be something inconsistent in Koyré's resolute rejection of the century of Buridan and Oresme as the real beginning of the new scientific outlook.[75] An ardent advocate of the priority of the conceptual element in science over the experimental or positive factors, Koyré refused to accept the well-known positivistic arguments in support of the newness of the science of Galileo. What constituted in Koyré's eyes the true scientific innovation of the generation of Galileo was rather a new mentality with respect to the investigation of nature. Koyré did not, however, consider in sufficient depth the fact that newness could only be established with reference to the old, and a substantial part of the old in this case was the Aristotelian world view with all the ramifications of its metaphysical and theological (pantheistic) underpinnings.

As an individual, the historian of science may or may not feel sympathetic toward what Koyré repeatedly called the "Christian God." In his role of historian of science the individual has, in theory at least, no such license. He must be a respecter of facts, not only those of nature, but the facts of history as well. The collective faith of the Middle Ages is a *fact* of history, and so is its enormous impact on the modern mind so proud of its science. It was mainly that faith which convinced medieval students of Aristotle that the sweep and exclusiveness of his account of the universe can very well be a pleasing but wholly misleading mirage.

It was the same faith which made them aware of the liberating possibility that the ultimate roots of the physical world may and do have characteristics which fly in the face of the most sacred assumptions of Aristotle about the cosmos. It was that faith which put them and kept them on new avenues of speculation which led inevitably to the stage where centuries later the slowly accumulating new considerations would all of a sudden fuse into one grandiose picture.

Aristotle had his share of critics long before the medievals, but the pantheistic necessitarianism of his synthesis had never received a broad and effective challenge before Christianity developed into an all-pervading cultural matrix during the Middle Ages. If, therefore, the primacy of a mental outlook over the bare facts of sensory evidence is worth defending when it comes to the analysis of the development of science, then the faith of the Middle Ages in a personal, transcendental, providential, and rational Creator should appear of paramount importance. It was a faith shared enthusiastically by most of those who functioned as teachers in a culture with truly novel features. It was a faith generating a sense of confidence, purpose, and guidance with respect to the fundamental issues about man's place in the universe. It was a faith fully aware of the rights of reason. If it insisted on going beyond the immediate range of reason it was because of an honest willingness to consider the credentials of the greatest factual claim made in human history, the claims of the Rabbi from Nazareth. Such a faith certainly represents a mental outlook and can be taken lightly by any professedly non-positivist historian of science only at the risk of grave inconsistency.

This inconsistency becomes glaring when viewed against specific passages in that landmark of early scientific literature, *Le Livre du ciel et du monde*, or Oresme's commentary on Aristotle's *On the Heavens*.[76] It throws a revealing light on the tacit presuppositions of some modern research on Oresme's scientific achievements that the vigour of his faith in the "Christian God" (to recall Koyré's somewhat tendencious parlance) is systematically played down. Such an unscholarly technique reaches its peak when Oresme is diligently described as a chief sceptic of a sceptical century, or a believer in the occult and the magic. In this latter respect he certainly was no worse than Copernicus, or Kepler, or a Galileo who kept casting horoscopes, or a Newton whose obscurantist proclivities are only beginning to come to full light. Yet, for all that, their crucial role in the scientific revolution is not minimized. As to Oresme's scepticism, it is at worst a mild anticipation of Nicolas of Cusa's "learned ignorance." Concerning his affirmations of the unlimited power of the Creator in the face of the pantheistic necessitarianism of the Aristotelian cosmos, they remain an integral and crucial part of his scientific thinking regardless of efforts to ignore or minimize them.

These passages show an acute awareness on Oresme's part to correct, challenge, and surmount in virtue of his faith in the Creator the Aristotelian explanation of the universe. With respect to Aristotle's insistence on the perfection of the universe Oresme emphasized that the perfection of the laws of nature was merely a modest reflection of the infinitely perfect attributes of the Creator.[77] Obviously then, Oresme had to reject out of hand the Aristotelian notion of an eternal, ungenerated, incorruptible heaven.[78] Oresme allowed incorruptibility to the celestial region only in a restricted sense of its being frictionless, a proviso which effectively started the road to a unitary discussion of the motion of earthly and heavenly bodies. About the alleged eternity of the heavens Oresme most categorically stated that "the heavens did have a beginning not by natural generation but rather by God's divine creation."[79] This sentence, which brings to a close his discussion of the question, is a worthy counterpart of the same section's starting phrase. In it Oresme admits that the eternity of the heavens, or of the universe, is not in itself a self-contradictory notion, but it is contrary to the revealed fact that "the divine power . . . created the world out of nothingness."[80] In view of the created character of the heavens Oresme refused to go along with Aristotle's insistence on its absolute changelessness and rejected the famous corollary of the Aristotelian claim of the eternity of the heavens, the endless recurrence of the same ideas and views. "This is not true," reads Oresme's terse rebuttal.[81]

This was not the only place where Oresme set the dogma of creation against the possibility of eternal recurrences. Years earlier he characterized it in his *Ad pauca respicientes* as a foolishness,[82] and declared it an error in philosophy and faith in his *De proportionibus proportionum*.[83] As a scientific argument he submitted that the periods of planets are most likely incommensurable, which means that they would never return at the same time to the same position, and as a result they would never restart an exact replica of their previous influence on the sublunary world. True, he later discarded[84] this argument by noting that perhaps the periods of planets were integer quotients of a very large period. He added, nevertheless, that such a period should be much larger than 36,000 years or the period of the precession of the equinoxes, the length traditionally assigned to the Great Year. His last comments[85] on the matter are interesting for several reasons. First, he noted carefully that Aristotle rejected the idea of eternal recurrences in the form of periodical conflagrations of the whole world. Second, he also reported, by citing St. Jerome, that according to Origen these conflagrations would go on endlessly. One should guard himself against seeing a tacit approval on Oresme's part of the idea which he failed to stricture in a particular instance. For Oresme had already stated that these periodic annihilations of the world would require a specific exercise of the infinite power of God for which he clearly saw no indication or purpose in line with his favourite dictum: God and nature do nothing in vain.

The difference between Augustine and Oresme in facing the idea of eternal recurrences is certainly remarkable. Augustine's attitude still reflects tension and urgency. His age was still steeped in the astrological morass of pagan antiquity and he felt keenly its pervasive cultural pressure. Oresme's calmness reflects the robust confidence of an overwhelmingly Christian ambience for which the once-and-for-all process of human and cosmic existence was almost as natural a conviction as the air one breathed. This is not to suggest that astrology in Oresme's time had greatly weakened in exerting its debilitating influence. In a booklet[86] Oresme felt impelled to call his royal patron's attention to the futility of reading the future from the position of stars and planets. But on the level of scientific accounts of the world the question of eternal recurrences had already yielded to a new era of thinking rooted in the liberating influence of the dogma about creation by an absolutely sovereign Creator, the sole source of human and cosmic destiny.

To see in some particulars this liberating influence, it is enough to take a close look at the context of Oresme's last remark on the idea of the Great Year. The context shows Oresme discussing the question of the plurality of worlds. For the necessitarian pantheism of Aristotle such a possibility was an anathema. For the Christian Oresme the Creator's omnipotence guaranteed such a possibility, and he probed into it with obvious delight. His description of worlds enclosed within one another is a fascinating cosmological speculation. It evidences in several ways the special intellectual benefits which Oresme derived from his faith in the Creator. In answering the various objections to a set of worlds in which each world is smaller in the ratio of 2000 to 1 with respect to the world immediately surrounding it, Oresme parted with the strict Aristotelian definition of natural places and also with the idea implicit in the Aristotelian description of the world that the quantity of matter contained in the universe is determined by intrinsic necessity. Oresme clearly had no patience with such necessities, and the reason for this is given in his reply to the fifth objection against the plurality of worlds. According to the objection, which bears the heavy stamp of Aristotelian necessitarianism, several worlds would imply several gods. For Christian thinking such reasoning was patent absurdity, a verdict which came through strongly in Oresme's concise reply: "One sovereign God would govern all such worlds."[87]

Another example of the plurality of worlds analysed by Oresme was the juxta-position of worlds in a vast, infinite, empty space.[88] Aristotle, as is well known, rejected such a possibility as being contrary to his ideas about the absolute directions of motion and about absolute heaviness and lightness. Oresme's discussion shows once more the liberating influence of his faith in the Creator. Starting from the premise that God's infinite power cannot be limited to one world, Oresme insisted on the relativity of motion with respect to a given co-ordinate system[89] and on the essential relation of lightness and heaviness to such systems: "From this it follows clearly that, if God in His infinite power created a portion of earth and set it in the heavens where the stars are or beyond the heavens, this earth would have no tendency whatsoever to be moved toward the centre of our world. So it appears that the consequence stated above by Aristotle is not necessary. I say, rather, that if God created another world like our own, the earth and the other elements of this other world would be present there just as they are in our own world."[90] Fully aware of the momentous implications of his belief in the Creator, Oresme went to the heart of the matter as he recalled Aristotle's "one world, one God" argument standing for necessitarian pantheism. After stating that God was infinite in His immensity and, if several worlds existed, not one of them would be "outside" Him nor "outside" His power, Oresme declared: "Assuming that all the matter now existing or that has never existed is comprised in our world, nevertheless, in truth, God could create ex nihilo new matter and make another world. But Aristotle would not admit this."[91]

In a most instructive manner Oresme exploited the direct consequences of his firm attachment to the dogma of creation which set him poles apart from Aristotle's pantheistic necessitarianism. According to Aristotle it was the Prime Mover's perfection which imposed a strictly spherical shape on the world. Oresme, of course, saw in such reasoning a curtailment of the Creator's power, and argued that God conceivably could have added sections to the outermost sphere in the form of parts protruding into empty space, another anathema for Aristotle. But for a mind such as that of Oresme, deeply steeped in belief in the Creator, it was most natural to admit the possibility of empty space, either outside the stellar sphere or inside it. What he added is a remarkable anticipation of that famous section of the Scholium to Newton's *Principia* where the idea of infinite empty space is anchored in the Creator's infinity, and is also represented as the indispensable conceptual framework for the dynamics of the new classical physics: "Such a situation can surely be imagined and is definitely possible although it could not arise from purely natural causes, as Aristotle shows in his arguments in the fourth book of the *Physics*, which do not settle the matter conclusively, as we can easily see by what is said here. Thus, outside the heavens, there is an empty incorporeal space quite different from any other plenum or corporeal space, just as the extent of this time called eternity is of a different sort than temporal duration, even if the latter were perpetual, as has been stated earlier in this chapter. Now this space of which we are talking is infinite and indivisible, and is the immensity of God and God Himself. Just as the duration of God called eternity is infinite, indivisible, and God Himself, as already stated above."[92]

If drastic modifications were imposed by the doctrine of Creation on some static aspects of the Aristotelian world view, the situation was no different with respect to its dynamical features. The Christian dogma of creation asserted a creation in time, or in other words it allowed for the universe only a finite time-span as measured backward from the present to the moment of the coming into being of all out of nothing. Such a doctrine was in flagrant conflict with Aristotle's emphasis on the strict eternity of the world and in particular with the eternity of the celestial movements. Oresme was very explicit in noting that Aristotle's position was anchored in the premise that "it is impossible to make something out of nothing."[93] But Oresme did not stop at the generalities. His intention was to weigh carefully the particulars of Aristotle's doctrine: "I intend to enumerate them faithfully and to examine them diligently, pointing out the faults of some of them in the light of natural reason, to dispel any occasion for major errors in the Catholic faith."[94] This intuition he articulated with care.

The first detail he subjected to severe criticism was Aristotle's theory about the quantitative relationship between the quantity of motion produced by a given force. For Aristotle the topic formed part of the thesis that whatever motion had a beginning must also have an end, and consequently a beginning of the world would entail its eventually going out of existence. This is not the place to recall the various sophisticated arguments by which Oresme tried to undermine a contention characterized by him as something opposed to the Catholic faith.[95] It should be obvious that in order to counter effectively Aristotle's contention Oresme had to emphasize the soundness of his own thinking about the principal motion in the universe, the motion of the heavens. The novel point he submitted was that the motion of the heavens was something that had been started by the Creator and would go on forever in virtue of a quality of motion imparted to the stellar sphere by the creative power of God at the moment of the creation. In contrast to the pantheistic and perennial contact of the Prime Mover with the uppermost heavens, the Christian idea of the Creator implied His transcendence over the world in connection with His actual influence on any created being. Clearly, this transcendence of an active Creator could readily be safeguarded by formulating the idea of an imparting by God of a given quantity of motion (impetus) to the world once and for all. That such thinking about motion had strongly adumbrated the notion of momentum should seem natural. Further refinements in the concept of impetus led to the correct definition of momentum together with inertial motion on which rests the whole edifice of physics.

It was this crucial conceptual development which was impossible to achieve within the framework of the pantheistic necessitarianism of Aristotle's physics and cosmology. For Aristotle the process implied in the notion of impetus was a theological, metaphysical, and physical impossibility, and in that order. Set against the Aristotelian background the notion of impetus meant a miracle which, however, could be performed, assuming the existence of a personal, rational, omnipotent, and transcendent Creator. Furthermore, a staunch and broadly shared faith in such a Creator could effectively prepare minds to entertain such an idea, to look upon it as a most natural notion, and to develop all its consequences. This process took three hundred years, the time that separates Oresme's *Livre du ciel et du monde* from Newton's *Principia*. The finished product may seem far removed from its incipient shape. Yet, if one is to trace the antecedents of Newton's definitions of motion, momentum, and inertia, the line of investigation leads inevitably to Oresme's inquiring mind, guided by a firm profession of faith which brings to a close a passage so typical of his speculations: "And since we are on this subject, I want to say that, just as the immensity of God extends outside the world which is finite in extent . . . it is equally reasonable to maintain that the eternity of God exists or could exist before the world existed, which world is or may be finite in time and certainly had a beginning."[96]

Oresme himself was not for a moment reluctant to indicate the ultimate provenance of that consideration on which everything else in science rested. It consisted of the biblical revelation about an eternal God and a world created by Him in time.[97] He dwelt at length on scriptural phrases about the transcendence, eternity, and omnipotence of the Creator before he set forth his account of the start of heavenly motions. What he first noted was that the clocklike precision of the heavenly motions evidenced God's power and wisdom.[98] His reference to a clockwork in connection with the heavens was telling evidence of the trend of his mind. True, he did not dispense with the celestial intelligences of Aristotle. But the differences are basic between his and Aristotle's thinking on this point. With Oresme the intelligences do not make the heavenly bodies animate and intelligent. He shows little sympathy for Aristotle's panorganismic concept of the world in which everything is pervaded by the throbbings of a universal, cosmic life. Oresme's intelligences are not emanations of the Deity, but created servants who are given a mandate as to the exact amount of motion to be imparted to this or that celestial body. Consequently, the famous passage of Oresme about the celestial clockwork retains its innovating, mechanistic ring:

If we assume the heavens to be moved by intelligences, it is unnecessary that each one should be everywhere within or in every part of the particular heaven it moves; for, when God created the heavens, He put into them motive qualities and powers just as He put weight and resistance against these motive powers in earthly things. These powers and resistances are different in nature and in substance from any sensible thing or quality here below. The powers against the resistances are moderated in such a way, so tempered, and so harmonized that the movements are made without violence; thus, violence excepted, the situation is much like that of a man making a clock and letting it run and continue its own motion by itself. In this manner did God allow the heavens to be moved continually according to the proportions of the motive powers to the resistances and according to the established order [of regularity].[99]

Oresme left no doubt about his full awareness of the ultimately theological roots of the crucial difference between his thinking and Aristotle's account of the universe. Against Aristotle's pantheistic assertion that God should be thought of as a perpetuum mobile keeping the heavenly bodies perpetually in motion and making them sharers of the divine nature, Oresme stated: "The principal action of God is to know and to love Himself – this is eternal life; and this action is God Himself, who is the end or object of Himself. But God has another occupation with things outside and other than Himself, and, according to Aristotle, this activity is to move the heavens and to govern the world. Speaking in this way, God is the principal purpose and primary intention of this activity and of all things. But speaking of the purpose of the second intention, or less absolutely, God is in this activity as if it were His own purpose. Thus Aristotle calls the heavens a divine, and Averroes a spiritual body because they consider them to be animated by the intelligence, which is God, that moves them, and so the heavens are divine. This we refuted in Chapter Five."[100]

It is again in the same theological consciousness that Oresme picked to pieces the famous *tour de force* by which Aristotle tried to derive the world and its principal features as a necessary effect of the nature of God. Oresme evidently delighted in marshaling the utterances of the Scripture about the self-sufficiency of God with respect to the created realm. His prose soared as he wrote: "For if He had created a hundred thousand times the number of angels and saints that He did create, His goodness would not increase a single iota, nor would it decrease by so much as a single speck if He destroyed every creature; and it would be the same if He were to create 100,000 worlds or if he destroyed this world and every other creature; for by this action His immensity, which we noted in Chapter One, should be neither increased nor diminished one jot or tittle. Briefly, it is simply impossible that anything could be required to add to His perfection, which depends on nothing whatsoever."[101] With no less animation Oresme also rejected a foremost cosmological principle of the Aristotelian system: the necessarily motionless stay of the earth at the centre of the universe.

In this rebuttal Oresme made one of his several references[102] to the list of propositions condemned in 1277 by Étienne Tempier, bishop of Paris. That Oresme espoused enthusiastically a decree revoked as early as 1325, or about half a century before the composition of the *Livre du ciel et du monde*, indicates that for Oresme the legally defunct document was, in its fundamentals at least, the embodiment

R

of some essential features of Christian mentality. The decree itself, the mentality underlying its essentials, and Oresme's theologically conditioned thinking, have been the target of more than one disparaging remark on the part of some historians of science. The tenor of such remarks sharply contrasts with the undeniably beneficial impact of the Christian faith in the Creator on Oresme's scientific speculations. True, that faith, not being a *Deus ex machina*, did not reveal all the secrets of nature for him. In fact Oresme's reverence for the words of Scripture stopped him short from declaring as fact the daily rotation of the earth, which he ably supported by several considerations.[103] His traditionalism on this point should, however, appear insignificant when set beside those instances, briefly outlined in the preceding pages, which show his mind noticing new aspects about the physical world precisely in virtue of his faith in the Maker of heaven and earth as described in the biblical revelation.

Centring attention on Oresme and Buridan can only be justified as a practical necessity. Beside these two giants there were, during the fourteenth century, a number of lesser figures, whose thought shows much the same pattern, namely, the emergence under the promptings of their belief in the Creator of an insight which was to prove invaluable for the future of physical science. A classic case in this respect is a remark by Franciscus de Marchia[104] from his commentaries on the foremost medieval textbook on theology, the *Sentences* of Petrus Lombardus. The question at issue, the manner in which sacraments conveyed grace, could not have been more purely theological. Through God's power, so faith in the sacraments demanded, certain sacramental signs could retain their power to convey grace, for some time after such signs were produced. This meant that they carried an "impressed" force or effectiveness. It was then the theologian's duty to find analogous processes in nature, to make the mystery of faith partly intelligible. From the viewpoint of conceptual genesis this meant an effort to find in the processes of nature a given pattern which otherwise might not have been looked for. The pattern in question was the imparting of an impetus on a body that it might remain active for a while. In the case of Franciscus de Marchia, it was such an approach that prompted him to conclude that the motion of a projectile through air should be explained, not by a continuous contact effected by the medium (air) between the mover and the moved, as Aristotle claimed, but by the motion or impulse of a power left behind in the stone by the primary mover, say one's hand.[105]

It seems that similar gems still would be spotted in fourteenth-century manuscripts many of which are awaiting careful study. One such study[106] did in fact show that belief in the Creator's omnipotence was not only the exclusive source of the acceptance during the fourteenth century of the *possibility* of an infinite void, but that considerations based on that possibility were an indispensable help in Newton's time to make the transition to the acceptance of the *reality* of an infinite space: "Whether conceived dimensionally or not, the belief in God's omnipresence in an infinite void made that overpowering and difficult concept more readily acceptable to a number of significant and theologically oriented philosophic and scientific figures in the seventeenth and early eighteenth centuries. This belief and the arguments justifying it were a legacy from the late Middle

Ages."[107] The scientific legacy of the Middle Ages has, of course, many other aspects that cannot be treated here in detail. Among them are the efforts to treat quantitatively the change of intensities in most varied processes; also an impressive array of crucial technological inventions, such as the first mechanical clocks, that were immediately referred to as small replicas of the Creator's great clockwork, the universe. Analysis of the motivation behind these advances often indicates with impressive explicitness that they bear witness to the sparks generated by a firm and humble faith in the words spoken in the Beginning. The Light diffused there was also the illumination that made possible the sighting of new and reliable horizons about the universe.

NOTES

1. A critical edition of the *Quaestiones naturales* was published by M. Müller in *Beiträge zur Geschichte und Theologie des Mittelalters*, Vol. 31, Fasc. 2 (Münster in W.: Aschendorff, 1934). On Adelard's life and scientific work, see F. Bliemetzrieder, *Adelhard von Bath* (Munich: Max Huber, 1935).
2. *Ibid.*, p. 8 (Quaest. 4).
3. *Ibid.*
4. *Ibid.*
5. *Ibid.*, p. 32 (Quaest. 25).
6. *Ibid.*, p. 27 (Quaest. 23).
7. *Ibid.*, p. 68 (Quaest. 76).
8. *Ibid.*, p. 69 (Quaest. 76).
9. The most satisfactory modern, general survey of sciences during the Middle Ages is A. C. Crombie, *Medieval and Early Modern Science* (Garden City, N.Y.: Doubleday, 1959), which is the second, revised edition of his *From Augustine to Galileo: The History of Science A.D. 400–1650* (London: Falcon Press, 1952). Shorter and less incisive is Part III, "The Middle Ages," by G. Beaujean in R. Taton (ed.), *History of Science*, Vol. I, *Ancient and Medieval Science*, translated by A. J. Pomerans (New York: Basic Books, 1963). An enormous storehouse of information is the ten volumes of L. Thorndike, *History of Magic and Experimental Science* (New York: Columbia University Press, 1923–58), which covers the first seventeen centuries of our era. All modern discussions of medieval science stand in debt to Pierre Duhem's *Le système du monde: Histoire des doctrines cosmologiques de Platon à Copernic* (Paris: Hermann, 1913–55), in ten volumes. The complete absence of references to medieval science in *The Legacy of the Middle Ages*, edited by C. G. Crump and E. F. Jacob (Oxford: Clarendon Press, 1926), is a telling indication of the myopia that until recently dominated scholarship about scientific thought during the so-called Dark Ages.
10. *Bernardi Silvestris De mundi universitate libri duo sive megacosmus et microcosmus*, edited by Carl S. Barach and Johann Wrobel (Innsbruck: Wagner, 1876).
11. See chaps. 28 and 29 in vol. I, of Thorndike, *History of Magic and Experimental Science*, on Arabic occult science and Latin astrology in the ninth, tenth, and eleventh centuries.
12. Its text starts with the words, "De septem diebus et sex operum distinctionibus," edited with an introduction by M. Hauréau in *Notices et extraits des manuscrits de la Bibliothèque Nationale*, vol. 32 (Paris: Imprimerie Nationale, 1888), pp. 171–85.
13. *Ibid.*, p. 172.
14. *Ibid.*

15. *Notices et extraits des manuscrits de la Bibliothèque Nationale*, vol. 32, (Paris: Imprimerie Nationale, 1888), pp. 175–76.

16. *Ibid.*, p. 177.

17. *Ibid.*, p. 180.

18. As amply evidenced in A. C. Crombie's magisterial monograph, *Robert Grosseteste and the Origins of Experimental Science 1100–1700* (Oxford: Clarendon Press, 1953).

19. *Ibid.*, p. 110.

20. *Ibid.*, pp. 103 and 102.

21. The former in the British Museum, the latter in the Cambridge University Library; see Crombie, *Grosseteste*, p. 327.

22. In *Opera omnia* (Paris: apud L. Billaine, 1674), vol. I, pp. 593ff.

23. See Thorndike, *History of Magic and Experimental Science*, vol. II, pp. 338–71.

24. See on this Theodore O. Wedel, *The Medieval Attitude toward Astrology, particularly in England* (New Haven, Conn.: Yale University Press, 1920), a study which though factually informative presents the conflict of Christianity with astrology as a clash between faith and science!

25. *Ed. cit.*, pp. 707–14 (Chapters xvi, xvii, and xviii in Section II of Part I).

26. *Ibid.*, p. 708.

27. *Ibid.*, pp. 1053–56 (Chapter xx in Section II of Part III).

28. *Ibid.*, p. 1055.

29. *Ibid.* In his *De legibus* William of Auvergne rejected Albumasar's interpretation of religious history in terms of the successive ascendency of the various planets. In *Opera omnia*, vol. I, pp. 54–57.

30. The classic treatment is that by E. Gilson in his *History of Christian Philosophy in the Middle Ages* (New York: Random House, 1955), pp. 361–86, with extensive bibliography. The Muslim philosophical struggles, against which Aquinas' efforts should be viewed, are well presented in Majid Fakhry, *Islamic Occasionalism and its Critique by Averroes and Aquinas* (London: George Allen & Unwin, 1958), see especially chap. iv, "The Causal Dilemma and the Thomist Synthesis," pp. 139–207.

31. *The "Summa Theologica" of St. Thomas Aquinas: Third Part (Supplement) QQ LXXXVI–XCIC and Appendices*, literally translated by Fathers of the English Dominican Province (New York: Benziger Brothers, 1922), pp. 48–69. Aquinas' discussion of the eternity of the world from the viewpoint of the Creator in Q. 46 of Part I of the *Summa theologica* contains no reference to world cycles. The reverse is the case in the *Summa contra Gentiles*. There the discussion of the problem is very short in connection with the final condition of the world after judgment (Book IV, chap. 97), whereas explicit reference is made in Book II, chap. 38, to the "infinite number of solar revolutions" and to the infinite number of individuals generated, if such were the case.

32. *The "Summa Theologica,"* p. 51.

33. *Ibid.*, p. 54.

34. On Albertus Magnus's scientific programme and accomplishments see the twenty-two essays in *Angelicum*, vol. XXI (1944).

35. In *Opera omnia*, edited by S. C. A. Borgnet (Paris: L. Vivès, 1895), vol. 31, pp. 694–714.

36. A short, delightful, and still scholarly introduction to Roger Bacon's life and work is *Roger Bacon in Life and Legend*, by E. Westacott (New York: Philosophical Library, 1953). Bacon's science is the topic of the majority of essays in *Roger Bacon: Essays Contributed by Various Writers on the Occasion of the Seventh Centenary of his Birth*, collected and edited by A. G. Little (Oxford: Clarendon Press, 1914).

37. Subsequent references will be to *The 'Opus Majus' of Roger Bacon*, edited with an introduction and analytical tables by John H. Bridges (Oxford: Clarendon Press, 1897).

38. *Ibid.*, vol. I, pp. 33–34 (Part II, chap. i).

39. *Ibid.*, vol. I, pp. 243–44 (Part IV, chap. xvi).

40. *Ibid.*

41. *Ibid.*, vol. I, p. 402 (Part IV, chap. xvi).

42. *Ibid.*, vol. I, pp. 253–58 (Part IV, chap. xvi).

43. *Ibid.*, vol. I, p. 263 (Part IV, chap. xvi).

44. *Ibid.*, vol. I, pp. 165–66 (Part IV, chap. xiv).

45. *Ibid.*, vol. II, pp. 377–81 (Part VII, chap xxi).

46. *Ibid.*, vol. I, pp. 180–83 (Part IV, chap. xvi).

47. See on this P. Mandonnet, *Siger de Brabant et l'averroïsme latin au XIIIe siècle, Ire Partie, Étude critique, IIe Partie, Textes inédits* (2d rev. ed.; Louvain: Institut Supérieur de Philosophie de l'Université, 1908–11).

48. "That is the eternal light of Siger, who lecturing in the street of straw [Rue de Fouarre] deduced truths which brought him envy." *The Paradise of Dante Alighieri*, edited with translation and notes by A. J. Butler (London: Macmillan, 1891), pp. 138–39 (Canto X, 136).

49. *Siger de Brabant*, vol. II, pp. 139–40. A similarly resolute defence of the eternity of the world by Boèce of Dacia does not contain reference to eternal recurrences. See *Boetii de Dacia Tractatus de aeternitate mundi*, edited by G. Sajó (new rev. ed.; Berlin: Walter de Gruyter, 1964).

50. Given around 1271. See the text of "De aeternitate mundi contra murmurantes" in *Opuscula omnia necnon opera minora*, edited by J. Perrier (Paris: P. Lethielleux, 1949), pp. 52–69.

51. See *Siger de Brabant*, vol. II, pp. 175–91. A topically arranged list of the condemned propositions is available in English in R. Lerner and M. Mahdi, *Medieval Political Philosophy: A Sourcebook* (New York: The Free Press of Glencoe, 1963), pp. 335–54.

52. *Le système du monde*, vol. VI, p. 66. Duhem offered the last five volumes of his monumental opus as proof of the proposition that modern science "was born, so to speak, on March 7, 1277 from the decree issued by Monseigneur Étienne, bishop of Paris." By this Duhem meant that the decree decisively reinforced, mainly at the University of Paris, a train of thought leading ultimately to the formulation of a new (classical) physics.

53. *Science and the Modern World: Lowell Lectures, 1925* (New York: The Macmillan Company, 1925), pp. 17–18.

54. Most of them can be found in the three volumes of Duhem's *Études sur Léonard de Vinci, ceux qu'il a lus et ceux qui l'ont lu* (Paris: Hermann, 1906–13).

55. As can be seen from the long bibliographies bringing to a close both volumes of Crombie's *Medieval and Early Modern Science*. As the yearly bibliographies of *Isis* show, interest in medieval science has been steadily increasing during the last ten years.

56. Edited by E. A. Moody (Cambridge, Mass.: The Medieval Academy of America, 1942).

57. As Moody noted, Galileo read Albert's *Quaestiones de caelo et mundo* in the collection of medieval writings on physics published by G. Lockert in Paris in 1516 and 1518. In his early work, entitled *Sermones de motu gravium*, Galileo offered against Aristotle's explanation of projectile motion arguments that are identical almost word for word with the ones in Albert's work which in turn follows very closely Buridan. *Ibid.*, pp. xxi–xxii.

58. In the edition by Moody, p. 152 (Lib. II, quaest. 6).

59. *Ibid.*, p. 90 (Lib. I, quaest. 19).

60. *Ibid.*, pp. 184–92 (Lib. II, quaest. 14).

61. *Ibid.*, pp. 44–49 (Lib. I, quaest. 10).

62. *Ibid.*, p. 152 (Lib. II, quaest. 6).

63. *Ibid.*, p. 164 (Lib. II, quaest. 9).

64. *Ibid.*, p. 229 (Lib. II, quaest. 22).

65. The passage is from Buridan's commentary on Aristotle's *Physics*. The whole context in which projectile motion is discussed is given in English translation in M. Clagett, *The Science of Mechanics in the Middle Ages* (Madison: University of Wisconsin Press, 1959), p. 536.

66. For a summary of Oresme's life and work, see the introductory essay to Nicole Oresme, *De proportionibus proportionum and Ad pauca respicientes*, edited with introductions, English translations, and critical notes by Edward Grant (Madison: University of Wisconsin Press, 1966), pp. 3–10.

67. Edited with an introduction, translation and commentary by M. Clagett under the title,

Nicole Oresme and the Medieval Geometry of Qualities and Motions (Madison: University of Wisconsin Press, 1968).

68. The first of Duhem's encomiums of Oresme that stirred a long overdue interest in the science of the Bishop of Lisieux, appeared significantly enough under the title, "Un précurseur français de Copernic: Nicole Oresme," *Revue générale des sciences pures et appliquées*, 20 (1909):866–73.

69. "Ueber die Stellung von Nicolas Oresme in der Geschichte der Wissenschaften," *Archeion*, 11 (1929):xv–xxii; reprinted in *Philosophisches Jahrbuch*, 45 (1932):58–64; see also his *Geschichte der Naturphilosophie* (Berlin: Junker und Dünnhaupt Verlag, 1932), pp. 76–77.

70. "Nicole Oresme and the Medieval Origins of Modern Science," *Speculum*, 16 (1941): 167–85.

71. "Le vide et l'espace infini au XIVᵉ siècle," *Archives d'histoire doctrinale et littéraire du Moyen Age* 24 (1949):45–91; see also his review of Crombie's *Grosseteste* and *Augustine to Galileo*, "The Origins of Modern Science: A New Interpretation," *Diogenes*, 16 (1956):1–22.

72. Nicole Oresme and Medieval Scientific Thought," *Proceedings of the American Philosophical Society*, 108 (1964):298–309.

73. "Late Medieval Thought, Copernicus, and the Scientific Revolution," *Journal of the History of Ideas*, 23 (1962):197–220.

74. See notes 66 and 67.

75. See his articles quoted above.

76. Nicole Oresme, *Le Livre du ciel et du monde*, edited by Albert D. Menut and Alexander J. Denomy. Translated with an introduction by Albert D. Menut (Madison: University of Wisconsin Press, 1968).

77. *Ibid.*, p. 57 (Book I, chap. i).

78. *Ibid.*, pp. 81–85 (Book I, chap. vi).

79. *Ibid.*, p. 85 (Book I, chap. v).

80. *Ibid.*, p. 81 (Book I, chap. vi).

81. *Ibid.*, p. 87 (Book I, chap. vii).

82. *Ed. cit.*, p. 382 (Part I, line 11).

83. *Ed. cit.*, p. 306 (chap. iv, lines 606–09).

84. See E. Grant, "Nicole Oresme and the Commensurability or Incommensurability of the Celestial Motions," *Archive for the History of Exact Sciences*, 1 (1961):440–41.

85. In *Le Livre du ciel et du monde*, p. 167 (Book I, chap. xxiv).

86. The Latin text was edited with a translation by G. W. Coopland under the title *Nicole Oresme and the Astrologers: A Study of his Livre de divinacions* (Liverpool: University Press, 1952).

87. *Le Livre du ciel et du monde*, p. 171 (Book I, chap. xxiv).

88. *Ibid.*, p. 169.

89. *Ibid.*, p. 171.

90. *Ibid.*, p. 173.

91. *Ibid.*, p. 175.

92. *Ibid.*, p. 177.

93. *Ibid.*, p. 187 (Book I, chap. xxvii).

94. *Ibid.*

95. *Ibid.*, see esp. pp. 197–203 (Book I, chap. xxix).

96. *Ibid.*, p. 271 (Book II, chap. i).

97. *Ibid.*, pp. 269–73.

98. *Ibid.*, p. 283.

99. *Ibid.*, p. 289 (Book II, chap. ii).

100. *Ibid.*, p. 357 (Book II, chap. viii).

101. *Ibid.*, p. 365.

102. *Ibid.*, p. 369.

103. *Ibid.*, pp., 521–39 (Book II, chap. xxv).

104. The remark is part of a detailed discussion of projectile motion by Franciscus de Marchia. The text is given in full in A. Maier, *Zwei Grundprobleme der scholastischen Naturphilosophie:*

Das Problem der intensiven Grösse. Die Impetustheorie (2d ed.; Rome: Edizioni di Storia e Letteratura, 1951), pp. 166–80.

105. See especially pp. 168, 170, 173.

106. E. Grant, "Medieval and Seventeenth-Century Conceptions of an Infinite Void Space beyond the Cosmos," *Isis* 60 (1969):39–60.

107. *Ibid.*, p. 60.

CHAPTER ELEVEN

The Interlude of "Re-naissance"

When Oresme died in 1382, the Middle Ages were already on the wane. In the place of typically medieval attitudes there came preferences which form a clearly identifiable whole, the spirit of the Renaissance.[1] Since the last hundred years, or more specifically since the publication of classic studies by G. Voigt[2] and J. Burckhardt,[3] the age known as the Renaissance has become a central issue for historians of modern culture. The Voigt-Burckhardt thesis, that the Italian revival of classical antiquity during the fifteenth century forms the cradle of modern Western mentality and culture, had not, of course, gone unchallenged. It has even been asserted that the Renaissance "simply continued all that was vivacious and ornate in the Middle Ages."[4] Whitehead's claim that "Thomas Aquinas would have enjoyed" the Renaissance, this "last spurt of the Middle Ages."[5] is in itself not beyond the realm of possibility though it will hardly ever command wide agreement. But there has developed no major dispute about the principal features characterizing the outlook of the chief representatives of the Rinascimento.

Among them the first great figure was Petrarch. Whether he was truly the first modern man, and the first Alpinist to boot, is a matter of definition. His *Secretum*,[6] an emulation of St. Augustine's *Confessions*, gives an unmistakable insight into the deepest recesses of his aspirations and motivations. Whatever Petrarch's professed admiration for Augustine's spiritual ideals, he anchored his own programme of spiritual perfection in the dicta of ancient Greek moralists. The Bible and the Fathers were completely ignored by him in this respect. In fact, he showed no interest in theology. This should not suggest that the external world, as an object of systematic study, had any appeal to him. Secular learning found no real advocate in Petrarch. His distaste for law, which he studied for some seven years, was as intense, as was his dislike for medicine. What attracted him to the spirit of classical antiquity consisted in a self-centred estheticism and philosophical mysticism which dominated classical thought during much of its Hellenistic period.

In the new-fangled cultivation of classical antiquity, the external world could serve as the stage for the myths of Plato at best, or for the myths of the gods at worst, as can be seen in the fifteen books of the *De genealogia deorum gentilium* of Boccaccio, a trusted friend of Petrarch. In its final form, published in 1373, it became the sourcebook for all Humanists who for the next two centuries discoursed about the world in terms of allegories, in much the same, scientifically worthless, obscurantist manner, in which Porphyry, one of their antique idols, did in his *De antro nympharum*. Porphyry was a chief of Neoplatonists and whatever there was philosophical in the Rinascimento, it was largely a reaffirmation of Neoplatonism, and of its studied neglect of a reasoned investigation of the external world. This can best be seen in the programme pursued by Marsilio Ficino, the renowned leader of the Platonic Academy in Florence. The programme centred on demonstrating that Platonism and Christian faith carried the same message. The result was a return to Plotinus and to his emanationist world view.

Returns are also departures and this time the departure was from Christianity, from its fundamental tenets, such as the doctrine of Creation and Incarnation. For Ficino the true prophets were Plotinus, Porphyry, and Iamblichus, who as he claimed, brought to perfection the gold in Plato's thought. When they arose, Ficino wrote to Cardinal Bessarion, "that gold, as if refined in the crucible of the smelter, was freed, through their explicative labours, from all the dross, and shone with the most resplendent light."[7] But what truly got eliminated in that crucible was the belief in the fundamental goodness and instructiveness of the world of matter. It was indeed logical on Ficino's part to pay only lip service to the articles of faith about Incarnation and Creation, and to claim at the same time the Christian character of the religious systems advocated by Zoroaster, Hermes Trismegistus, Orpheus, Pythagoras, and Plato.[8] Such was a motley but telling list. Equally expressive was Ficino's presumption to serve up his extreme arbitrariness by which he handled the facts of religious history. He should have remembered that his vaunted prophets were the leaders of the last great stand made in late antiquity against Christianity. He should have also known about Augustine's regret for having commended Neoplatonism, precisely because it had no room for the doctrine that God created the universe in the beginning out of nothing, and that in the fullness of time the Word of God was made flesh. It is no accident that Ficino's only explicit discussion of creation runs to a mere nineteen lines in the two heavy folio volumes of his collected works.[9] Obviously, the best he could do was to skirt an issue which, had he faced it in all candor, would have forced him to admit that differences between Gospel and Plotinus were anything but negligible. The Christian creed's first article demanded historically, philosophically, and psychologically far more than a furtive remark that God created all matter in the first instant of creation.[10]

Modern admirers of Ficino's "soaring humanism," who could poke fun at the "nonconcept" of creation out of nothing, found no fault with the noncommittal faith with which Ficino espoused the dogma of creation.[11] Bias of this type is a deep shadow cast on the breadth and width of one's scholarship. A portrayal of Ficino's philosophy which makes no reference to Ficino's addiction to magic and astrology comes very close to being a caricature. At any rate, Ficino's espousal of the crowning tenet of astrological lore, the doctrine of the Great Year, could hardly have been more energetic and explicit. For him it was important, just as it was for Plotinus, to discuss the number of men generated within one Great Year. Ficino spoke, therefore, in a matter-of-fact style about the identical return of all forms as one Great Year followed another.[12] Ficino's beliefs included the transmigration of souls in an endless process of moving up and down the ladder of perfection. The Renaissance he represented was a return to the various aspects of the antique belief in endless re-naissances. His acquiescence was a far cry from the utter revulsion felt in this respect by Augustine of Hippo, once a great admirer of Platonic and Neoplatonic thought.

For a Christian of the stature of Augustine a matching of Christ and Apollo was out of the question, whereas Ficino devoted lengthy lucubrations to what he called the "nuptial number" in Plato's *Republic*. Ficino's verbosity was only in part due to the fact that Plato himself, as Ficino put it, "gives in multiple, nay almost innumerable, variations such a number so that its measure could accommodate the whole life of the world, namely, its restoration from deluge to deluge, or in other words, in each Great Year."[13] Out of the various mystical and cosmic interpretations of numbers which he found in Plato's works, which he was the first to translate in full, Ficino was particularly attracted to 1728, or the product $12 \times 12 \times 12$. He saw in that number a major period for cosmic recurrences, but he showed no concern about the fact that no planetary cycle or its multiple corresponded to that number. In Ficino's science numbers preceded physical reality and the latter could readily be ignored if the cabbalistic interpretation of a given number seemed satisfactory. Thus, Ficino simply declared as proof that in the figure 1728, 1 referred to the firmament, the immensity of which could only be measured in thousands, 7 expressed the influence of the seven planets, whereas 28 took care through the lunar cycle of the events to happen in the sublunary world under the influence of the planets.

The planets themselves were merely the tools and expressions of an arcane doctrine of numbers. "Plato selects," wrote Ficino approvingly, "especially those numbers that correspond to the universe and express its resonances, in order to show that events in the sublunary world depend on the firmament through certain numbers and measures, particularly when the numbers coincide with the cycles. Thus, Plato arrives at some fatal numbers that are integers and supreme, and demonstrates that once they are fulfilled the totality of things slowly begins to turn into the opposite direction."[14] The cosmological supremacy of numbers was also a basic belief with Plotinus and Proclus, as Ficino hastened to add.[15] He was also eager to treat the triad of Trinity in the same context in which he discussed the respective influence of odd and even numbers on fertility, character, and political fortunes. Clearly, his developments on Plato's "nuptial numbers" served not the wedding of faith and reason but the coupling of vague, cosmic aspirations with astrology and magic. There was indeed something magical if not demonic in the manner in which Ficino put a biblical seal on his revelling in sundry periods and recurrences. For Ficino, who very likely engaged in the celebration of some demonic magic reconstructed from the writings of Hermes Trismegistus,[16] brought to a close his discussion by quoting the hallowed phrase from the Book of Wisdom about God, who disposed everything according to number, weight, and order.[17] This blatant abuse of intended meaning was well matched by Ficino's ignorance of the numbers and patterns that formed the backbone of scientific astronomy. His astronomical knowledge remained within the vague generalities that fill Plotinus' *Ennead* about the circular motion of heavenly bodies. Like his idol, Plotinus, Ficino too had his "astronomical vision" centred on the claim that souls could reach their appropriate or circular motion only after their liberation from bodily bondage.[18]

The real bondage was not that of the soul but of Neoplatonism and of its astrological fatalism out of which at least one member of Ficino's circle, Giovanni Pico della Mirandola, sought a resolute escape. The elemental urge to do so came only after he learned that learning was not necessarily knowing. In the way of learning he had covered much in a short time. He started the study of canon law at fifteen, and ten years later he was a master of philosophy and theology. He also got a good exposure to Hebrew and Greek. Added to this was an expertise in Italian and Latin poetry. During his visits to Florence he found a friend in Ficino who introduced him to Plato, Plotinus, and the Hermetic literature. Young Pico's intellectual cup must have "runneth over," as after his one-year stay in Paris he decided to challenge the whole commonwealth of intellect by positing 900 theses for disputation. His famed discourse "On the Dignity of Man" was to preface the debate which never materialized. Four out of the 900 theses were found to be heretical and Pico had to retire from public life. Through Ficino's good services he was given by Lorenzo dei Medici a villa at Fiesole.

What Pico wrote during his last five years at Fiesole contrasted sharply with his erstwhile confidence that all differences could be reconciled with enough good will, persistence, and talent. In his discourse "On the Dignity of Man" there is hardly a trace of the limits that could stand in the way of one's effort to become a truly universal man. For such a man, as Pico depicted him, every philosophy and every creed was a treasure chest of practically equal worth. Cabbala and magic were as instructive as the most highly regarded branches of theology and philosophy. Pico was ready to make sallies up and down the whole gamut of knowledge, speculation, and mystical longings: "We shall at one time be descending, tearing apart, like Osiris, the one into many by a titanic force; and we shall at another time be ascending and gathering into one the many, like the members of Osiris, by an Appollonian force; until finally we come to rest in the bosom of the Father, who is at the top of the ladder, and are consumed by a theological happiness."[19]

In that theology Zoroaster was as welcome as Moses, Apollo as well as Christ, the Chaldeans as well as the Church Fathers. Pico's long list of the respective merits of various philosophical and theological sects reached its high point with the Neoplatonists: "And if you turn to the Platonists, to go over a few of them: in Porphyry you will be pleased by an abundance of materials and a complex religion. In Iamblichus you will feel awe at a more hidden philosophy and at the mysteries of the barbarians. In Plotinus there is no one thing in particular for you to wonder at, for he offers himself to our wonder in every part."[20] In Plotinus' teaching Pico particularly singled out his praises for the magician who "marries earth to heaven, that is, lower things to the qualities and virtues of higher things."[21] By this Pico meant "natural magic" which in his words drives its practitioners to sing with the Psalmist: "The heavens are full, all the earth is full of the majesty of Thy glory."[22]

The injudiciousness of this bold galloping over the universe of knowledge can be seen at once if the praise accorded by Pico to Roger Bacon is contrasted with the sustained attack on the volatile friar in Pico's critique of astrology, the *Disputationes adversus astrologiam divinatricem*[23] completed in 1494, the year he died at the youthful age of thirty-one. The *Disputationes* is a spirited effort to break through the clutches of astrological fatalism and secure thereby the dignity of man rooted in his moral freedom. Six years, for only six years separate Pico's famous discourse and his massive refutation of astrology, are not much, especially if they span one's late twenties. But for Pico those six years were a lifetime of maturing and of gaining a true insight. The dimensions of Pico's attack on astrology are also a reflection of the heavy influence astrology still had in his time. The major stepping stones and thrusts of that attack show that Pico found his strongest inspiration against the disease of astrology in a rediscovery of the uniqueness of the Christian interpretation of human life and cosmic existence.

Pico's repeated insistence that a Christian cannot at the same time be a Chaldean[24] marks not only a crucial departure from the syncretistic Neoplatonist Christianity of the "Dignity of Man." It also flies in the face of the astounding claim that Pico conquered astrology through a purely humanistic faith in the autonomy of the creative powers of man which "excludes the possibility of any determination from without, be it 'material' or 'spiritual'."[25] Pico acquired a far deeper insight into the dimensions of human freedom than did some modern advocates of rationalistic humanism. For him the human freedom became anchored in the creative power of God, who alone could create human volition and all the choices of which that volition is capable, and yet keep its freedom inviolate. Pico's discussion of the human freedom is not an anticipation of rationalistic dicta and ethics. Much rather it is an echo of the Thomistic analysis of the freedom of will as related to the absolute sovereignty of the Creator. "Certainly," declared Pico, "God not only foresaw our volitions, but also foreordained them . . . so that not only the outcome of our plannings, but even our thoughts, deliberations, resolutions . . . not only as it is written, the hearts of kings, but the hearts of all of us are in the hands of God, who alone governs all."[26]

Only when this biblical and theological tone of Pico's analysis of man's freedom is kept in mind, can one explain why he devoted numerous chapters to showing that astrology could be of no help whatever to Christian religion. His special targets were Roger Bacon and Pierre d'Ailly,[27] but what he attacked really was the attitude of those who leaned over backward lest they should find some fault with paganism, old and modern. He pointedly asked the advocates of an accord between astrology and theology whether they had never read Saint Paul's sharply pointed question: "Can there be anything common between light and darkness or any association between Christ and Belial?"[28]

It was not merely some minor detail in Bacon's and d'Ailly's efforts to vindicate Christianity in terms of planetary cycles that Pico found fault with. He unerringly sensed that individual freedom and collective or historical destiny could only remain meaningful if it was anchored in the biblical doctrine of the Creator's sovereignty and reasonability. It was relatively easy for him to show that the alleged periods of 360 and 1460 years based on the revolutions of Jupiter and Saturn did not fit the major turning points of political and religious history.[29] It took, however, a penetrating insight to go to the central issue and put the astrological interpreters of history on the spot: "Let them therefore tell me, admitting for the sake of argument the reality of these great periods, whether such great periods started anew at the very first moment of the creation of the world."[30] The heart of Pico's stratagem could not have been clearer. It consisted in the warning that the Christian belief in creation was incompatible with astrological periods and eternal recurrences.

Pico's case with astrology contains an indirect though very instructive moral for the fate and fortunes of the scientific enterprise during the Renaissance. The Platonic

Academy to which he belonged was certainly the most representative and novel aspect of the rebirth of letters during the fifteenth century apart from the cultivation of poetry in the vernacular. The Academy had little if anything to do with the scientific endeavour. Nor was the atmosphere generated within the Academy germane to scientific reasoning. The only time it produced a perspective favourable to scientific thought came when Pico took up the cudgels against astrology. It was not Neoplatonism that gave him the spurs, but the vistas and guidelines contained in the dogma of Creation. Actually, nothing could have been more natural for Pico than to become a captive of the mental labyrinths of Ficino's Academy. Its Neoplatonism reimposed the same stifling confines in which ancient Greek thought once became trapped, and claimed its new victims with startling effectiveness. The best illustration of this is Francesco Cattani da Diacceto[31] who, as one of his biographers put it, "was a most devoted disciple of Ficino and absorbed avidly not only all the precepts of his Platonic ideology but also carried it over most minutiously into his own comportment, so that he might be known as a true Platonist and a disciple of Marsilio Ficino."[32]

Diacceto's main association with Ficino coincided with the years when Pico, though living near Ficino, was already pursuing a course leading steadily away from the master's fundamental tenets. The result was the finest critique of astrology and of its world view written during the Renaissance. Diacceto, the faithful disciple, produced the last notable reaffirmations of the Platonic idea of the Great Year. He did so first in his work on *De pulchro*,[33] a treatise which was more metaphysical than esthetical, in close analogy to Plotinus' work on the same topic. The doctrine of the Great Year was for Diacceto not an ancient tradition to be respectfully recalled but a tenet to be firmly defended in its integrity which implied the identical return of all beings and events. To make his point crystal clear, he quoted from Virgil's Fourth Eclogue the lines stating how the mighty Achilles would fight the same battles over and over again under the walls of Troy. This was the picturesque counterpart of what Diacceto called "the recondite decree of the Academy" according to which "everything shall at one time return to the same condition."[34] In Diacceto's explanation the doctrine was a logical sequel to such Platonist tenets as the finite number of ideas, the role of the celestial region in grafting them on ordinary matter, and the perennial endurance of all in spite of changes and transformations. "For if we believe Plotinus," Diacceto remarked, "nothing can be completely corrupted. Everything would rather endure in at least a rudimentary condition, even in the case of an apparently complete dissolution."[35] The actual execution of this tenet was assured by the revolution of the heavens: "Great is indeed the name of that year through the completion of which the heavens invariably resume their original position."[36] And so would everything recur on earth exactly in the same manner in which a runner, to recall Diacceto's comparison, would, in making a thousand rounds in a stadium, pass the very same starting line as many times. In the same context he also referred to the faithful return of the sun in every twelve months to the same section of the zodiac. Clearly, one had to concentrate on the tireless march of the heavens to overcome the sense of weariness of running forever in circles.

In defending the principle of identical returns after one Great Year, Diacceto cared to mention only one group of opponents. They were some Peripateticians, who took exception not to the Great Year but to the return in numerically identical forms. About Christian philosophers, theologians, Church Fathers, Diacceto said not a word. Nor was there any reference to the dogma of Creation in his "Epistle on the Great Year" addressed to Bindaccio Recasoli, which is more historical in character.[37] In it Diacceto disclosed his burning desire to gain a clear idea of a doctrine so central to Plato's teaching. The starting point, tellingly enough, was the coming and going of departed souls as suggested by Socrates in the *Meno*. Souls, Diacceto recalled Socrates' dictum, never die but go perennially through various cycles of rebirth, purification, decadence, and renewal. He added in the same breath that, according to Platonic and Pythagorean doctrine, the more noble was one's conduct on earth, the less frequent were his soul's returns to the bondage of the body. Diacceto made much of the section in the *Phaedrus* where Socrates claimed that a soul did not return to the place from whence it came for ten thousand years. It remained less clear how a soul, which had chosen the life of a philosopher, would gain its wings within three successive periods of a thousand years and speed away to its (final) freedom. But Diacceto was not detained by such inconsistencies. He boldly pressed forward to his goal, the assertion that emphasis on the figure twelve in the *Republic* meant to suggest that in every twelve thousand years everything would undergo an identical rebirth. In support of this he recalled that in the *Timaeus* the figure of a dodecahedron was attributed to the firmament and that Plato was fond in many other respects of the number twelve. "From this it should appear," stated Diacceto, "that in Plato's belief that number had a supreme authority in ruling, directing, and bringing to a completion the universe."[38] Since, according to Diacceto, "this number of twelve was the cause and author of the harmony of the whole world," he could greet his discovery of it in Plato's writings as a reason for immense satisfaction.

Enthusiasm for Plato was one thing, casting one's mental lot with Neoplatonism was another. The former did not start with Ficino. Grosseteste and other leading figures of medieval science had already found in the Platonic philosophy of numbers fruitful inspiration, but their Christian faith in the Creator enabled them to see the difference between genuine gold and that of fools. The Plato which proved itself scientifically productive during the Renaissance was the continuation of the Platonism of medieval thought which carefully avoided the syncretistic morass exuding from Ficino's thought and from the Neoplatonism of his Academy. Diacceto's Neoplatonist paganism came from the very core of that morass and it certainly was not a negligible aspect of his thinking.[39] What it represented was rather the very matrix out of which emerged an interpretation of the world which became increasingly alien to considerations that constituted the real progress of science during the Renaissance.

The valuable contributions to the scientific understanding of the world were made during the Renaissance by thinkers whose faith in the intelligibility of the world was rooted in a sincere attachment to the first article of the Christian creed. Most of them were also admirers of Plato but their faith saved them from becoming prisoners of the immanentism and apriorism of his interpretation of the world. Their line opens with Nicolas of Cusa (Cusanus) and with his classic *De docta ignorantia* written in 1440.[40] There the future cardinal made a resolute departure from the closed Aristotelian universe, argued on behalf of innumerable centres in an unlimitedly large universe, discarded the absolute motionlessness of the earth, proposed the basic similarity of all cosmic bodies, and claimed unhesitatingly the presence of living beings everywhere in the universe. Any of these startling propositions could have secured for Cusanus a lasting place in the history of science. He carved for himself, however, a special niche by presenting not only a whole array of such propositions, but by putting them forward as part of a tightly knit system of thought.

The tone of that reasoning was distinctly Platonic. Yet, whenever Cusanus spoke of Plato and Platonists, he did so with an eye on the doctrine of Creation. As he defended the method of explanation based on the analysis of numbers and geometrical figures, he referred to Plato and Pythagoras who like the chief Christian philosophers, Augustine and Boethius, "have not hesitated to assert that number was the essential exemplar in the mind of the Creator of all things to be created."[41] At the same time he unhesitatingly laid bare the essential differences between the Platonist theory of the origination of the world and the Christan dogma of creation. This is not to suggest that he did not try to put in the best possible light Plato's teaching which he defended against Aristotle's criticism.[42] But Plato shared with all ancient philosophers the dictum *ex nihilo nihil fit* and precisely because of this, so Cusanus argued, Plato's account of the ideas failed to do justice to the absolute distinctness of God from the world. The Platonist theory about the eternal images of forms had, therefore, to be discounted. To this conclusion Cusanus added: "The efficient and formal and final cause of all is God, who in the one Word creates all things however different they be; and every creature owes its existence to this creative act of God and for that reason is finite. Between it and the Creator there is an infinity. God alone is Absolute, all else is finite."[43]

Once this was firmly laid down, there could be no harm in making ample use of such distinctly Platonic notions as the world-soul and the world-organism to convey the marvellous harmony and interrelation among the constituent parts of the cosmos. Cusanus borrowed from mathematics and geometry the analogies by which he illustrated his central theme, the relation between the One (Creator) and the many (created beings), between the infinite and the finite.[44] This was a most appropriate procedure as the First Cause of Cusanus was the one "who created all things in number, weight and measure." He placed this biblical reference in a chapter which brought to a close the second main section of his book, devoted to the topic of how the world could be unlimitedly large, lack a centre, and still be harmoniously ordered.[45] It was the same biblical doctrine about the Creator that also generated in him the vivid awareness of the intrinsic limitations of the human mind. Man's share was a "learned ignorance" which Cusanus pointedly anchored in the scriptural text about seeing "in a mirror and in a dark manner."[46]

The only exception to this limitation was the Incarnate Word in whose mystery Cusanus saw the bridge between the One and the many. The whole Third Book of the *De docta ignorantia* is a moving reflection on Christ both in his singularity and in his plenitude, the Church. The sincerity and depth of the Christian conviction that animates those pages made it impossible to turn Cusanus into a "progressive figure" in the rationalistic sense of the expression. More reprehensibly, the Christian resources of his scientific methodology and attainments were readily ignored or belittled whenever interpretation rushed ahead of the task of carefully setting forth the record. In Cusanus' case, the record bespeaks the vigour of the biblical vision of a creation once and for all, and of one single ultimate goal for the cosmos and mankind in the mystery of final resurrection. Thus, Cusanus could refer to the moment of Incarnation as having taken place after "many cycles of ages,"[47] without giving the slightest support to the doctrine of cycles. His vision of the cosmos and history was aligned along a straight path which the Creator's hand traced firmly into the realm of existence.

The reflection of that firmness in Cusanus' lifelong labours is something that here can only be mentioned briefly. His efforts to channel the course of history toward the ideals of peace and unity stand in stark contrast with the counsel of one who is never absent in any portrayal of the Renaissance, his namesake, Niccolo Machiavelli. The famed author of the infamous *Il Principe* illustrates the point of this book only in a negative way. Machiavelli's brazen advocacy of "the end justifies the means"[48] reflects a paganism for which he was already strictured in his own time. Less offensive to his contemporaries appeared the cynical vein of his view of history. Cynicism raises its ugly head in the absence of confidence in lasting ideals, and it is the proper goal that ultimately disappears when it can be pursued by any means. There is, indeed, a distinct absence of truly inspiring and permanent goals in Machiavelli's sundry reflections on history which intersperse his lengthy account of the history of Florence.[49] One of these reflections is a classic description of history as an ever recurring treadmill. The reflection opens, it is worth noting, the second half of his *Florentine History* which covers the developments in Renaissance Florence from 1434 to the death of Lorenzo the Magnificent in 1492: "In the changes that they make, countries are wont to pass from order to disorder, and from disorder again to order; for as Nature never suffers the things of this world to come to a stay, so soon as they reach their ultimate perfection, there being nothing higher to which they can mount, they must needs descend; and, in like manner, when, in their downward course they have reached, through their disorders, their lowest point of degradation, since then they can descend no lower, they must needs rise. And always in this way from good we descend to evil, and from evil mount to good. For valour begets tranquillity, tranquillity ease, ease disorder, and disorder ruin. And conversely out of ruin springs order, from order valour, and thence glory and good fortune."[50]

Rather typically, the improvement was ascribed by Machiavelli to the ascendency of strong-willed military leaders, whereas he singled out philosophers and the pursuit of culture as the harbingers of decay: "It has been noted by the discerning that letters follow after arms, and that in all countries and cities captains come before philosophers. For when good and well-disciplined arms have brought victories, and victories peace, the vigour of warlike minds can be corrupted by no more specious ease than that of letters; nor can ease find entrance into well-constituted States by any more seductive or more dangerous snare."[51] Machiavelli, therefore, approvingly recalled the action of Cato who upon seeing the impact of the philosophers Diogenes and Carneades, envoys from Athens, on the Roman youth, had a law passed which banned the admittance of philosophers to Rome. This failure to see in culture a positive and lasting goal to be pursued was a logical consequence of a thinking for which history was a fatalistic cycle: "It is from these causes," wrote Machiavelli, "that countries come to ruin; which being reached, men, taught wisdom by suffering, return again, as has been said, to order, unless indeed they remain strangled by some irresistible force."[52] The Florence of Lorenzo the Magnificent in which Machiavelli grew up was no exception to this rule. In fact, the concluding paragraph of the *Florentine History* recounts some heavenly signs, such as the destruction of the pinnacle of the Church of Santa Reparata by lightning, taken as omens of the downturn of the fortunes of Florence.[53]

This overtly astrological buttressing by Machiavelli of his interpretation of history was a pattern present also in other Renaissance supporters of the cyclic nature of history.[54] Revealingly, none of them had been part of the growth of the scientific movement during the sixteenth century. As Bacon was later to observe, their cyclic theory of history sapped precisely that confidence in steady progress which the scientific enterprise needed if it was to prosper.[55] Long before Bacon made this comment based in part on the wisdom of hindsight, a small book, the *Utopia*[56] of Thomas More, the finest Humanist of Renaissance times, proclaimed the basic principles which Bacon outlined in detail in his *New Atlantis*. Wisely enough, More implied in the very title of his book the basic unattainability of the ideal set forth there. Yet, its very writing evidenced the confidence that it was worth working toward at least a partial realization of that ideal.

The identification of the deepest source of that confidence should be no problem with a Christian of More's stature. The "natural" religion of the Utopians is not an idealized reconstruction from some historical precedent, but a studied "naturalization" of Christian faith and morals. This should be clear from the firm resolve by which More's Utopians held fast to such tenets as the immortality of the soul and its eternal reward or punishment. By this latter point More set a great store. The hope rooted in eternal life was for him the only sound foundation on which society and its cultural pursuit could securely be based.[57] Distinctly Christian is also the set of principal features which More put forward to describe the faith of the Utopians in the Creator of all.[58] He is one, eternal, far above the reach of human mind, everywhere present in the universe, but as More carefully adds with his Christian instinct, the omnipresence in question is not a physical diffusion, but an influence of power. The same instinct is also evident in More's emphasis on the image of the Creator as the Father of all and the source of all beginnings, changes, and ends. Existence, personal, social, and cosmic, was for the Utopians, a once-and-for-all process, designed and directed by a personal, supreme Benevolence. Not surprisingly, More also took pains to note that the universe could not be carried along by chance.[59]

More's philosophical perspectives took on very distinct contours as he tied his Utopians' boundless thirst for knowledge to their belief in the Creator. More's noble pagans are a far cry from the Gentiles whom Isaiah and Paul chastized for their obstinate refusal to recognize from the lawfulness of nature its Lawmaker. The investigation of nature was for the Utopians "an act of worship acceptable

s

to God."[60] This had to be so, if the laws of nature bespoke the Lawgiver of nature: "When . . . they explore the secrets of nature, they appear to themselves not only to get great pleasure in doing so but also to win the highest approbation of the Author and Maker of nature. They presume that, like all other artificers, He has set forth the visible mechanism of the world as a spectacle for man, whom alone he has made capable of appreciating such a wonderful thing. Therefore He prefers a careful and diligent beholder and admirer of His work to one who like an unreasoning brute beast passes by so great and so wonderful a spectacle stupidly and stolidly."[61]

This religious motivation is to be kept in mind if one is to understand the supreme honour which learning enjoyed among the various pursuits and crafts available for the Utopians. The finest privileges were accorded in Utopia to those capable both morally and intellectually of most intense studies.[62] These studies ran the whole gamut of learning with special emphasis on natural sciences and medicine. In speaking of "Physicke" More had in mind not only the art of healing.[63] His list of various Greek scientific authors studied by the Utopians and his description of the various branches of natural science cultivated among them show that his ideal society aimed at far more than at keeping itself physically healthy. Still, More's repeated insistence on securing the highest possible measure of health for society should ring exceedingly progressive. The same holds true of his ideas about a planned equilibrium between the food needed by a settlement and its site, of his ideas about the establishment of an optimum size for all social units, and last but not least of his ideas about a national defence strong but not more than adequate to keep belligerency and bloodshed at a minimum.

Equally advanced was that liberality of spirit in which More wanted the sciences to be cultivated. Clearly, if the divine mind was so superior to man's intellect, no definitiveness could be accorded to any specific explanation of the world as a whole. The arrogance of Aristotle's cosmology and physics had no sway on More who singled out questions about the origin and nature of the heavens as being the most uncertain in philosophy.[64] More's stricture of astrology was given in the same context in which he praised his Utopians for their versatility in devising "instruments in different shapes, by which they have most exactly comprehended the movements and positions of the sun and moon and all the other stars which are visible in their horizon."[65] Here More conjured up the vision of future progress. He did the same when he spoke about the ability of the Utopians to "forecast rains, winds, and all the other changes in weather by definite signs which they have ascertained by long practice."[66] Again, More showed his grasp of the future as he described the eagerness of the Utopians to learn the arts of paper making, printing, and navigation.

It hardly makes a favourable reflection on the standard historiography of science that More's *Utopia* is regularly omitted in the listings of the major steps that constitute the advance of scientific enterprise and spirit during the Renaissance.[67] The neglect is strange not only because of the enlightening contents of the *Utopia*, but also in view of its extraordinary influence. No single book written by an English author before the nineteenth century was read as avidly as was More's small masterpiece.[68] It represented the primary articulation of ideas that sparked the scientific movement three generations later into an irresistible, gigantic forward march. The sixteenth-century literary record in England is a clear rebuttal of the cliché which ascribes without any qualification to Francis Bacon the first formulation of the spirit of the new scientific endeavour. That spirit, as was shown in a careful study of astronomy in Elizabethan England, was "the publicly avowed creed of the English scientific workers throughout the latter half of the sixteenth century."[69] More's *Utopia* is in fact the first of its kind[70] which is not only free of patent obscurantism and absurdities, but in which one also finds in a reasoned fashion the chief components of that creed: a judicious criticism of ancient literary and scientific authorities, the perception of the invaluable potentialities of science for social improvement, the emphasis on a sustained observation of phenomena, and the voicing of an optimistic faith in progress through learning, science, crafts, and also through socio-economical planning.

More importantly, the *Utopia* put forward a faith in science which, had it been carefully cultivated in subsequent centuries, might have forestalled the development of science into a Mephistophelic enterprise.[71] Indeed, if there is anything lampooned and deplored in the *Utopia*, it is pride, moral as well as intellectual. The science and progress which the *Utopia* advocated were strictly subject to limits set by moral considerations. Science, its tools and plannings, were proposed in the *Utopia* with qualifications, in the absence of which the art of inventions inevitably becomes, as history proved, an infatuation heedless of consequences that are not perfectly obvious immediately. The last three hundred years of the scientific movement, and especially its very last phase, served ample evidence that much of the present threat of science and technology to society cries for a remedy which has much in common with the *Utopia*'s Christian world view, moral firmness, liberality of mind, and unquestionable humility. The same qualities were also needed to welcome without fear and suspicion the startling scientific novelties that were proposed as More's own century was running its course through some of history's major reorientations.

The foremost of these radical innovations was the heliocentric system of Copernicus. It was the result of a lifelong resolve and inspiration that owed little if anything to the literary Renaissance, let alone to its unabashed paganism or to its "revolutionary" spirit. By the time Copernicus arrived in Italy, his commitment to the heliocentric system seems to have been firmly established.[72] His debt to tradition was, however, enormous. His thorough conservatism[73] remained unnoticed until very recently because of the persistence of a carefully cultivated smokescreen. It was spread by a historiography born under the auspices of an Enlightenment which was more intent on propaganda than on respect for facts, especially for those of history. Copernicus showed no particular embarrassment over answering in terms of Aristotelian physics the difficulties posed by his advocacy of the earth's motion. The firmness of his mind was rooted in considerations that had little to do with observational evidence. Copernicus himself was not a prolific observer and the new system assured no greater accuracy in predicting the position of planets than did the Ptolemaic system through its medieval improvements. He was eager to point out that Pythagoras and others had already anticipated his contention. But his references to classical antiquity were merely tributes paid to the fashion of his time, as were the encomiums heaped by him on the sun. He certainly did not follow the obscurantist segment of the Renaissance in its emulation of solar worship of old. What was akin in his scientific attitude to a worshiping stance consisted in his faith in the ordering of the world along patterns of geometry. In this respect he did not differ from Plato or Ptolemy. Nor was anything essentially new in his insistence that the ordering of the world had to show a maximum measure of simplicity.

Simplicity was an avowed ideal for ancient Greek astronomers, still they found it impossible to abandon for a higher measure of it the evidence of senses and geocentrism. In his *Almagest* Ptolemy called it ridiculous and absurd to remove the earth from the centre of the universe.[74] Copernicus, as Galileo later pointed out,[75] had to commit a rape of his senses in putting forward the heliocentric ordering of planets. Galileo could hardly contain himself in praising the faith of Copernicus in the simplicity of nature. Such a faith, as Galileo explained in another context,[76] rested on the Christian faith in the Creator, whose nature demanded that His handiwork should reflect His own perfect simplicity. Curiously, what Galileo found so important to emphasize, many a historian of science preferred to belittle.[77] This should appear all the more puzzling as Copernicus himself voiced with eloquence the importance of that faith for his own work. The simplest ordering of the planets, according to Copernicus, was "the sure scheme for the movements of the machinery of the world." This had to be so, as the machinery in question "has been built for us by the Best and Most Orderly Workman of all."[78] Although not a child of the modern age of meticulous self-analysis, Copernicus revealed enough of the genesis of his confidence in the view that the startlingly new and physically so paradoxical scheme was reality itself: "For who, after applying himself to things which he sees established in the best order and directed by divine ruling, would not through diligent contemplation of them and through a certain habituation be awakened to that which is best and would not wonder at the Artificer of all things, in Whom is all happiness and every good? For the divine Psalmist surely did not say gratuitously that he took pleasure in the workings of God and rejoiced in the works of His hands, unless by means of these things as by some sort of vehicle we are transported to the contemplation of the highest Good."[79]

The triumph of faith is rarely rapid, and the first fortunes of Copernicanism illustrate this all too well. The very same Bible which inspired the faith in the Creator also contained passages that seemed to be incompatible with the notion of a moving earth. But this was not the most difficult of hurdles to be faced by the theory of Copernicus, who brushed aside the question of the Bible. He very likely did so because of his awareness of a long-standing theological tradition about the respective merits of scientific demonstrations and scriptural phrases that could very well be figurative. Moreover, as the Spanish theologian, Diego de Zuñiga (Didacus de Stunica), pointed out in 1576, there were Scripture passages which could effectively be used against the idea of an absolutely immobile earth.[80] The zeal by which the Reformation espoused the Bible certainly weakened the willingness to make the most of such an approach. A Protestant in a Protestant land, Tycho Brahe was one of those for whom certain passages of the Bible about the immobility of the earth loomed forbiddingly large.[81]

The crucial obstacle in the way of Copernicanism was, in addition to an increasingly stiff theological attitude, the fact that the Newtonian science of motion and gravitation was still more than a century away. Even Galileo, armed for some time with the law of free fall, had to rest content with the position that no experiment performed on the earth could prove or disprove its daily and annual motions.[82] Pre-Galilean physics was simply no match to the task although the study of motion made some progress during the Renaissance. The list of those who formulated the most significant speculations in this respect is short but very instructive. The line starts once more with Nicolas of Cusa, and especially with his *De staticis experimentis*.[83] There he called the balance the chief of all measuring instruments and illustrated the point with many examples. Among them the most striking concerned the analysis of the fall of bodies. Cusanus suggested that by weighing the water collected in a clepsydra during the fall of a body from different heights, the time of fall could be inferred.[84] The emphasis on the balance harked back to Archimedes, although in Cusanus' case it was not only the heritage of Archimedes, but even more so, faith in the Creator who arranged everything in weight, measure and number, that served as a spark for the exploration of the intricacies of nature by geometry.[85] The same should also be noted about Leonardo da Vinci, whose notebooks reveal not only astonishing flights of fancy, blueprints of wondrous machines, anticipations of some laws of Newtonian physics, but also a devout reverence for the Creator, an evangelical piety, and a revulsion against magic, astrology, and necromancy.[86]

Leonardo was strongly influenced by Cusanus. Leonardo, in turn, made no major impact on Tartaglia and Benedetti,[87] the sole notable contributors to the science of motion during the sixteenth century. Of the two, Benedetti's was by far the more soaring mind. He proposed nothing less than the establishment of a mathematical philosophy of nature. He saw the immediate foundation of this plan in the statics of Archimedes. Accordingly, Benedetti blamed Aristotle's neglect of mathematics for his denial of the plurality of worlds, of the possibility of an infinite space, and of the mutability of the heavens. Benedetti had a clear vision of the theological reasons underlying the main errors of Aristotle's physics and cosmology as he praised the Stagirite for recognizing that the world could not be a product of chance.[88] It should also be noted that Benedetti's effort to base the analysis of motion exclusively on mathematics (geometry) owed nothing to Archimedes. In fact, it was by heeding in full Archimedes' precepts that Stevin, the greatest engineer of the sixteenth century, dismissed the application of geometry to the dynamics of motion. Happily for science, Galileo's admiration for Archimedes eschewed such consistency. Benedetti, whom the young Galileo followed at times almost verbatim in his *De motu*, failed, as is well known, to anticipate the insights of the mature Galileo. This was in a great part due to the fact that Benedetti could not purge his thinking completely from the remnants of the animistic, or organismic view of the world. He believed with Aristotle that motion was a "change of state," a notion which Aristotle tied most explicitly to the alleged purposefulness of every process in nature. In addition to that, there was Benedetti's flirtation with astrology[89] which almost invariably imposed the organismic viewpoint.

A generation or two after Benedetti, Galileo still was unique in his account of the world free of animism, while casting at the same time horoscopes for his princely patrons. Such a feat was well-nigh impossible during the century of Benedetti that saw a powerful comeback of a thoroughly animistic conception of the world. The principal propellent of that comeback was the translation and printing of works ascribed to Hermes Trismegistus. Characteristically, the chief promoter of this venture was none other than Ficino, who was also very interested in cabbala and magic.[90] The extraordinary flourishing of alchemy during the sixteenth century bespeaks a similar ascendency of animistic obscurantism. One would look in vain, for instance, in the writings of Paracelsus even for a modicum of the clarity that transpires from the analogy of machinery. For Paracelsus, for alchemy, for cabbala, and for magic the world was not a machine but a huge living entity. To be sure, the suppleness of the living is a wonderful feature to study, but undue emphasis on it certainly pre-empts the possibility of a quantitative analysis which the analogy of a machine so readily invites. Animism also encourages an explanation which starts with empathy, intuition, and identification, and ends in quasi-mystical conjectures and a willful cultivation of inconsistencies.

The barrenness of Paracelsus' writings in scientific values is a direct outcome of this logic. As for Tycho Brahe, his half-way house solution of the planetary problem is also a reflection on his "cosmic physics" that blandly ignored the problems of motion. In that "physics" the sun, the earth, the planets, and the four elements stood in the same animistic correlation that was believed to exist among the various organs of the human body.[91] That Tycho did not get wholly trapped in the morass of that "physics" was due to his commitment to observational and mathematical astronomy, in which he saw a most exquisite service to the Creator.[92] When such a faith was lacking, the safeguards of reason also vanished rapidly. The classic case is that of Giordano Bruno. His vision of an infinite universe composed of an infinitely large number of "worlds" or planetary systems, is mentioned on the pages of a legion of modern works, whose authors seem to demonstrate only their second-hand knowledge of Bruno, or their own bent on propagating a dubious brand of rationalism.

The real Bruno of the record and the real nature of his infinite universe is a far cry from his portrayals presented in most histories of science and of Western culture. Instead of being a champion of reason and a voice of reputable progress, Bruno was the hapless captive of Hermetic and cabbalistic tradition[93] and of a mind revelling in the denial of rational, clearly identifiable patterns. His scorn for practically every tenet of Christian faith was motivated by the same consideration that made him renounce the great aim of Copernicus, a mathematically and observationally cogent representation of the universe. The consideration was an idea of pantheistic infinity in which the infinite number of entities were forever subject to a flux of unfathomable transmutations. With this consideration, faith in creation, Incarnation, sacraments, redemption, and resurrection were as incompatible as was exact science. Bruno denounced mathematics whenever the occasion arose,[94] and in his very first cosmological work, *La Cena de le Ceneri*, made no secret of the fact that he wanted to see nature through his own eyes and not through the eyes of Copernicus. According to him, Copernicus failed to go sufficiently beyond

Ptolemy, Hipparchus, and Eudoxus, because he was more enamored of the study of mathematics than of the study of nature.[95]

Bruno's advocacy of Copernicus was a largely utilitarian attitude. Copernicus was for him a welcome ally only so long as heliocentrism helped discredit the closed world of Aristotle. Once Aristotle was out of the way, so were Copernicus and science. The world of Bruno rested not on order but on the fate of necessity. It is on this theme that culminate the three best known cosmological works of Bruno, all published in 1584, *La Cena de le Ceneri*, *De la causa, principio et uno*,[96] and *De l'infinito universo et mondi*.[97] The same is also true of his major works of ethical philosophy, the *Spaccio de la bestia trionfante*[98] and the *De gl'heroici furori*.[99] That his thinking remained in the clutches of the same theme is well attested in his last and longest work, *De immenso et innumerabilibus; seu de universo et mundis libri octo*,[100] published in 1591, a year before he started a seven-year long captivity which ended in his tragic burning at the hands of the Inquisition.

The only praise for Aristotle in the *Cena de la Ceneri* came as a reference to what Bruno called the "halting" and "confused" recognition by Aristotle of great geological cycles in his *Meteorology*.[101] Bruno, of course, gave himself credit for perceiving in all clarity the universal validity of the perennial, cyclic transmutation of everything into everything. In an animated passage preceding the one on Aristotle, Teofilo (Bruno) gives the following reply to the question about the deepest reason of the earth's motion: "The reason of such a motion is the renovation and rebirth of that body which cannot be perennial by its very disposition; just as the things that cannot be perennial as individuals . . . perpetuate themselves as a species; the substances, that cannot perpetuate themselves under the very same appearance, go on changing their faces completely, because . . . any thing must be subject in all its parts to all forms, so that in all its parts . . . it might become all, be all, if not at the same time and instant of eternity, at least at different times in various instants of eternity, successively and vicissitudinously . . . Therefore, since death and dissolution are not proper to that entire mass of which this globe, this star [the earth] consists, and since annihilation is not proper either to its entire nature, from time to time, in certain order, it gets renovated by transforming, changing, altering all its parts . . . And we ourselves, and our things, go and come, pass and return; there is no thing [part] of ours which would not become alien [to us], and there is no alien thing which would not become ours . . . Thus, all things in their kind have all the vicissitudes of dominion and servitude, of happiness and unhappiness, of that state which is called life, and of that which is called death, of light and darkness, of good and bad. And there is no thing to which it would be naturally convenient to be eternal, except to the substance which is matter; to which it is no less convenient that it should be in continuous transmutation."[102] This fundamental, infinite substance was called by Bruno the "Great Individuum," and in the same animistic vein he spoke of planets and stars as huge animals that wander by instinct through infinite spaces.[103]

A passionate thinker like Bruno could hardly observe a cool detachment as he explored the deepest consequences of this principle of perennial, universal permutation. Long before him the Sophists argued on the same basis that everything was, therefore, a purely accidental configuration. Whether it was true or not that the

Sophists, as Bruno claimed, greatly feared the prospect of inevitable dissolution should not detain us. A crucial truth was, however, revealed in the passionate manner in which Bruno sought to vindicate immortality which he desired with an elemental force. Against the madness of the Sophists, he wrote in his *De la causa, principio et uno*, "nature cries out in a loud voice, assuring us that neither bodies nor souls should fear death, because matter as much as form are the most constant principles."[104] The sonnet that followed, mirrored what Bruno really was, an agitated, often frenzied poet:

> O race, atremble with fear, with the icy terror of dying,
> Wherefore dread ye the Styx, vain names, and the forms of shadows
> Idle subjects for parts, and dangers of the word that exist not?
> Whether the funeral pile shall consume our bodies with fire,
> Or old age wasting away, think not that we can suffer evil.
> Souls are not subject to death, but former dwellings abandoned
> Rise to shelter eternal, where they may inhabit forever.
> Thus do all things suffer change, but nothing ever shall perish.[105]

In his *De l'infinito universo et mondi* the problem of individual and cosmic death was dissolved in the same manner into the mirage of the "Great Individuum," or infinite nature, the eternal substance of all: "When we consider more profoundly the being and substance of that universe in which we are immutably set, we shall discover that neither we ourselves nor any substance doth suffer death; for nothing is in fact diminished in its substance, but all things wandering through infinite space undergo change of aspect."[106] Bruno even tried to inject the notion of purpose and good into a framework of absolute necessity: "And since we are all subject to a perfect Power, we should not believe, suppose or hope otherwise, than that even as all issueth from good, so too all is good, through good, toward good; from good, by good means, toward a good end."[107] The Power in question was the inexorable, eternal flux and reflux of atoms[108] and Bruno made no secret of his indebtedness on this point to Lucretius, Epicurus, and Democritus. He failed, on the other hand, to live up to the consistency of Democritus who rejected all qualities and goals as mere illusions in a universe of atoms moving at random. Bruno was more consistent, as will be seen shortly, when he preferred to refrain from any comment on what was a primary aspect of at least the sublunary part of Aristotle's finite world, namely, its cyclic destructions and rebirths: "I leave aside that if there should come to pass the destruction of a world followed by the renewal thereof, then the production therein of animals alike perfect and imperfect would occur without an original act of generation, by the mere force and innate vigour of Nature."[109]

What Bruno could not leave aside was the existential agony deriving from the absence of a real target and resting point in his universe of perpetual flux. He acted the heroic in his *De gl'heroici furori*, but the genuine voice of his humanity was not to be stifled. It was one thing to contemplate the circular motion of the world-soul, it was another to face the eternal "oscillations" of one's soul between the extremes of an almost divine condition and that of a most repulsive beast.[110] For as Bruno explicitly stated in his *Spaccio de la bestia trionfante*, the course of transmigration was ultimately ruled by a "Fatal Justice" and not by merit and de-merit. The "Fate of Mutation," to recall another expression of Bruno, left no logical room for personal responsibilities. Nor was it responsible reasoning when Bruno tried to prove his point from the analysis of human faces: "Since we see in the faces of many in the human species, expressions, voices, gestures, affects, and inclinations, some equine, others porcine, asinine, aquiline, and bovine, so we are to believe that in them there is a vital principle through which, by virtue of the proximate past or proximate future mutations of bodies, they have been or are about to be pigs, horses, asses, eagles, or whatever else they indicate, unless by habit of continence, of study, of contemplation, and of other virtues or vices they change and dispose themselves otherwise."[111]

Beneath this weird reasoning lurked the spectre of a treadmill. The *De gl'heroici furori* offers as a conclusion the comparison of the fate of a snake wriggling helplessly in snow with that of a child engulfed in flames. Each would prefer the other's condition as if indeed a shift of fates would solve anything. "The same fate vexes, and the same [fate] torments both the one and the other – that is immeasurably without mercy and unto death."[112] But in fact not even death was forthcoming. In Bruno's universe the fundamental truth is expressed in Tansillo's words: "All times to me are full of woe; All things time takes from me, And gives me naught, not even death."[113] No wonder that the last sonnet ends with the line: "Clear is our evil fate – all hopes resign."[114]

The *De gl'heroici furori* comes to a close by a remark which is pathetic, not heroic, dejected, not enthusiastic: "let us go and . . . seek to untie this knot – if possible."[115] Bruno himself looked for solution not in science, not in philosophy, but in numero-logy and in a form of ancient Egyptian religion which really existed only in his own mind. Both represented the very denial of cogent reasoning. The numerology was the very antithesis of what numbers stand for: distinctness and clarity. Bruno's poem, *De monade*, is entirely made up of speculations of the type: "The number five, born from the first even and from the first uneven, will in turn be good and bad, just as the five senses, the five fingers, are the vehicles of good and bad."[116] The Egyptian religion, as imagined by Bruno, was also worlds removed from that rationality the lack of which Bruno bitterly deplored in Christianity as well as in Judaism and Mohammedanism. The Hermetic tradition also contained the idea of great cyclic conflagrations and rejuvenations of the world. Bruno himself reported in detail the words ascribed to his oracle, Trismegistus, who conjured up as a cure for the present corruption the coming of a traumatic transformation for all: "But do not doubt, Asclepius, for after these things have occurred, the lord and father God, governor of the world, the omnipotent provider, by a deluge of water or of fire, of diseases or of pestilences or of other ministers of his com-passionate justice, will doubtlessly then put an end to such a blot, recalling the world to its ancient countenance."[117]

In the long run, however, Bruno realized that the doctrine of the Great Year even in its Egyptian, or Hermetic form, contradicted the very core of his world view. For what Bruno ultimately wanted was to become a carefree ripple on infinite waves agitated forever without any aim, any direction, any pattern. The Great Year, the eternal recurrence of all, was a pattern, and it had therefore to be discredited. Bruno's muddled arguments fill two chapters in his *De immenso et innumerabilibus*.[118] It is doubtful that they convinced anyone. They were not even enlightening on the history of the question. Seventeenth-century authors had to turn to a dissertation written by Lipsius in the 1590's for information on ancient views on the Great Year.[119] But what throws the sharpest light on Bruno's obscurantist references to non-concentric spheres and irregular lines is the totally different manner in which Kepler dismissed in 1596 the doctrine of the Great Year in his *Mysterium cosmographicum*.[120] For Kepler it was enough to recall in a few lines Oresme's classic argument about the incommensurability of planetary periods.

The inner logic, which in Bruno's case deprived the concept of the infinite universe of sound scientific value, can also be seen at work in the cosmological speculations of Jean Bodin. He earned the admiration of his contemporaries by his political theory in which the state figured as a sort of ultimate entity. The deterministic feature of Bodin's political thought was clearly in evidence not only in his emphasis on obedience to all laws of the state, but also in the extent to which he tried to make the state the function of physical environment. Religion too received a similar "rationalist" streamlining in Bodin's hands. In his *Colloquium heptaplomeres*,[121] which for many years had to circulate as a manuscript, he declared the distinctive features of the major religions to be mere customs. At the same time he emphasized the general validity of the Decalogue though not as a set of laws revealed by God. Actually, there was no room for a personal God or Creator in Bodin's universe. The obsequious references to the Creator in his cosmological work, *Universae naturae theatrum*,[122] were a mere lip service as can very well be seen in his definition of the act of creation.[123] In an age of censorship he had, of course, to be cautious, and he knew how to advocate rank heresies in a roundabout way. A principal one of these was his defence of the eternity of the world.[124] He even gave his vote to the Pythagorean doctrine of metempsychosis, after laying down the tactical proviso that it could not be accepted in its original crude form.[125] He kept its less crude form his own secret.

It should not, therefore, be surprising that Bodin was a firm supporter of astrology and also advocated the doctrine of the Great Year.[126] True, he poured scorn on the Hindus who, according to him, believed in cosmic cycles of 700,000 years. Bodin also dismissed the Chaldeans' contention about 470,000 years. He had, however, only praise for the Egyptians who reckoned, so Bodin claimed, the history of the world in units of 48,000 years. The chief merit of this figure was its closeness to 49,000 which Bodin equated with the precession of the equinoxes and with the period of the cyclic rebirths of the universe. One wonders what he meant as he wrote: "In that creation and re-creation of worlds God's eternal power and goodness is evidenced."[127] He was patently more consistent when he emphasized the ironclad determinism deriving from the "inflexibility of pattern" of the motion of the stellar sphere.

That Bodin failed to join the Copernicans was forgivable, but this can hardly be said about his haughty dismissal of Copernicus. The heliocentric theory can, he wrote, "easily be refuted, because of its lack of proofs, although up to now nobody set himself to the task seriously."[128] Such was a strange claim that gave away Bodin as a "man of science," as much as did his contention that Eudoxus and

Archimedes were supporters of the heliocentric theory. Whatever Bodin's respect for facts, whether for those of nature or for those of scientific history, he certainly manhandled both realms. Caught in the trap of inexorable, eternal returns, he chose to base on a priori laws the architecture of the cosmos. For this he found the corner-stone in the earth-sun distance which he stated to be 576 earth diameters. The same figure, the square of 24, corresponded, according to him, also to the sum of right angles that could be constructed in the five perfect bodies.[129] Apart from the dubious merits of such reasoning, Bodin's intentions hardly justified bringing all this to a close by quoting the Book of Wisdom about the Creator who disposed everything according to weight, measure, and number.

It sheds no favourable light on a not too distant phase of the historiography of science that Bodin's dabbling in cosmology could be described as revealing "amazing ideas of geology, physics, and astronomy."[130] As in the case of Bruno, here too only a quick look is needed at Kepler's *Mysterium cosmographicum*. There Kepler made an effort which looked rather similar to Bodin's "layout" of the universe. In essence Kepler's approach, as will be seen later, had ingredients that were wholly lacking with Bodin.

That difference also gives in a nutshell the gist of the last phase of that travail through which the scientific enterprise had to go before coming to a full birth during the seventeenth century. The successful outcome of that travail was ulti-mately determined by the fact that faith in the Creator and recognition of the fundamental features of scientific knowledge found natural allies in one another. In the opposite camp there reigned the antiscientific obscurantism of Ficino, Diacceto, Paracelsus, Bruno, and of many smaller figures, such as Palingenius, Telesius, and Campanella. The allegedly crucial contribution to science of some of them constitutes one of the myths still beclouding the historiography of science. Pathetic is at best any effort which tries to prove, for instance, Bruno's influence on the outstanding figures of seventeenth-century science. It was in one such effort that the astounding claim was made that Bruno "received from Lucretius the vision of the dignity of the human soul."[131] Needless to say, all such efforts also overlook both Kepler's revulsion for Bruno and the fact that Galileo never advocated an infinite universe, let alone a universe as void of all-embracing patterns as Bruno postulated.

The strict infinity of the universe was not accepted by Descartes, Guericke, Boyle, Hooke, and Huygens, to name only a few of the leading seventeenth-century scientists. The first major scientific discourse on the infinity of the universe came only with the *Principia* of Newton, and especially with the Scholium added to it in 1713. It was a step in which encouragement from Bruno's writings or memory had no part. Those whom Bruno really inspired were the chief represen-tatives of German Idealism and Naturphilosophie. Their pantheism, antiscientific lucubrations, and infatuation with the idea of eternal recurrence will form another chapter of this story of the inner logic of some fundamental standpoints.

The names of Nicolas of Cusa, Leonardo da Vinci, Copernicus, Tycho, Benedetti, and Kepler, to which one may add many others, represent not only the story of what was best in science during the Renaissance. Those names also attest the inspiration and safeguard which faith in the Creator provided for scientific endeavour during an age that witnessed a hardly concealed desire on the part of many to bring about a "re-naissance" of classical paganism. Man's longing for science, buttressed by the Christian faith in the Creator, prevented that desire from becoming more than an interlude. The desire itself was in part caused by that shift which consisted in a heavy preference for feelings and in a weariness of the rationality of scholasticism. In many cases the process ended in rank subjectivism. The proclivities of Kepler, the last of the giants of Renaissance science, illustrate the issue on hand in a graphic manner. He might have developed into the crudest of "sleepwalkers" had it not been for that faith in a rational Creator.[132] It saved him not only from a disastrous measure of involvement with astrology to which he was also attracted by his psychic traits. The same faith also inspired him to superhuman efforts to fit theory to the data of observations. The data were the given facts of the celestial world, existing by a sovereign act and planning of the Creator. The theory had two aspects, one which allowed no tampering with, and another that was the task of the scientist to establish. The former aspect derived from the belief that the work of a rational Creator was structured along the lines of geometry. Kepler's lifelong convictions on this point were well expressed in his greatest work, the *Harmonice mundi*, published in 1619. There he apotheosized geometry as something co-eternal with the divine Mind. On the ground that no real distinction could exist in God Kepler spoke of geometry as "God Himself." Geometry, he wrote, "supplied God with a model for the creation of the world and was implanted into human nature along with God's image and not through man's visual perception and experience."[133]

Such expressions were only a hair's breadth away from pantheism, but Kepler, a deeply convinced Christian, was not to be swayed by the glitter of its apriorism. He was in fact most explicit about the intellectual safeguards which Christian faith provided for scientific speculation. It was in connection with the apriorism of the world view of antiquity that Kepler wrote: "Christian religion has put up some fences around false speculation which is on the wrong track, in order that error may not rush headlong but may become in other respects harmless in itself."[134] Equally revealing was what Kepler said in the same context about the "strength of mind" based on "the highest confidence in the visible works of God" that he needed to go on with his work.[135] When Kepler interspersed his reflections on scientific method with biblical quotations on the wisdom, power, and glory of the Creator evidenced by His Creation, he voiced convictions that were most genuinely Christian. His enthusiasm for geometry and his superlatives for Pythagoras did not constitute a compartment severed from his deeply experienced Christianity. His was a Christian God and not, as it was claimed, a "Lutheran God" of the Bible shadowed by a "Pythagorean God" of nature.[136] Kepler's notion of the Creator reflected the best of Christian theological tradition, which enabled him to state that the Scriptures "never intended to inform men about the things of nature" and that the first chapter of Genesis taught "only the supernatural origin of all things."[137] There is, therefore, no need to fall back on some psychological device, namely, the alleged unity in Kepler's vision of the abstract and the concrete, to account for the synthesis he achieved. It was in the Creator of human destiny *and* of nature *and* of geometry that he found the measure of rationality his work presupposed. Faith in the same Creator supplied him with encouragement and perseverance which his work demanded in perhaps an even greater measure.

This last remark was not made in view of the hardships that were his lot. Faith in the Creator was also a recognition of the contingency of the universe, but it had to be a hard lesson to learn that this contingency might play havoc with most cherished theories and hallowed experiences. It turned out that Kepler's youthful vision of 1595 about the divine arrangement of the planetary system in terms of the five regular solids was but a pleasant illusion.[138] Reality did not support his belief that the orbits of the planets had to be circles. Thirteen years later and after an exhausting effort on recalculations, he surrendered to the conclusion that the orbits were elliptical. But ellipses too were geometry which provided the finest mental communing with God, and in that consideration his mind found rest. His finding and his two other laws became the cornerstone around which the celestial dynamics of Newton was later constructed.

The impact which the doctrine of creation had on Kepler's mind is most palpably shown by his efforts to prove that the world was created in time. For Kepler the time elapsed from the moment of creation equalled the few thousand years of biblical chronology. In the last chapter of the *Mysterium cosmographicum* he meticulously set the date of creation at 3977 B.C., April 27, Sunday, at 11 a.m., Prussian local time.[139] Twenty-five years later, in his *Epitome astronomiae copernicanae*, he claimed to have established on purely astronomical grounds "that it is impossible that the world should not have been created at a fixed beginning in time."[140] The specifics were rather transparent and unconvincing. It was clear that Kepler tried to find a conspicuous constellation of planets for the epoch suggested by biblical chronology. The end of the world too was supposed to be signified by a similar constellation. Such were no doubt harmless exercises in computations that were to scandalize a more "enlightened" posterity engrossed with the vision of an immense geological past. Those armed with the wisdom of facile hindsight were forgetful of the most enlightening lesson in scientific history. To be sure, the scientific edifice that began to take shape in virtue of the belief that the world was the handiwork of the Father and Creator of all was not perfection itself. It remained to the seventeenth century to provide the crossbeams of synthesis for that edifice. But the edifice already in existence represented a structure that could be meaningfully and consistently improved and in a sense to be brought to completion. Therein lay the factor that made a world of difference and ultimately a different world. In that world, science became the implementation of the age-old drive sparked by the hallowed injunction: "Fill the earth and conquer it."

NOTES

1. Although books on the Renaissance are legion, appreciation of Renaissance science has until rather recently been considerably weakened by the claim of the Encyclopedists that everything was dark before Galileo. The section "Renaissance" in R. Taton (ed.) *History of Science*, vol. II, *The Beginnings of Modern Science from 1450 to 1800*, translated by A. J.

Pomerans (New York: Basic Books, 1964) is probably the best general survey, partly because it was co-authored by A. Koyré. Another recent work that should be mentioned is M. Boas, *The Scientific Renaissance 1450–1630* (New York: Harper & Brothers, 1962). It should be significant that for historians of science the Renaissance starts around 1450 whereas the Rinascimento got under way much earlier and was a thing of the past by 1600. This time difference between scientific and literary Renaissance is one of the evidences about a profound difference between the two. This difference received a fine analysis in G. C. Sellery's *The Renaissance: Its Nature and Origins* (Madison: University of Wisconsin Press, 1950). On the other hand, none of the studies in *Renaissance Essays*, edited by P. O. Kristeller and P. P. Wiener (New York: Harper & Row, 1968), tried to trace the scientific bareness of the literary (and philosophical) Renaissance to its infatuation with Neoplatonism and to its flirtation with the idea of the Great Year.

2. *Die Wiederbelebung des klassischen Althertums, oder das erste Jahrhundert des Humanismus* (Berlin: G. Reimer, 1859).

3. *Die Kultur der Renaissance in Italien* (Leipzig: Seemann, 1860) and many subsequent editions and English translations.

4. G. Santayana, *Genteel Tradition at Bay* (New York: C. Scribner's Sons, 1931), p. 9.

5. A. N. Whitehead, "The Problem of Reconstruction," *Atlantic Monthly*, 169 (1942): 173.

6. See *Petrarch's Secret; or, The Soul's Conflict with Passion. Three Dialogues between Himself and S. Augustine*, translated from the Latin by W. H. Draper (London: Chatto & Windus, 1911).

7. In Marsilio Ficino, *Opera omnia* (1561), reprint edition by Bottega Erasmo (Torino, 1959), vol. I, tom. I, p. 616.

8. In his epistle "Quod divina providentia statuit antiqua renovari," a reply to Ioannes Pannonius, in *Opera*, vol. I, tom. I, p. 871.

9. *Opera*, vol. I, tom. I, pp. 492–93.

10. *Ibid.*, p. 492.

11. See, for instance, *The Philosophy of Marsilio Ficino* by P. O. Kristeller, translated into English by Virginia Conant (New York: Columbia University Press, 1943), p. 46.

12. "In Plotinum," in *Opera*, vol. I, tom. II, p. 1767.

13. "Expositio circa numerum nuptialem in octavo de Republica," in *Opera*, vol. II, tom. I, pp. 1414–25; quotation is from p. 1414.

14. *Ibid.*, p. 1416.

15. *Ibid.*

16. See on this D. P. Walker, *Spiritual and Demonic Magic from Ficino to Campanella* (London: The Warburg Institute, 1958), pp. 45–53.

17. "Expositio . . .", in *Opera*, p. 1425.

18. "De immortalitate animorum," Lib. IV, cap. 2, and Lib. XVI, cap. 7; in *Opera*, vol. I, tom. I, pp. 134 and 380.

19. *On the Dignity of Man*, translated by Charles G. Wallis (Indianapolis: The Library of Liberal Arts, 1965), p. 10.

20. *Ibid.*, pp. 22–23.

21. *Ibid.*, p. 28.

22. *Ibid.*, p. 29.

23. References will be to the edition by E. Garin (Florence: Vallecchi, 1946).

24. See, for instance, p. 82.

25. The claim was made by E. Cassirer in his *The Individual and the Cosmos in Renaissance Philosophy*, translated with an Introduction by Mario Domandi (New York: Barnes & Noble Inc., 1963), p. 119.

26. *Disputationes*, p. 454 (Book IV, chap. 4).

27. On Roger Bacon, see pp. 226–28 above. Pierre d'Ailly's (c. 1350–c. 1420) defence of astrology was more qualified. In his *Tractatus contra astronomos* (1410) he exempted Christian religion from the deterministic influence which was exerted by the planetary

periods on all other religions and secular institutions. The tract is available in *Joannis Gersonii Opera omnia* (Antwerp: Sumptibus Societatis, 1706), vol. I, cols. 778–804.

28. *Disputationes*, p. 118 (Book I, chap. 4).

29. *Ibid.*, pp. 526–39 (Book V, chap. 2).

30. *Ibid.*, p. 530.

31. Born in 1466 into a noble family in Florence, Diacceto was a faithful supporter of the Medicis; early in his career as a philosopher he associated himself with Ficino and taught various parts of the philosophical curriculum at the University of Florence until his death in 1522. The best modern account of his life and work is by P. O. Kristeller, "Francesco de Diacceto and Florentine Platonism in the Sixteenth Century," in *Miscellanea Giovanni Mercati*, vol. IV, *Letteratura classica e umanistica* (Città del Vaticano: Biblioteca Apostolica Vaticana, 1946), pp. 260–304. Diacceto's collected works were published under the title, *Opera omnia*, in 1563 (Basel: per Henricum Petri et Petrum Pernam).

32. The biographer in question is Frosino Lapini whose account of Diacceto's life introduces the *Opera omnia*. For quotation see p. 6.

33. Book II, chap. 6, in *Opera omnia*, pp. 51–54.

34. *Ibid.*, p. 52.

35. *Ibid.*

36. *Ibid.*

37. *Ibid.*, pp. 341–43.

38. *Ibid.*, p. 343.

39. Contrary to Kristeller's presentation of Diacceto's thought. A subtly revealing aspect of Diacceto's paganism was his ambivalent protestation of orthodoxy as the theological faculty of the University of Paris took some of his writings under scrutiny. In his "Apologia contra Parisienses philosophos pro Platone" (1509), addressed to Germain Ganay, Bishop of Cahors (*Opera omnia*, pp. 332–37), Diacceto admitted that the doctrine of a creation in time was one of the few points where Platonism conflicted with Christianity, but he did not consider the matter serious. He upheld the principle of double truth, which he justified on the ground that the Averroists of his time resorted to the same expediency.

40. References will be to its English translation by G. Heron published under the title, *Of Learned Ignorance*, with an Introduction by D. J. B. Hawkins (London: Routledge & Kegan Paul, 1954).

41. *Ibid.*, p. 26 (Book I, chap. 11).

42. *Ibid.*, p. 101 (Book II, chap. 9).

43. *Ibid.*, p. 103.

44. *Ibid.*, p. 140 (Book III, chap. 4). One of these compared the relation between divine nature and human nature to the one between a circle and a polygon inscribed in it.

45. *Ibid.*, p. 119 (Book II, chap. 13).

46. *Ibid.*, p. 25 (Book I, chap. 11).

47. *Ibid.*, p. 137 (Book III, chap. 3).

48. *The Prince*, translated by L. Ricci, revised by E. R. P. Vincent, with an Introduction by C. Gauss (New York: The New American Library, 1952), p. 94.

49. *The Florentine History*, translated by N. H. Thomson (London: Archibald Constable, 1906).

50. *Ibid.*, vol. II, p. 1.

51. *Ibid.*

52. *Ibid.*, p. 2.

53. *Ibid.*, p. 284.

54. For details, see H. Weisinger, "Ideas of History during the Renaissance," in *Journal of the History of Ideas* 6 (1945): 415–35, where excerpts from the works of Gabriel Harvey, Louis Le Ray, and Pierre Charron illustrate the popularity of the cyclic theory of history during the Renaissance.

55. See his Essay, "Of Vicissitude of Things," in *The Works of Francis Bacon*, edited by J. Spedding, R. L. Ellis, and D. D. Heath (Boston: Taggard & Thompson, 1863), vol. XII, p. 275.

56. First published in Louvain in 1516. References will be to the modern English translation facing the original Latin in *The Complete Works of St. Thomas More*, Vol. 4, edited by Edward Surtz and J. H. Hexter (New Haven, Conn.: Yale University Press, 1965).

57. *Ibid.*, pp. 163 and 223.

58. *Ibid.*, p. 207.

59. *Ibid.*, p. 221.

60. *Ibid.*, p. 225.

61. *Ibid.*, p. 183.

62. *Ibid.*, p. 127.

63. *Ibid.*, p. 183. In the first English translation of the *Utopia* from 1551 "re medica" is rendered with "Physicke," a word then mainly referring to medical science. But as J. H. Lupton, the great nineteenth-century More scholar had already noted, in the context More was speaking of science in general, much in the sense of the modern meaning of the word "physics." See his *The Utopia of Sir Thomas More*, with introduction and notes (Oxford: Clarendon Press, 1895), p. 217.

64. *Ibid.*, p. 161.

65. *Ibid.*, pp. 159–61.

66. *Ibid.*, p. 161.

67. The significance of More's utterances on scientific endeavour is also neglected in most monographs on the *Utopia*; see for instance, R. S. Johnson, *More's Utopia: Ideal and Illusion* (New Haven, Conn.: Yale University Press, 1969), R. Ames, *Citizen Thomas More and his Utopia* (Princeton, N.J.: Princeton University Press, 1949), or the archetype of More's Marxist interpretations, Karl Kautsky's *Thomas More and his Utopia* (New York: Russel & Russell, 1959), originally published in German in 1888. Again, the question of More's impact on science has not been taken up in any of the essays published in the first 22 issues (1963–69) of *Moreana* (Organe de l'Association Amici Thomae Mori, Angers, France). The situation is only slightly better in this respect in P. Hogrefe, *The Sir Thomas More Circle: A Program of Ideas and their Impact on Secular Drama* (Urbana: University of Illinois Press, 1959) and in E. Surtz, *More's Utopia* (Cambridge, Mass.: Harvard University Press, 1957). A short but emphatic presentation of the *Utopia*'s significance in scientific history is given in R. P. Adams, "The Social Responsibilities of Science in *Utopia, New Atlantis* and After," *Journal of the History of Ideas* 10 (1949): 374–98.

68. A convenient illustration of this is the massive documentation by R. W. Gibson, *St. Thomas More: A Preliminary Bibliography of his Works and of Moreana to the Year 1750* (New Haven, Conn.: Yale University Press, 1961).

69. F. R. Johnson, *Astronomical Thought in Renaissance England: A Study of the English Scientific Writings from 1500 to 1645* (Baltimore: The Johns Hopkins Press, 1937), p. 296.

70. The scientific ingredient in the various "Utopias" published from classical times to the mid-17th century is the topic of the monograph by N. Eurich, *Science in Utopia: A Mighty Design* (Cambridge, Mass.: Harvard University Press, 1967).

71. The absence of this Mephistophelic trait in More's scientific program was aptly noted in B. Willey, *The Seventeenth Century Background* (London: Chatto, 1934), p. 38.

72. See on this the findings of the great Polish Copernicus scholar, L. A. Birkenmajer, reported by his son A. Birkenmajer, "Comment Copernic a-t-il conçu et réalisé son œuvre?", *Organon: Revue Internationale* (Warsaw), 1 (1936): 123 and 126.

73. Concisely set forth by H. Butterfield in his *The Origins of Modern Science* (London: Bell, 1949), Chap. II, "The Conservatism of Copernicus".

74. Book I, chap. 7.

75. In the *Dialogue concerning the Two Chief World Systems – Ptolemaic and Copernican*, translated by Stillman Drake (Berkeley: University of California Press, 1962), pp. 328, 334, and 339.

76. *Ibid.*, pp. 102–104.

77. A more recent example is the article, "Late Medieval Thought, Copernicus, and the Scientific Revolution," by R. Grant, in *Journal of the History of Ideas*, 23 (1962): 197–220.

78. In the Dedication to Pope Paul III by Copernicus of his *On the Revolutions of the Heavenly*

Spheres, translated by C. G. Wallis, in *Great Books of the Western World*, Vol. 16 (Chicago: Encyclopedia Britannica, 1952), p. 508.

79. *Ibid.*, p. 510 (Introduction to Book I).

80. *Didaci a Stunica Salmanticensis Eremitae Augustiniani in Job commentaria* . . . (Toledo: per Ioannem Rodericum, 1584), pp. 205–06. The English translation of the passage was given prominently in the *Mathematical Collections and Translations in Two Parts. From the Original Copies of Galileus and Other Famous Modern Authors* by Tho: Salisbury, Esq.; (London: Printed by William Leybourn, for George Sawbridge, 1667), pp. 468–70.

81. *Astronomiae instauratae progymnasmata*, in *Tychonis Brahe Danis Opera omnia*, edited by I. L. E. Dreyer (Copenhagen: Libraria Gyldendaliana, 1916), vol. III, p. 175.

82. As is well known, Galileo contradicted himself when in the Fourth Day of his *Dialogue* he presented the tides as a combined effect of the earth's daily rotation and orbital motion around the sun.

83. In *Nicolai de Cusa Opera omnia* (Leipzig: Felix Meiner, 1937), vol. V, pp. 120–39.

84. *Ibid.*, p. 121.

85. Cusanus quotes the passage from the Book of Wisdom as he argues that study of the weights of various bodies can effectively lead "to a truer grasp of the secret of things and that the same method would also yield much knowledge." *Ibid.*, p. 120.

86. See *The Notebooks of Leonardo de Vinci*, arranged, rendered into English and introduced by E. MacCurdy (Garden City, N.Y.: Garden City Publishing Company, 1941), pp. 81–87, and E. MacCurdy, *The Mind of Leonardo da Vinci* (New York: Dodd, Mead & Co., 1928), pp. 213–29. The efforts to present Leonardo as a freethinker started with Vasari's *Lives* and characterize still today the majority of studies on Leonardo. Less offensive though still not fully objective is the position that credits Archimedes' influence for everything valuable in Leonardo, to the exclusion of medieval scientists whom Leonardo carefully studied. Examples of this are the essays and discussions commemorating the fifth centenary of Leonardo's birth, published under the title, *Leonard de Vinci et l'expérience scientifique au XVIᵉ siècle, Paris, 4–7 juillet 1952* (Paris: Centre National de la Recherche Scientifique, 1953). Koyré's well-known contention that the revival of Platonism was the crucial factor in the birth of classical physics, also implied a studied neglect of the role which Christian faith in the Creator played in that birth.

87. See, for instance, *Mechanics in Sixteenth-Century Italy: Selections from Tartaglia, Benedetti, Guido Ubaldo & Galileo*, translated and annotated by Stillman Drake and I. E. Drabkin (Madison: University of Wisconsin Press, 1969).

88. See his essay addressed to Hyeronimus Condrumerius, "Quod recte Arist. senserit coelum casu non esse productum," in *Io, Baptistae Benedicti. . . Diversorum speculationum mathematicarum et physicarum liber* (Turin: Haer. Nic. Bevilacquae, 1585), p. 412.

89. If, indeed, he was the author of *De coelo et elementis liber. Joanne Benedicto Tiernaviense auctore. Ad Franciscum Vrbinatum Ducem Sereniss.* (Ferrara: excudebat Victorius Baldinus, 1591, 48ff in 12º.

90. Only by ignoring such details is it justified to praise "the spectacular humanistic movement and . . . the personally more novel and original literary Platonism of the Florentines." The quotation is from "The Development of Scientific Method in the School of Padua," (1940), by J. H. Randall Jr., reprinted in his *The School of Padua and the Emergence of Modern Science* (Turin: Editrice Antenore, 1961). p. 19.

91. See, for instance, a short section of his inaugural lecture of his course on astronomy, delivered in September 1574, in *Opera omnia*, vol. I, pp. 155–56.

92. See the introductory part of the same lecture, *ibid.*, pp. 145–46.

93. As conclusively shown by Frances A. Yates, *Giordano Bruno and the Hermetic Tradition* (Chicago: University of Chicago Press, 1964).

94. As carefully pointed out in Paul-Henri Michel, *La cosmologie de Giordano Bruno* (Paris: Hermann, 1962), pp. 3, 32–36.

95. See the critical edition by G. Aquilecchia ([Turin]: Giulio Einaudi, 1955), pp. 90 and 92 (Dial. I).

96. References will be to the English translation by S. Greenburg in his *The Infinite in Giordano*

T

Bruno with a Translation of his Dialogue "Concerning the Cause, Principle, and One" (New York: King's Crown Press, 1950).

97. References will be to the English translation by Dorothea W. Singer in her *Giordano Bruno: His Life and Thought, with Annotated Translation of His Work "On the Infinite Universe and Worlds"* (New York: Henry Schuman, 1950).

98. References will be to the English translation by A. D. Imerti under the title, *The Expulsion of the Triumphant Beast* (New Brunswick, N.J.: Rutgers University Press, 1964).

99. References will be to the English translation by L. Williams under the title, *The Heroic Enthusiasts* (London: George Redway, 1887).

100. In *Jordani Bruni Nolani Opera latine conscripta* (Naples-Florence: Morano-LeMonnier, 1879–91), vol. I, parts I and II.

101. *Ed. cit.*, pp. 217–18 (Dial. V).

102. *Ibid.*, pp. 216–17.

103. *Ibid.*, p. 192 (Dial. IV).

104. *Transl. cit.*, p. 119 (Dial. II).

105. *Ibid.*, p. 120.

106. *Transl. cit.*, p. 244 (Introductory Epistle).

107. *Ibid.*

108. *Ibid.*, p. 285 (Dial. II).

109. *Ibid.*, p. 376 (Dial. V).

110. *Transl. cit.*, pp. 117–18 (Dial. IV).

111. *Transl. cit.*, p. 78 (Explanatory Epistle).

112. *Transl. cit.* p. 169 (Dial. V).

113. *Ibid.*, p. 168.

114. *Ibid.*, p. 170.

115. *Ibid.*

116. *De monade*, chap. 6., in *Opera*, vol. I, part II, p. 405. Bruno, in putting forth his interpretation of the number 5, and of the figure of the pentagon, refers, revealingly enough, to ideas entertained by Ficino. It was that "more secret type of mathematics and physics," claimed in the full title of the work, which had nothing to do with real science.

117. *The Expulsion of the Triumphant Beast*, p. 242 (Dial. III).

118. Lib. III, cap. vi and vii, in *Opera*, vol. III, pp. 361–72.

119. Dissertations xx–xxiii in Book II of his *Physiologiae stoicorum libri tres* in *Iusti Lipsi Opera omnia* (Vesaliae: Typis Andreae ab Hoogenhuysen, 1675), vol. IV, pp. 950–64.

120. See chap. xxiii, "De initio et fine mundi astronomico et anno platonico," in *Gesammelte Werke* (München: C. H. Beck, 1938), vol. I, pp. 78–80.

121. See the modern edition by R. Chauviré of the French translation, *Colloque de Jean Bodin des secrets cachez des choses sublimes entre sept sçavans qui sont de différens sentimens* (Paris: L. Tenin, 1914).

122. Frankfurt: apud heredes A. Wecheli, C. Marnium, 1597. The subtitle states that in the work "the efficient and final causes of all things are considered and discussed through the uninterrupted series of five books."

123. *Ibid.*, p. 52.

124. *Ibid.*, pp. 36–52.

125. *Ibid.*, pp. 536–38.

126. *Ibid.*, pp. 557–60.

127. *Ibid.*, p. 558.

128. *Ibid.*, p. 580.

129. *Ibid.*, p. 596.

130. D. Stimson, *The Gradual Acceptance of the Copernican Theory of the Universe* (Hanover, N.H., 1917), p. 45.

131. D. W. Singer, *Giordano Bruno*, p. 51.

132. A point which is overlooked with several other important aspects of Kepler's science in A. Koestler, *The Sleepwalkers: A History of Man's Changing Vision of the Universe* (New York: Macmillan, 1959).

133. Book IV, chap. 1, in *Werke*, vol. VI, p. 223.
134. Introduction to Book IV of *Epitome astronomiae copernicanae*, in *Werke*, vol. VII, p. 254.
135. *Ibid.*, p. 252.
136. G. Holton, "Johannes Kepler's Universe: Its Physics and Metaphysics," *American Journal of Physics*, 24 (1956): 51.
137. In his letter of March 28, 1605 to Herwart von Hohenburg, in *Werke* vol. XV, p. 182.
138. He first spoke of it in his letter of Aug. 2, 1595 to Maestlin (*Werke*, vol. XIII, p. 28) in which he also disclosed his deep commotion on being seized with the idea on July 20, after his mind had been in the grips of the problem "since the day of Pentecost."
139. *Werke*, vol. I, p. 78.
140. *Werke*, vol. VII, p. 254.

CHAPTER TWELVE

The Creator's Handiwork

For most phases of scientific history the selection of a starting point is a perplexing task. But Galileo is certainly an inevitable choice for the honour of being the first full-blooded representative of modern scientific mentality. This should be especially clear when Galileo's scientific portrait is drawn with Kepler's figure looming in the background. While the latter gave only tantalizing inklings of modernity, the former kept only occasional ties with pre-modern mentality. Tenuous were, indeed, the ties that connected Galileo with his greatest scientific contemporary, Kepler, who pursued with equal zeal astronomy and astrology. Professional jealousy no doubt played a part in Galileo's less than appreciative attitude toward Kepler's achievements. This made Galileo a loser in a truly ironical sense. His crusading advocacy of the Copernican system could have been greatly strengthened had he kept a more open mind for Kepler's contributions to science which contained such priceless gems as the three laws of planetary motion. Still, in Galileo's professed distaste for the heady mysticism and animism exuding from not a few pages of Kepler's writings there was more than the specious excuse of ill-concealed envy. Galileo might have had a blind spot for mysticism in general, but he unerringly sensed that the world view of science must be free of animistic and purposive features.

About two thousand years earlier the pre-Socratic *physikoi* were groping with somewhat similar insights. The sweeping manner in which they all banished purposiveness as a reliable form of experience indicated that they lacked some intangible though indispensable safeguards to maintain balance in reasoning about man and the universe. The coldly mechanistic interpretation of existence, cosmic and human, as conjured up by the Ionians and the atomists, had an equally unbalanced reaction in the philosophical revolution inaugurated by Socrates. As a result, the spirit of purposiveness came to pervade everything, but as it were, with a tropical warmth which suffocated the climate of a cold, purely quantitative investigation of the external world. The reversal of this trend was a painfully slow process that had already been at work for about three centuries before Galileo. The whole drama of Galileo's career is a major proof of how much agony was implied in his great step, the resolute parting with categories smacking of purpose, or of good and evil, in the scientific study of the physical world.

The structuring of Galileo's great testament of scientific philosophy, the *Dialogue concerning the Two Chief World Systems*,[1] shows his full awareness of the need of a new state of mind if science was to be cultivated in an effective manner. It was hardly to Galileo's credit that he failed to give his forerunners their due. His slighting of his great scientific contemporaries also matched his silence on a several-century-long scientific tradition to which his debt was, as his early writings show, simply enormous. But there was undoubtedly a fine measure of originality in the incisiveness as he spelled out some main features of that new state of mind. The foremost of them consisted in transcending the anthropomorphic categories of purpose in the light of the Creator's infinite wisdom and often inscrutable ways. In the *Dialogue* the discussion of the earth's diurnal and annual motion comes only after a thorough analysis of what makes the universe truly perfect, or rather purposive, but on a much higher than purely anthropocentric level. The conceptual edifice to be demolished was Aristotle's universe steeped in pantheistic and organismic purposiveness. In its place had to come a notion of the universe permitting a science of nature whose conclusions "are true and necessary and have nothing to do with human will."[2] This phrase of Salviati, who represents Galileo himself in the *Dialogue*, is from that phase of the First Day's discussion where the argument about the perfection of the universe comes to a head. To discredit the Aristotelian notion of that perfection resting on the dichotomy between the terrestrial and the heavenly regions, Salviati calls for nothing less than for a "reform of the human mind."[3] His remark is aimed at Simplicio, whose ultimate defence of the traditional world view consists in the teleological and implicitly theological claim that all celestial bodies "are ordained to have no other use than that of service to the earth."[4] The truth of such a claim implied a drastic curtailment for a scientific investigation of the universe: with respect to the celestial bodies nothing else was to be ascertained except their motion and light.[5]

The *Dialogue*'s reply to such a claim is both sarcastic and illuminating. The sarcasm serves to bring out the barrenness of the Aristotelian notion of the perfection of the heavenly regions: "Take away," argues Salviati, "this purpose of serving the earth, and the innumerable host of celestial bodies is left useless and superfluous, since they have not and cannot have any reciprocal activities among themselves, all of them being inalterable, immutable, and invariant . . . I do not see how the influence of the moon or sun in causing generations on the earth would differ from placing a marble statue beside a woman and expecting children from such a union."[6] The illuminating part of the reply derives from a new appraisal of the notion of change. Whereas in the theological outlook of the Socratic, Platonic, and Aristotelian presuppositions change was frowned upon, within the light of the Christian doctrine of Creation change and diversity were as many evidences of the Creator's infinite richness. Thus, to Simplicio, who is more Aristotelian than Christian, the idea of a moon populated by living beings seems not only mythical but impious.[7] But to Salviati the contrary conclusion is the one in keeping with the attributes of the "Maker and Director" of the universe. He conjures up the image of a moon, and by inference of a universe, populated by a great variety of beings, "acting and moving in it, perhaps in a very different way from ours, seeing and admiring the grandeur and beauty of the universe and of its Maker and Director and continually singing encomiums in His praise. I mean, in a word, doing what is so frequently decreed in the Holy Scriptures; namely, a perpetual occupation of all creatures in praising God."[8]

An equally momentous consequence of a world picture reflecting the infinite richness of the Creator concerned, according to Salviati, the need for experimentation, if man was to learn something reliable about the universe. Against a Creator scattering His richness all across the cosmic spaces, man's mind could not appear qualified to fathom the construction of the universe in the distinctly a priori manner outlined in Aristotle's *On the Heavens*. There pantheism sapped the willingness to experiment; here experimentation, laborious but confident, was the only way for man to decipher the structure of at least some parts of the Creator's immense handiwork. According to the *Dialogue*, a reliable investigation of nature presupposed men convinced of the fact that "they know only the tiniest portion of what is knowable, exhaust themselves in waking and studying, and mortify themselves with experiments and observations."[9] Such a rugged determination rested on the assumption that the human mind was patterned after nature and not vice versa, or to quote Sagredo: "nature first made things in her own way, and then made human reason skilful enough to be able to understand, but only by hard work, some part of her secrets."[10]

Nature, here, stood for God, not of course in a naturalistic sense, but in the sense made possible by the belief that nature was the work and a faithful symbol of a most reasonable Supreme Being. Therefore nature, in analogy to her Maker, could only be steady and permeated by the same law and reason everywhere. From permanence and universality of the world order followed, for instance, that the same laws of motion were postulated for the earth and the celestial bodies.[11] It also followed that regularly occurring phenomena, such as tides, baffling as they might appear, should not be assigned a miraculous cause.[12] The most important consequence of the permanence and universality of the world order anchored in the Christian notion of the Creator was the ability of the human mind to investigate that order. Such was an inevitable consequence if both nature and the human mind were products of one and the same Creator. As to the human mind Galileo most emphatically stated that it was a "work of God's and one of the most excellent."[13] The rapid survey of man's various intellectual achievements, which closed the First Day, served indeed for Galileo as proof of precisely such a theologically oriented conclusion.

The intimate relatedness of the human mind to the Creator's was analyzed by Galileo with great interest. In particular he made detailed comments on the mathematical or geometrical simplicity and truth as evident to both God and man. In perceiving the simplicity of geometrical truths, man's mind participated in a knowledge that differed from the divine only in the sense that God knows *all* geometrical truths and knows them *always*, whereas man perceives only some of them and only by a step-by-step process.[14] But once the divine sublimity of geometrical propositions was claimed, their study demanded a singlemindedness which allowed for no hesitation in proclaiming their truth whatever the appearances to the contrary. For if nature and mind were works of the same God, they both had to reflect the very simplicity of divine truths of which the geometrical ones seemed the most palpable. This is why Galileo praised repeatedly[15] the *faith* of Copernicus in the simplicity of the heliocentric arrangement of the planets. As it was most emphatically proclaimed in Galileo's rare but significant reference to scientific history, the new science had its origin in a faith intimately tied to the belief in the Creator. Or as Salviati reminded the theologizing Simplicio: "Copernicus admires the arrangement of the parts of the universe because of God's having placed the great luminary which must give off its mighty splendor to the whole temple right in the centre of it, and not off to one side."[16]

If anything could be truly alien to a picture of the world anchored in the planning and wisdom of a divine Craftsman, Architect, and Maker, to recall Galileo's favourite references to the Creator of all, it was the vision of a universe going through endless and senseless births and deaths. The Galilean universe resting in the Creator's hands implied a stability contingent on the Creator's steady will and wisdom, with which redundant cosmic repetitions were patently incompatible. To the total lack of reference by Galileo to eternal cycles one can only assign one cause: the lucid atmosphere of the *Dialogue* was not to be spoiled by any useless recall of a most devious and murky world picture, utterly antagonistic to the light of reason. This meant that the freedom of speculating about the universe, which Galileo poignantly advocated in a marginal note in his own copy of the *Dialogue*, had its limits. Charged with introducing novelties he noted with justified bitterness that worse disorders resulted "when minds created by God" were "compelled to submit slavishly to an outside will," when "people devoid of whatsoever competence" were made "judges over experts" and were "granted authority to treat them as they please," and, most importantly, could urge them "to deny their senses and subject them to the whim of others."[17]

Galileo could not be unmindful that he too advocated the denial of the evidence of one's senses when he praised Copernicus' scientific faith, which he fully shared also in its theological ramifications. Freedom in thinking implied no justification, according to Galileo, for departing from the very faith in the Creator which justified scientific pursuit itself. And ultimately faith it was, and not experimental evidence, when it came to the truth of the heliocentric arrangement of the universe. As another handwritten remark of Galileo in his own copy of the *Dialogue* stated in utter candor: "Take note, theologians, that in your desire to make matters of faith out of propositions relating to the fixity of sun and earth you run the risk of eventually having to condemn as heretics those who would declare the earth to stand still and the sun to change position – *eventually*, I say, at such a time as it might be physically or logically proved that the earth moves and the sun stands still."[18] (Italics added.)

Eventually, experimental and observational evidences were marshalled in support of the heliocentric arrangement though not as soon as Galileo expected. Meanwhile faith, and in particular faith in the Creator, was needed to provide firm, logical foundation for the scientific pursuit. This is very clearly evidenced in the reflections on scientific methodology by Descartes and Bacon, the other main heralds of the new age of science. Of the two, Descartes was by far the more systematic. The determination with which he pursued his goal, a mechanistic account of the universe, stamped the three centuries of classical physics with the Cartesian spirit and motto of "matter and motion." But for Descartes this merely represented the superstructure. He was equally concerned with the ultimate foundations of science. What made scientific inquiry possible, according to Descartes, was that the notion and reality of God as Creator secured a double feature for the totality of created entities. One of these was their participation in eternal reasons on which rested the notion of physical law, the postulate of the homogeneity of the universe, of its consistency and harmony. The other was the contingency of nature and this seemed to be most palpable for Descartes in the countless particular characteristics of physical things and processes. The specifications of these particulars and the demonstration of a general law at work in them entailed for Descartes the overriding need for experimentation and the justification of his insistence to part with the mentality of the Schoolmen.

How crucial such considerations appeared to Descartes is well evidenced in his three major discussions of the new scientific method. The first of these, *Le Monde ou Traité de la lumière*,[19] was on the point of being published, but Descartes, on hearing about Galileo's condemnation in 1632, suddenly changed plans. Long before the incomplete text of the work was printed posthumously in 1664, the world of the learned had already been given a glimpse of its main thrust in the *Discourse on the Method*[20] which, as the rest of the title indicated, aimed at establishing rules of "rightly conducting the reason and seeking for truth in the sciences."[21] If Descartes' longing for truth was all-absorbing, so was his distaste for doubts. But he was equally determined not to be trapped. Great as was for him the appeal of clear and distinct notions, especially those of geometry, he felt wary about them unless they could be anchored in an unassailable source of truth. This source was none other than God. Only an utter persuasion about His existence could, Descartes insisted, remove all doubts even about the most clearly perceived ideas and notions. For him it was the basis of all truth to hold that "all the things that we very clearly and very distinctly conceive of are true, is certain only because God is or exists, and that He is a Perfect Being, and that all that is in us issues from Him."[22] On what all this ultimately implied he could not be more explicit: "If we did not know that all that is in us of reality and truth proceeds from a perfect and infinite Being, however clear and distinct were our ideas, we should not have any reason to assure ourselves that they had the perfection of being true."[23]

To be sure, Descartes got carried away time and time again by his eagerness to remove doubts. He tied too many "clear and distinct ideas" to the Creator's divine and necessary nature. He provided more than one reason for being considered a chief advocate of an a priori physics where some "clear and distinct ideas" about a world structured on the clarity of Euclidean geometry could generate the rest of physics. He often made the boastful claim that the formation of the physical world from the original chaos would follow phases which he confidently specified.[24] He even wrote that the knowledge of his rules would enable one "to have, in terms of the scholastic parlance, a priori demonstrations of everything that can be produced in this new world."[25]

The new world was a hypothetical, new creation by God and the main burden of *Le Monde* and of *Les principes de la philosophie*[26] consisted in showing that the features of that "new world" would be essentially identical with the actual one, because God's nature set certain inevitable patterns for the interaction of material particles. The three principal of these patterns or rules stated that (1) any particle of matter remains in the same condition until coming into contact with another; (2) no body when colliding with another can give it any amount of motion without losing an equal amount at the same time; (3) when a body moves, all its parts tend to continue their movements along a straight line.[27] Descartes believed that in addition to these rules he needed only the demonstrations of geometry which were, in his eyes, equally anchored "in most certain and evident demonstrations, according to which, as God Himself taught us, He disposed all things in number, weight, and measure."[28]

This biblical ring is only one of the indications that Descartes' innovation aimed at bringing out in full force the best of an already vigorous intellectual tradition. The other revealing sign of this is Descartes' emphatic insistence that his crucial break with the followers of Aristotle concerned the concept of motion. "The notion, of which they speak, is so different from the one I conceive of, that it may easily turn out that what is true of one, is not so of the other."[29] The new concept was the rectilinear inertial movement and Descartes most explicitly derived it from the act of Creation. As he stated it in explaining the first two rules: "These two rules evidently follow from that alone, that God is immutable, and that acting always in the same manner, He produces always the same effect . . . For assuming that He put a *certain quantity of motion* into the entirety of matter in general, from the first instance that He created it, one has to admit that He conserves the same quantity of movement, or else one cannot believe that He always acts in the same manner."[30] (Italics added.)

If Descartes had not been aware that Buridan and Oresme long before him had said much the same thing, as an avid critique of the Schoolmen he could hardly be ignorant of the fact that his unhesitating claims to originality were subject to a qualification or two. While he seemed to be oblivious of this perhaps secondary point, his consciousness was very explicit on a more fundamental issue. He fully realized that if he were to avoid introducing necessitarianism into God as a Creator, his sweeping ratiocination had to be carried out, to quote his words, "without doing outrage to the miracle of Creation."[31] The sincere awe which he always felt for its uniqueness played the role of an all-important safeguard in his thinking. For the one who confidently wrote that his "entire physics is nothing but geometry,"[32] was at pains to specify that those who required of him geometrical demonstrations in some points of physics expected him "to do the impossible."[33] He frankly admitted that he succeeded in making other sciences only "almost similar to" but not completely identical with geometry and mathematics.[34] According to him the reason for this lay in the infinite richness of the phenomena both in their juxtaposed and in their intrinsic variety. It was the sole privilege of the Creator of all to have a perfectly lucid and exhaustive knowledge of any or all physical phenomena. Man had to make extensive observations and experimentations to establish some true patterns involved in their mutual interactions: "Then when I wished to descend to those which are more particular, so many objects of various kinds presented themselves to me, that I did not think it was possible for the human mind to distinguish the forms or species of bodies which are on the earth from an infinitude of others which might have been so if it had been the will of God to place them there, or consequently to apply them to our use, if it were not that we arrive at the causes by the effects, and avail ourselves of many particular experiments."[35]

This is not the place to review in detail Descartes' numerous references to the need of experimenting, or the "countless experiments" he claimed he had performed, or his famous call to all the learned to communicate their experiments to him, the sole possessor of the unfailing method of interpreting them correctly.[36] Undoubtedly, he had been familiar with the activities and ideas of contemporary scientists, though like many men of science of those times, Descartes grudged to acknowledge others' eminence.[37] This is well illustrated in Descartes' ostensibly generous praise of Bacon.[38] It merely served to undercut the intellectual stature of the Lord Chancellor, the renowned apostle of the inductive method. The developments of scientific reflections fully justified Descartes' claim that collection of the facts of nature could only be fruitful if it was done judiciously, namely, with some carefully formed set of assumptions. Descartes was also closer to the truth with respect to the idea of an *experimentum crucis*[39] by which Bacon set great store. In fact, no scientist of any stature ever followed Bacon's endless specifications as to how the scientific method ought to be practiced. Nor did his voluminous writings offer any original scientific observation. In addition, Bacon displayed more often than not a baffling lack of perception about what was truly valuable in the work of the scientists of his time, to say nothing of their immediate predecessors.

For all that, Bacon's place in the history of science is firmly established. He succeeded in articulating the intellectual temper of an age which grew increasingly preoccupied with reflections on science. Much of what Bacon said about the general philosophy or foundations of science was in the air, and the great popularity of his publications was due precisely to the fact that the learned among his readers found their own thoughts mirrored or further articulated therein. This is well to keep in mind if one is to grasp Bacon's view on the questions why Greek science bogged down in a blind-alley, and why a new and effective undertaking of the scientific enterprise became possible and promising. Bacon's reply to these questions rested on the enormous disparity which the doctrine of Creation introduced between the pagan and the Christian outlook on the world. The first version of Bacon's thoughts on this point dates from almost the beginning of a career devoted to a thorough reform of learning. The *Twoo Bookes of Francis Bacon of the Proficience and Advancement of Learning Divine and Human*, as reads the original title page of a work published in 1605 and simply referred to as the *Advancement of Learning*,[40] contains an expressive parallel between the respective impact on thinking by classical pantheism and Christian faith. In discussing the difference between the supernatural and natural knowledge of God, Bacon not only voiced the traditional biblical view that nature as a work of God was a powerful evidence of a Creator, but he also gave a most instructive account of the theological structuring of the classical world picture: "For as all works do shew forth the power and skill of the workman, and not his image; so it is of the works of God; which do shew the omnipotency and wisdom of the maker, but not his image; and therefore therein the heathen opinion differeth from the sacred truth; for they supposed the world to be the image of God, and man to be an extract or compendious image of the world."[41]

The intimate connection between pantheism and the inordinate trust of the Socratic school in divining the various forms of cosmic purposiveness is stated with full explicitness by Bacon. The outcome of that connection was, to quote his words, "the great arrest and prejudice of further discovery."[42] Final causes, which otherwise represented in Bacon's eyes a most worthy target of inquiry, turned into "remoras and hinderances to stay and slug the ship from further sailing, and have brought this to pass, that the search of the Physical Causes hath been neglected and passed in silence."[43] The ship was the ship of science and it had to run aground on the shallows of pantheistic presumptions about man's introspective capacity to penetrate the inner workings of a universe made not by human hands. No wonder that Bacon felt greater sympathy with Democritus' approach to nature than with that of Plato and Aristotle. But one wonders how some recent students of Bacon could find in this a revival of materialism by Bacon,[44] who took pains to warn about the difference between the philosophical materialism of Democritus and the usefulness of his ateleological approach to the particular phenomena of the material world: "And therefore the natural philosophy of Democritus and some others, who did not suppose a mind or reason in the frame of things, but attributed *the form thereof able to maintain itself to infinite essays or proofs of nature*, which they term *fortune*, seemeth to me . . . in particularities of physical causes more real and better enquired than that of Aristotle and Plato."[45]

How dear such considerations had been to Bacon can be seen from the fact that he kept elaborating them in his later works. In his *On Principles and Origins according to the Fables of Cupid and Coelum*,[46] written about 1609, Bacon presented the doctrine of Creation as a unique vantage point for the understanding of nature. If ancient philosophers erred in various degrees about the primordial formation of the universe and its limited life-span, it was because, as Bacon put it, "Creation out of nothing they cannot endure."[47] At the same time Bacon's stance also exemplified the dangers which could come from an overly enthusiastic acceptance of the usefulness of the doctrine of Creation for scientific thinking. While Bacon was right in tracing some patent aberrations of the cosmologies of both the atomists and of the Socratic school to their assumptions about a "self-existing" matter, he was reluctant to assign such potencies to matter which would predetermine the evolution of material forms "in a long course of ages."[48] Bacon grossly overstated the cause of creation when he insisted in the same context that the actual configuration of the world before the fall "was the best of which matter (as it had been created) was susceptible."[49] Leibniz himself was not to go that far.

The latter claims were not stressed in the Latin version of the *Advancement of Learning* which he prepared in 1623 as part of his final comprehensive synthesis aimed at giving wings to scientific inquiry. The Latin version[50] contains a number of phrases which add further emphasis to some salient points of the original English text. Thus, Bacon paid more detailed attention to the theological reasons for the stillbirth of Greek science. Plato, but especially Aristotle, to recall Bacon's indictment of both, "left out the fountain of final causes, namely God, and substituted Nature for God."[51] In turn, a pantheistic notion of God was bound to prompt an onrush with preconceived ideas about the universe, although Bacon added that Aristotle "took final causes themselves rather as the lover of logic than of theology."[52] Also, the absence of theology could in a sense still be "theological" and Bacon illustrated with the case of Aristotle the workings of the inner logic of a pantheistic denial of God: "Aristotle, when he had made nature pregnant with final causes, laying it down that 'Nature does nothing in vain, and always effects her will when free from impediments,' and many other things of the same kind, had no further need of a God."[53] This pseudo-theological engrossment with final causes represented, as Bacon put it, "excursions and irruptions into the limits of physical causes," and this in turn "bred a waste and solitude in that tract."[54]

Once more Bacon not only vindicated the soundness of searching for final causes within the framework provided by the Christian idea of God, but he also emphasized the instructional value of the search for physical causes toward a deeper knowledge of God the Creator: "So far are physical causes from withdrawing men from God and Providence, that contrariwise, those philosophers who have been occupied in searching them out can find no issue but by resorting to God and Providence at the last."[55] In stating this Bacon pregnantly articulated a principal aspect of seventeenth-century scientific work: discoveries of scientific laws represented as many new evidences of the Creator, the Author of every law of nature. From Bacon through Boyle to Newton such was an outlook most firmly adhered to by all scientists of some stature. This unanimity indicated the robustness of that intellectual atmosphere which after several centuries of maturing had finally come into its own. Its self-confidence can be best gauged by the fact that the ancient, cyclic view of an organo-pantheistic universe ceased to loom large on the intellectual horizon. Its fall into discredit seemed by the mid-seventeenth century to be almost complete. Only occasional references to it suggested that it was not completely forgotten; but as a wholly vanquished enemy it could be left safely aside. Thus, Bacon, so attentive to many small details of classical myths and notions, made only a cursory reference to it: he described the Platonic or Great Year in so far as it implied an eternal recurrence of all, as "the fume of those that conceive the celestial bodies have more accurate influences upon these things below than indeed they have."[56]

The only notable figure of seventeenth-century science who returned repeatedly to the question of the Great Year was Father Marin Mersenne,[57] possibly the most selfless servant the scientific community had ever had. His disarming generosity, his tireless eagerness to learn and to communicate, earned him the trust, respect, and gratitude of most of those who played a part in ushering in, what Whitehead aptly called, the century of genius. Mersenne's motivation in carrying for decades an almost superhuman workload of scientific correspondence and publications derived from that evangelical candor and zeal which his order, the Minimes, aimed to achieve. In Mersenne, there was the rare combination of theologian, scientist and, above all, of a virtuous Christian. It should not, therefore, be surprising that Mersenne wrote more extensively than others in his century about the relation of science and theology, and he did so with a consistently high level of judiciousness. In the clarity of the new mechanistic science he saw a powerful antidote to crude and refined forms of obscurantism alike. He fought with equal zeal the astrological tradition and the pantheistic animism of which in his time Robert Fludd, whom he used to call "the evil magician," was the chief advocate. As Mersenne saw it correctly, the belief in the Creator was incompatible with both.

In his monumental discussion of the biblical account of creation, the *Quaestiones celeberrimae in Genesim*, Mersenne delved into every possible ramification of the dogma of creation. He did so in the conviction, which he expressly stated, that no phrase of greater portent was ever formulated than the one declaring that 'In the beginning God created the heaven and the earth.'[58] His discussion of the Great Year[59] shows that previous positions in the matter were well remembered at least by scholars of Mersenne's caliber. Among authors from the Renaissance he singled out Ficino and Diacceto, and explicitly mentioned Oresme's arguments based on the incommensurability of planetary cycles given in his *De proportionibus proportionum*. From Christian antiquity Augustine and Macrobius were recalled. As chief supporters of the idea of eternal recurrences Mersenne referred to Albumasar, Ptolemy, Plato, and the Pythagoreans. That no reference to the Great Year occurs in Mersenne's *L'impiété des déistes, athées et libertins de ce temps*[60] shows the sudden disrepute into which the doctrine had fallen. Although it was a natural consequence of a belief in a pantheistic and eternal world, by Mersenne's time it could only do disservice to those who tried to promote a distinctly non-Christian view of the universe. They were relatively few.

The same consideration seems to apply to the silence of Pascal and Boyle on the Great Year. As the scientific enterprise grew more robust with each passing decade, so did the chimera of eternal returns approach the vanishing point. Thus, both Pascal and Boyle referred to the Stoics, but the former did so only to take them to task for their moral presumptuousness,[61] while the latter mentioned their view on cosmic conflagration as something borrowed from Israel.[62] Not even in discussing the pre-existence of souls did Boyle make as much as an indirect hint to the idea of eternal returns. Boyle's reticence on the topic might also have been occasioned by the fact that his chief target, Hobbes' materialism, contained no pointers toward a cyclic notion of the universe.

What had to be asserted against Hobbes' contentions was the utter dependence of the realm of matter on the Creator's Fiat. Boyle did this by presenting Hobbes' views as an unpardonable rehash of the errors of ancient philosophers about the eternity of matter. Boyle admitted that intimations of an actual beginning of the world might have occurred to some pre-Socratics, but he added the following remark on the all-important point of a creation out of nothing *and* in time: "But whether or no mere natural reason can reach so sublime a truth, yet it seems not, that it did actually, where it was not excited by revelation-discovery. For though many of the ancient philosophers believed the world to have had a beginning, yet they all took it for granted, that matter had none; nor does any of them, that I know of, seem to have so much as imagined, that any substance could be produced out of nothing. Those that ascribe much more to God than Aristotle, make him to have given form only, not matter to the world, and to have but contrived the pre-existent matter into this orderly system we call the universe."[63]

For Boyle the doctrine and belief in the Creator represented the very foundation of sound reasoning about the world. This also implied for him a limited time-span for the world in its present, natural form. About the actual length of the world's past existence, he preferred the short or what he believed to be the biblical time-scale of a few thousand years. This he contrasted with the speculations of "some extravagant ambitious people" whose accounts "reached up to forty thousand or fifty thousand years."[64] Among these he singled out the "fabulous Chaldeans" without, however, referring to the idea of eternal returns tagged as "Chaldean" throughout so many centuries. His silence on this point could only indicate contempt for the scientific merit of an idea which he must have considered utterly absurd, if a cosmic past of some fifty thousand years had already appeared to him extravagant.

Boyle must have certainly been familiar with the past history of the idea of eternal returns. The England of his time witnessed the ascendency of the Cambridge Platonists, among them Henry More. The latter's *Immortality of the Soul* had a special chapter for the Stoics' claim about a Great Year ending invariably in an all-consuming fire.[65] More's principal interest lay not, of course, in the cosmological aspects of the question. He merely tried to show the correlation between some repulsive aspects of the Stoics' ethics and their cosmology.[66] Characteristically, the cosmology of eternal returns was dismissed out of hand in the much more scientifically oriented *Three Physico-Theological Discourses*[67] by John Ray, member of the Royal Society. The work, a vindication of the main points of the biblical account of the origin and destiny of the cosmos, was originally published in 1692 under the title, "Miscellaneous Discourses concerning the Dissolution and Changes of the World," but it concerned itself exclusively with three points: the creation of the world, the deluge, and the fiery end of the world. It was in connection with this last point that Ray discussed in a special chapter[68] the idea of the Great Year mainly as the Stoics held it, though he also pointed out the ubiquitous presence of the idea in the antique world. As Ray put it, the "ancient Gauls, Chaldeans, and Indians" could not get the idea from the Stoics.[69] But the surprisingly numerous references by Ray to various antique sources served only two purposes. The first was to discredit the Stoics' ethics on the basis of their weird cosmology; the second consisted in bringing out the credibility of the biblical doctrine of final conflagration as something which was held, however incorrectly, by all peoples and cultures. Ray wasted no time on the idea of cosmic periodicity. In his eyes it was not worth refuting. The assent commanded by the rationality of the once-and-for-all cosmic purpose was, for all practical intent, overwhelming.

This also explains Boyle's and others' adherence to the biblical time scale of geological and cosmic past. It was as much motivated by an almost complete lack of evidence of a drastically longer past, as by piety, which in Boyle's case was all too conspicuous. Still, he stopped short of considering the biblical time scale as an unquestionable truth. Much similar was the attitude of other prominent men of science, among them Newton, for whom speculation on the chronology of the whole past history had become an all-absorbing interest. Although Newton's and other chronologists' style was overassertive more often than not, they rarely proposed their chronological schemes as ultimate verities, and much less were those schemes accepted as such by most contemporaries. It indeed took the naiveté, if not tendenciousness, of some latter-day cosmologists, historians, and popularizers of science to gloat over Archbishop Ussher's meticulous setting of the "beginning of time . . . upon the entrance of the night preceding the twenty third day of October in the year of the Julian Calendar, 710," or 4004 B.C.[70]

Those who equipped with the wisdom of hindsight greet Ussher's and others' undeniable eagerness with condescending smiles, are apt to overlook the utmost respect for the biblical faith in creation which underlies those efforts. As Director of the Mint, Newton devoted much of his spare time to studies of world chronology, largely because his thinking was dominated by the idea of Yahweh, the Creator and Supreme Lord of all.[71] Newton greeted with obvious satisfaction Bentley's efforts to vindicate the role of the Creator against atheism and deism on the basis of the main conclusions of the *Principia*. On his part, Newton most explicitly endorsed the notion of a Creation once and for all as the only sound framework of natural philosophy. This he did as he added the Scholium to the second (1713) edition of the *Principia* and as he kept enriching the subsequent editions of the *Opticks* with further Queries. The universe for Newton was a clockwork constructed by the Creator out of basic components created out of nothing. It was also the Creator who kept, according to Newton, this world mechanism in good repair by preventing it from a too early unwinding through His repeated interventions. Again, it was the Creator who secured the ultimate transition of this world into its final transformed condition at the end of time.

If there was anything alien to Newton's mind, it was the "succession of worlds" in the sense of cyclical, let alone of perpetual returns.[72] In 1706, when the Latin version of the *Opticks* appeared, Newton added to the original sixteen Queries seven new ones, among them the longest and most famous of all Queries, listed from the second edition of the original English (1717) on as Query 31. In that Query Newton most explicitly stated about the basic building blocks of the Universe, the "solid, massy, hard, impenetrable, moveable Particles," that "it became Him who created them to set them in order."[73] In the same Query Newton categorically rejected the idea of an autonomous nature arising out of chaos by mere chance or "by the mere Laws of Nature; though being once form'd, it may continue by those Laws for many Ages."[74] That the present course of Nature could not go on indefinitely was traced by Newton to two causes: friction in the ether and the mutual disturbance of the planets. Both these causes demanded, according to Newton, that the Creator should restore by periodic interventions the original "quantity" of motion and orbits of planets. In 1713, in a paragraph added to the second edition of the *Principia*, Newton speculated about a third cause, the gradual loss of light and heat of the sun and stars. As a replenishing factor of the sun's (and stars') energy he offered the periodic fall of comets into those celestial bodies.[75] This he repeated in substance in March, 1725, in a private conversation with John Conduitt.[76] But there is no evidence whatever that Newton rescinded his strictures of Descartes' contention that by gradually losing their heat and light suns would turn into comets and these into planets. In his third letter to Bentley, Newton labeled as absurd the belief in "the growth of new systems [of planets] out of old ones, without the mediation of a divine power."[77]

To suggest that such statements on Newton's part were "possibly expressions more of his hostility to Cartesianism than to cosmogony,"[78] seems arbitrary and presupposes a drastic belittling of Newton's religious convictions. Unorthodox as he was with respect to the doctrine of Trinity, his espousing in the Scholium of the biblical idea of the utter contingency of the world on Yahweh the Creator, represented a fundamental, unwavering conviction on his part: "This Being governs all things, not as the soul of the world, but as Lord over all; and on account of his dominion he is wont to be called *Lord God* παντοκράτωρ, or *Universal Ruler*; for *God* is a relative word, and has a respect to servants; and *Deity* is the dominion of God not over his own body, as those imagine who fancy God to be the soul of the world, but over servants. The Supreme God is a being eternal, infinite, absolutely perfect; but a being, however perfect, without dominion, cannot be said to be Lord God; for we say, my God, your God, the God of *Israel*, the God of Gods, and Lord of Lords; but we do not say, my Eternal, your Eternal, the Eternal of *Israel*, the Eternal of Gods; we do not say, my Infinite, or my Perfect; these are titles which have no respect to servants."[79]

Newton's warning against imagining God as a soul of nature reveals his awareness of the pantheistic world view, which dominated ancient thinking and which represented a natural option for those ready to part with Christianity. Some of the logical implications of a pantheistic world view were spelled out in detail by Bentley in his famous series of Sermons,[80] the first Boyle-lectures, in which the educated public received the first glimpse of the theological bearing of a number of conclusions in the *Principia*. As Bentley aptly noted, the atheistic and pantheistic positions were tied up with the eternity of the world which Bentley opposed on various grounds. One of his arguments consisted in stating that had the world existed eternally, all inventions would have long ago been perfected leaving nothing to discover for the very age which proudly looked upon itself as the age of discovery. The typical atheistic rejoinder to this was to take refuge in the occurrence of global catastrophies such as deluges, of which there must have occurred an infinite number, if, as Bentley remarked, eternity was taken consistently.[81]

The infinity of successive geological phases and human generations presented a welcome target for Bentley's wit. But in addition to mild sarcasm he displayed both historical and scientific insight. On the one hand, he showed in detail the intimate connection of eternal cosmic repetitions with Chaldean, Egyptian, Greek, and Roman astrology.[82] On the other hand, he noted that infinite successions of geological epochs and human generations implied an infinite resource of matter. But it was precisely the idea of the infinity of matter that he opposed convincingly both in the seventh sermon of the series and shortly later in his correspondence[83] with Newton on the scientific points made in the *Principia*.

By questioning the infinity of the world Bentley attacked one of the most powerful prejudices that ever got hold of scientific thinking. The gravitational and optical paradoxes of an infinite universe had been ignored, dismissed, and belittled with baffling superficiality from at least the closing decades of the seventeenth century until rather recently.[84] In part because of a justified reaction to the Aristotelian universe, preference for infinity in space had grown into an irrepressible intellectual fashion and on its coat-tails came the option for infinity along time, or eternity. Leibniz, whose sincere attachment to the Christian faith cannot be doubted, illustrates well this transformation, as far as eternity was concerned. In his essay, "On the Ultimate Origination of the Universe,"[85] he described the universe as being "engaged in a perpetual and spontaneous progress."[86] By perpetual he meant not continuous but strictly eternal. Recorded history of a few thousand years was only "an infinitesimally small part of the eternity which stretches out beyond measure."[87] Reversals in that grandiose march toward the ultimate in perfection allowed for some partial reversals, but only to give new and greater impetus to progress. It could easily be seen, however, that the actual world, the earth, to be specific, hardly reflected the best of conditions which should have long since materialized had the world been eternal. Leibniz parried the objection by noting that progress was a continuous line that could be divided into an infinite number of infinitesimally small parts: "the continuum being infinitely divisible, there will always remain in the unfathomable depth of the universe some somnolent elements which are still to be awakened, developed, and improved – in a word, promoted to higher culture. This is why the end of progress can never be attained."[88]

The passage is revealing not only of the momentary weakness of a genius capable of being trapped in a sophism which already Zeno had some trouble selling. The image of a universe which Leibniz conjured up stretched to its very extremes the Christian concept of the universe. Leibniz nowhere denied the creation in time, that is, "in the beginning," but he clearly let the idea fade into the background. This was all the more significant as he showed otherwise an acute awareness of the role to be played by the notion of the Creator in the justification of scientific pursuit. His criticism[89] of Descartes' *Principia* is a proof of how seriously he took the matter, namely, the proper analysis of the actual physical laws and of the properties characterizing the true Creator. Two of Leibniz's achievements, the enunciation of the conservation of "vis viva" (later energy) and the foundation of infinitesimal calculus, were used by him to show that the actual physical world was indeed the best and that it had a goal. By insisting on this latter point Leibniz tried to overcome the cold, soulless mechanism inherent in the Cartesian and Newtonian universe. But a lasting sense of purpose was far more inherently tied to the biblical outlook than Leibniz suspected. He could hardly have guessed that his ideas and his philosophical optimism paved the way to deism and to the faith of the Enlightenment in an infinite progress in an infinite world.

The inner logic of this new view of the universe could easily remain hidden so long as the universe was practically the solar system. Leibniz's controversy with Samuel Clarke, who represented Newton, revealed its true portent only in retrospect. Nor was it noticed that the concept of a cyclic universe was slipping in through the back door when the young Kant published anonymously his *Universal Natural History and Theory of the Heavens*.[90] Its lasting scientific merit consisted in Kant's bold description of the nebulous patches as systems of stars comparable to our Milky Way. This gave a truly universal significance to Thomas Wright's somewhat fumbling analysis of that whitish band, the Milky Way, stretching all across the night sky. As Kant outlined his own world picture in the seventh chapter of the Second Part of his work, the world was created by God in the form of a chaos comprising an infinite amount of matter spread out in an infinite space. In the absolute centre of it[91] matter soon began to condense under the influence of gravitational attraction. The subsequent formation of the first galaxy, or world, was then followed by the formation of other galaxies around it as the process of condensation spread outward, resulting in larger and larger systems of galaxies. At the same time Kant also insisted that whatever has ever reached the highest point of its development must also decay. He therefore claimed that our present and fully developed system of stars was bounded on both sides by matter in a state of chaos. On the side away from the centre matter was still in its prime chaotic state, while on the side toward the centre matter consisted of the ashes of already decayed systems. These ashes were, however, supposed to undergo again a new process of condensation and world-building. Thus, the whole universe of Kant

U

could be pictured as a spherical wave spreading continually from the centre, the peaks of that wave representing those spherical layers of the universe where matter existed in fully developed forms, or planetary and stellar systems, which in turn were all populated by living beings.

The presence of living beings everywhere in the universe[92] followed, for Kant, from his emphasis on the principle of plenitude based on God's infinity. To be sure, Kant carefully exempted human souls from the cyclic treadmill pervading his spatially infinite universe which was eternal only with respect to the future but not to the past. Following man's death his soul was forever united with God while spherical layers of the universe developed, decayed, and rose again from the cosmic ashes like "Phoenix," as Kant put it,[93] at regular intervals or periods. The length of those periods Kant did not dare to specify, but he was very conscious of the novel feature of his propositions: "According to that law [of periodic development and decay] the heavenly bodies that perish first, are those which are situated nearest the centre of the universe, even as production and formation did begin near this centre; and from that region deterioration and destruction *gradually spread* to further distances till they come to bury all the world that has *finished its period* through a gradual decline of its movements, in a single chaos at last. On the other hand, Nature unceasingly occupies herself at the opposite boundary of the developed world, in forming worlds out of the raw material of the scattered elements; and thus, while she grows old on one side near the centre, she is young on the other, and is fruitful in new productions."[94] (Italics added.) Thus, every region in the universe was to witness throughout eternity an endless recurrence of the same phases of a circular physical process. For as Kant confidently asked: "Can we not believe that Nature, which was capable of developing herself out of chaos into a regular order and into an arranged system, is likewise capable of re-arranging herself again as easily out of the new chaos into which the diminution of her motion has plunged her, and to renew the former combination?"[95] Clearly, the prospect of a cosmic perpetuum mobile gave him no second thoughts, as befitted any zealous scientific amateur blissfully building his own cosmos.

Kant's speculations made no considerable ripple. That his cosmology was essentially cyclic remained unnoticed not only in his time but also by more recent interpreters of his cosmology. Lambert, the most prominent contemporary of Kant among German scientists, wrote his famous *Cosmologische Briefe*[96] with no knowledge of Kant's work published six years earlier.[97] Unlike Kant's universe, Lambert's cosmos was finite and basically static, in the sense that Lambert paid no detailed attention to such problems as evolution and decay of planetary systems, stars, and galaxies. Lambert's brief reference to the Platonic Year[98] was rather innocuous. In his hierarchical arrangement of galaxies and of some five hundred higher orders of systems of galaxies everything rotated around an immensely massive, dark, central body. It was, therefore, natural for Lambert to ask the question about the period of rotation of our sun around that mythical centre. Was it equal to the Platonic Year or only a mere fraction of it? To this question Lambert did not profess to know the answer though he did not hesitate to speak positively about equally problematic points of his hierarchical world order.

With Lambert the hierarchical world order reflected a design pointing to a

Supreme Architect assuring stability and purpose for the cosmic clockwork. Those for whom the notion of God, much less its Christian version, had no appeal, turned their attention instinctively to traditionally "atheistic" conceptions of the universe. Thus, Hume reached back to the old Epicurean hypothesis, "the most absurd system that has yet been proposed," as he put it, but obviously with tongue in cheek. The whole gist of his *Dialogues concerning Natural Religion*[99] was to show that religious faith, natural or supernatural, was an attitude that rested on emotional and not on rational justifications. The "atheological" world view that alone suited reason had, therefore, to be such in which design and purpose had no right to intrude. As Hume correctly guessed, the Epicurean or atomistic concept of a world, which represented in each moment a chance configuration of atoms, perfectly fitted the foregoing specification. That design and purpose were wholly alien to this world model could be demonstrated palpably, according to Hume, by assuming that the number of atoms was finite in the world. In such a world the haphazard, endless recurrence of any and all possible configurations of atoms was an inevitable result and this meant for him the complete elimination of design and purpose.

It was on that foundation that Hume tried to elaborate what he called "a new hypothesis of cosmogony that is not absolutely absurd and improbable."[100] By this he seemed to suggest that it was most probable and anything but absurd. At any rate, he was impressively consistent with his own intellectual proclivities though not with the facts of scientific history. He argued that the perenniality of motive forces in matter could be maintained without the assumption of a First Cause. He claimed that the assumption of "an unknown voluntary agent is mere hypothesis and hypothesis attended with no advantages. The beginning of motion in matter itself is as conceivable *a priori* as its communication from mind and intelligence."[101] If Hume's postulate of a cosmic perpetuum mobile had already revealed something of the liberties which an advocate of scepticism could take with reasoning, even more powerfully did this come through as he outlined the second main point of his cosmology. The surprisingly large measure of order and stability shown by nature was, so Hume claimed, merely an appearance. In a perennial change of configurations involving an extremely large number of minute components, the over-all features would change very slowly and this in turn would create the illusion of permanency of pattern and order. "Every individual," wrote Hume, "is perpetually changing, and every part of every individual; and yet the whole remains, in appearance, the same. May we not hope for such a position or rather be assured of it from the eternal revolutions of unguided matter; and may not this account for all the appearing wisdom and contrivance which is in the universe?"[102] If the order was only apparent, what was reality? Hume could not be more outspoken in giving his reply: "The universe goes on for many ages in a continued succession of chaos and disorder."[103]

By eliminating all order, all permanence, all law and causality except the effectiveness of haphazard chance from the ontological structure of the universe, Hume felt he had proved his main contention: not only the revealed but also the natural religion was untenable on rational grounds. All religions, systems, and their advocates, Hume insisted in a closing remark on his cosmology, "prepare a complete triumph for the sceptic."[104] On a radical scepticism no dent could, of course, be made by any objection, not even by the stubborn facts of nature which consistently indicate a very limited and uniform selection out of infinite possibilities on the part of nature in flat rebuttal of the philosophy of the haphazard. Hume's triumphant conclusion was equally belied by the facts of scientific history. He remained utterly oblivious of the fact that the only viable birth of science was accomplished by a most reasoned reliance on that faith which for Hume was acceptable only if one were willing to abdicate reason.

Significantly, the principal creative scientists were not Humeans either then or later. Many of them kept their faith in an objectively ordered nature contingent on an ultimate and Supreme Cause in the metaphysical sense. A good case in point is Laplace who in the early nineteenth century offered some famous though not really sound considerations on the possible development of solar systems from gaseous condensations. This he did in a Note to his *Exposition du système du monde*.[105] In part the book was a semipopular exposition of his classic studies which he carried out in collaboration with Lagrange about the various features of the solar system, especially its stability. Neither his forbiddingly technical *Mémoires*, nor the *Exposition*, contained any reference to God.[106] As Laplace aptly noted, no particular scientific law or phenomenon was an "ultimate" in the sense that beyond it one could not assume a more comprehensive law or phenomenon, but immediately and only the First Cause.[107] It was not the scientist at grips with a particular quantitative correlation that needed the "hypothesis" of God, but the scientist as a thinker in search of a comprehensive or fully consistent explanation.

Like any other scientific conclusion, the one reached by Laplace and Lagrange on the stability of the solar system could be used and abused. A somewhat far-fetched though still defensible extrapolation of it appeared in *The Monthly Review* in 1795 from the pen of the anonymous reviewer of the contents of the fifth (1794) volume of the *Transaction of the Royal Irish Academy*. In connection with Thomas Garnet's "Observations on Rain-gages," the reviewer conjured up[108] a "meteorology of the future" based on the exact knowledge of the influence on the earth's climate by various periodic processes in nature. Among these the periodicity of planetary perturbations was singled out by the reviewer as the most encompassing one. Referring to Lagrange's conclusion about "certain vast cycles, on the return of which the same motions are perpetually renewed" in the motion of planets, the reviewer concluded that "all the events within the immeasurable circuit of the universe are the successive evolution of an extended series, which, at the returns of some vast period, repeats its eternal round during the endless flux of time."[109] Once the exact influence of this great and many smaller cycles was known men could "conjecture with tolerable precision the succeeding changes of the weather."[110]

Wishful thinking as these expectations could appear, they were distinctly modest and reasonable when compared with those efforts which had nothing in common with Laplace's cautious remarks on the interpretation of some conclusions of scientific research. Possibly the chief of these heedless interpreters of Laplace was Condorcet, a most dedicated apostle of unlimited progress on the basis of a resolute and exclusive application of sciences and mathematics. His *Memoirs on Public Instruction* and his *Sketch for a Historical Picture of the Progress of the Human Mind* were extremely influential in spreading the belief that the world machine was to run for all practical purposes in much the same state for eternity, securing thereby the physical framework for unlimited progress.[111] It was along these lines that he tried to legislate the future of mankind in his *Memoirs on Public Instruction*, while his *Sketch* interpreted the past in the same sense. Subsequent events of history quickly dissipated his sanguine vision of the future. His reading of the intellectual past was in turn discredited by modern historical research unfettered by the dogmas of the Enlightenment. It is no longer the mark of undisputed scholarship to repeat with Condorcet that "the triumph of Christianity was the signal for the complete decadence of philosophy and the sciences."[112]

Contrary to Condorcet's expectations his brand of rationalism failed to make notable converts precisely there where the most abundant harvest should have taken place, the world of exact scientists. The leading physicists and astronomers of the nineteenth century were conspicuous by their firm adherence to a Christian or at least to a theistic interpretation of the universe. Questions about the earth's geological past that were heatedly debated between the Neptunists and the Vul- canists during the first decades of the century had only tenuous ties with cosmology. The main cosmological interests of astronomy during the nineteenth century turned not about the origin but rather about the actual shape of the stellar realm, and in particular about the question of whether there were outside the Milky Way other independent galaxies or "island universes." The younger Herschel, Struve, Nichol, Mädler, Littrow, Lord Rosse and others who tried to fathom this question during the first half of the century had all been animated by the convictions of the older Herschel who, in the words of his son, "was a sincere believer and a worshipper of a benevolent, intelligent, and superintending Deity whose glory he conceived himself to be legitimately forwarding by investigating the magnificent structure of the Universe."[113]

It should not, therefore, be surprising that the younger Herschel raised a firm caveat when the idea of eternal recurrence reappeared in slightly veiled form in Humboldt's famous *Kosmos*. In his almost book-length review of it[114] Herschel took issue at the very outset with the two main notions of Humboldt's scientific philosophy. One was the contention that there were no qualitative changes in the world but only a mechanical push and pull pervading and dominating every- thing. The other was, as Herschel quoted Humboldt's words, "the ancient belief that the forces inherent in matter, and those which regulate the moral world, exert their action under the government of a primordial necessity, and in recurring courses of greater or less period. It is this necessity, this occult but permanent connection, this periodical recurrence in the progressive development of forms, of phenomena and of events, which constitute nature."[115] As Herschel aptly intimated, it was incongruous on Humboldt's part to add that nature by repeating itself through endless cycles was "obedient to the first imparted impulse of the Creator."[116] The notion of a Creator was hardly compatible with an outlook in which nature, "moral as well as physical" was "a piece of mechanism, which wound up and set going, has been abandoned to itself, to evolve its changes in variously superposed periods, without choice or option, according to the combinations of an occult wheelwork."[117]

Herschel's emphatic proviso aimed at what seemed to hark back to occult qualities in Humboldt's speculations. Herschel had no quarrel with the idea of a wheel-work universe operating as a clock. By then this notion had been the hallowed shibboleth of science for well over two hundred years. But the world of science soon had to learn that the cosmic clockwork was not likely to run forever. The realization of this began to be felt shortly after the decade had come to a close which witnessed the enunciation of the principle of the conservation of energy by Mayer, Joule, Helmholtz, and others. While Moleschott, Vogt, and Büchner tried to forge from the principle an unassailable foundation for their "scientific" materialism, perceptive scientists discovered a most far-reaching aspect of energy transformations. Upon hitting in 1848 on an almost forgotten paper by Carnot, W. Thomson, the future Lord Kelvin, realized that something was always lost in the transformation of heat into mechanical energy or vice versa. Although Joule, for one, urged the abandonment of Carnot's principle, Kelvin saw that such a step would entail "innumerable other difficulties."[118]

As it turned out, Joule's fears were unfounded. The generalization of Carnot's principle did not contradict the principle of conservation of energy, but it certainly shattered the idea of a world machine running forever. The cosmological implication of the loss of a part of the utilizable energy in every physical process was spelled out by Kelvin as early as 1852.[119] Two years later Helmholtz himself appraised Carnot's principle "as a universal law of nature" which radiated light "into the distant nights of the beginning and of the end of the history of the universe."[120] In 1865 Clausius summed up the Second Law of thermodynamics in the now famous statement: "The entropy of the universe tends towards a maximum."[121] After that only a few years passed before two theologically minded Scottish physicists, B. Stewart and P. G. Tait, concluded that the law of entropy proved it absolutely certain that the minimum and maximum entropy of the universe represented its beginning and end.[122]

The cogency of such reasoning could be readily challenged in an age that still blissfully believed in the infinity of the universe in spite of the gravitational and optical (Olbers') paradox. Kelvin insisted with baffling naiveté that it was "impossible to conceive a limit to the extent of matter in the universe; and therefore science points rather to an endless progress, through endless space, of action involving the transformation of potential energy into palpable motion and thence into heat, than to a single finite mechanism, running down like a clock, and stopping forever."[123] Maxwell's position was less optimistic. In 1870 he noted at the meeting of the British Association in Liverpool that the maximum cosmic entropy represented a limit beyond which science could not reach. The entropy brought certain ideas home for the late-nineteenth-century scientist with greater surprise, he added, than "any observer of the course of scientific thought in former times would have had reason to expect."[124]

The measure of surprise can be sensed in the somewhat frantic efforts that tried to dissipate the sombre theological clouds cast over scientific cosmology. Some of these efforts were patently hapless, while others were sophisticated though clearly tendencious. All, however, implied the notion of a universe capable of restoring in endless cycles the energy dissipated across the endless expanse of space. As early as 1852 W. J. M. Rankine, one of the founders of thermodynamics, speculated under the impact of Kelvin's statement made in the same year about immense, concave ether-walls in the universe. To these Rankine assigned the role of refocusing the dissipated energy into small volumes of space making thereby the formation of new stars and new planetary systems possible. According to Rankine's conception, the universe consisted of cosmic compartments in any of which either the reconcentration or the dissipation of energy was going on at any given time. "The world," to quote his words, "as now created, may possibly be provided within itself with the means of reconcentrating its physical energies, and renewing its activity and life."[125] Here 'possibly' seemed to mean 'really'.

Whatever one may think of Rankine's speculations, he at least faced with frankness a real problem instead of trying to talk it away or give it the silent treatment. This latter procedure was taken by John Tyndall, Faraday's successor at the Royal Institution. The most startling feature of his voluminous monograph, Heat a Mode of Motion, which in the 1860's went through four editions, was its author's silence about the Second Law of thermodynamics.[126] Those familiar with the monism of Tyndall's famous address[127] given before the British Association meeting in Belfast in 1874, can easily surmise the key to Tyndall's cultivated silence. For the one who saw in the First Law of thermodynamics the very foundation of an eternal universe, the Second Law must have been a bitter pill, not to be swallowed if possible. Tyndall clearly tried to chase away the ghost of entropy as he described in his book on heat a universe derived from only the First Law: "The energy of Nature is a constant quantity, and the utmost man can do in the pursuit of physical truth, or in the applications of physical knowledge, is to shift the constituents of the never-varying total, sacrificing one if he would produce another. The law of conservation rigidly excludes both creation and annihilation. Waves may change to ripples, and ripples to waves – magnitude may be substituted for number, and number for magnitude – asteroids may aggregate to suns, suns may invest their energy in florae and faunae, and florae and faunae may melt in air – the flux of power is eternally the same. It rolls in music through the ages, while the manifestations of physical life as well as the display of physical phenomena are but the modulations of its rhythm."[128]

The smallness of the last word was in inverse proportion to its significance in Tyndall's cosmological views. This was recognized by a no less literate advocate of scientific monism at that time than Herbert Spencer. He noted with satisfaction that "after having for some years supposed myself alone in the belief that all motion is rhythmical, I discovered that my friend, Professor Tyndall, also held this doctrine."[129] The word doctrine could not have been better chosen. For Spencer the doctrine of the rhythmic nature of every motion was a fundamental tenet in the sense of a metaphysical or religious doctrine. His First Principles, which fittingly enough was the first volume in his System of Synthetic Philosophy, is a repetitious contention on behalf of the cyclic nature of every process including the universe as a whole: "Rhythm is a necessary characteristic of all motion. Given co-existence everywhere of antagonist forces – a postulate which, as we have seen, is necessitated by the form of our experience – and rhythm is an inevitable corollary from the persistence of force."[130]

The only saving grace in Spencer's rambling cosmological discourse was his assertion that "infinity was inconceivable," but he applied this dictum to the universe only to note that equally inconceivable was a motion "which never had a commencement in some pre-existing source of power."[131] This chain of physical causality formed in his eyes a series returning into itself and turning the whole universe into an all-encompassing cyclic entity:

> The universally co-existent forces of attraction and repulsion, which, as we have seen, necessitate rhythm in all minor changes throughout the Universe, also necessitate rhythm in the totality of its changes – produce now an immeasurable period during which the repulsive forces predominating, cause universal concentration – and then an immeasurable period during which the repulsive forces predominating, cause universal diffusion – alternate eras of Evolution and Dissolution. And thus there is suggested the conception of a past during which there have been successive Evolutions analogous to that which is now going on; and a future during which successive other such Evolutions may go on – ever the same in principle but never the same in concrete result.[132]

A little reflection could have shown Spencer that given an infinite universe in space and time the same concrete configurations were bound to reoccur including another Spencer, nay an infinite number of Spencers, writing the same books in the infinitely numerous planetary systems in an eternal universe. But the one who impetuously jumped at the age of seventeen into so-called "railroad engineering" and refused afterwards any systematic training, could grandly pass over some problems created by the exactness of mathematical physics. The poet and rhetorician in Spencer got the upper hand time and again. This is well illustrated in the manner in which Spencer somewhat dreamily speculated about concave boundaries of the ether in cosmic spaces so that the dissipated energy might be reradiated to keep the cycles going.[133] Even more naive was his solution of the problem in the absence of such boundaries. A sidereal system might be restored, he opined, by energy radiated into its expanse from neighbouring sidereal systems. He could not help feeling that those ether-walls gave a tenuous support at best to his reasoning about the universe. It was a reasoning that looked down on anything at variance with it, whether in science or in history. Only utter smugness could inspire Spencer's words about faith in Creator and creation: "If we analyze early superstitions, or that faith in magic which was general in later times and even still survives among the uncultured, we find one of its postulates to be, that by some potent spell Matter can be called out of non-entity, and can be made non-existent. If men did not believe this in the strict sense of the word (which would imply that the process of creation or annihilation was clearly represented in consciousness), they still believed that they believed it; and how nearly, in their confused thoughts, the one was equivalent to the other, is shown by their conduct."[134]

Such misrepresentation of cultural history matched the hollow optimism which Spencer zealously preached. He held high the prospect of the state of greatest happiness, the grand state of equilibrium of contending physical forces in a given sidereal system. Yet, he had to admit that it was bound to end in dissolution and in the repetitions of the same forever. Whether it was that prospect that created modern scientific civilization in which he prided himself was a question of which he remained unaware. His friend, Tyndall, offered in his Belfast Address as ultimate human destiny the prospect of the melting "of you and I, like streaks of morning cloud, . . . into the infinite azure of the past."[135] More verifiable was Tyndall's warning given to his large and distinguished audience that the evolutionary and cyclic concept of the universe was steadily rising in its attractiveness.[136]

The assent which Rankine's concave ether-walls could hardly elicit was more spontaneously forthcoming when decades later Boltzmann rested much the same contention on the laws of statistical thermodynamics. It appears unlikely that Boltzmann simply wanted to echo Rankine, but the fundamental choices in cosmology are too narrow to prevent unwitting though highly revealing repeat performances, such as are contained in Boltzmann's words: "The laws of probability calculus imply that, if only we imagine the world to be large enough, there will always occur here and there regions of dimensions of the visible sky with a highly improbable state of distribution."[137] In a somewhat bolder vein Rankine had suggested that some distant points of light in the sky might be the evidences of fresh energy concentrations in the foci of concave ether-walls.[138]

What seemed to give special weight to Boltzmann's reasoning was the fact that he was a chief architect of statistical thermodynamics. Still, something was missing in Boltzmann's argument aimed at exorcising the theological perspective from cosmology. He perhaps banished God, but hardly that demon which was first conjured up by another chief figure of statistical thermodynamics and an idol of Boltzmann, James Clerk Maxwell. In his *Theory of Heat* Maxwell described a being "whose faculties are so sharpened that he can follow every molecule on its course."[139] He placed that truly demon-like figure at the stopcock connecting two vessels to illustrate the point that a "miraculous" intervention was needed to achieve the practically impossible: the reconcentration of high-energy molecules in one of the vessels. If the case of two small glass containers was fraught with "miracles," what then about the infinite number of molecules darting about in immensely large volumes of space with unimaginable wide ranges of speed? Clearly, the demon was a reminder, in the style of Maxwell's gentle sarcasm, about a direction which pointed toward wishful thinking and not toward fruitful scientific speculations.

The hint could not, however, be well taken by those who saw in the spectre of the finiteness of the world in time a threat to scientific thinking. Their eagerness to find a way out of the dilemma well illustrated the fact that scepticism would ultimately invade rationalistic thought when narrow-minded rationalism is defended against overwhelming evidence. A case in point was Mach's struggle against the cosmological bearing of the Second Law of thermodynamics. According to him it was logically defective to extend the law of entropy to the entirety of the universe as science could only investigate a limited number of phenomena. He insisted that as far as scientific reasoning went there was a certain gap between the parts and the whole of the universe and he summarized this point in several variations. "The universe," he wrote, "is like a machine in which the motion of certain parts is determined by the motion of others, only nothing is determined about the motion of the whole machine." The restatement of this in terms of a cosmic timetable meant that if one part of the universe served as a clock for the other part "we have nothing left over to which we could refer the universe as to a clock."[140] He was still to be shocked by radioactivity.

What Mach really offered here was a somewhat ambiguous exercise in logic. Furthermore, he was patently inconsistent as in other respects he firmly maintained the applicability of the laws of physics to the whole universe. It is to him that modern relativistic cosmology owes the notion that the inertia of matter might be due to the interaction of *all* matter in the universe. It was, of course, his unshaken belief that the universe was infinite, and therein lay the cause of his baffling oversight of that elementary rule of logic which forbids the interchange of proofs and claims. He should have known better. His lifelong silence about the gravitational and optical paradoxes inherent in an infinite, homogeneous universe constitutes a most puzzling phenomenon in view of his irrepressibly philosophical bent of mind. As an observer and critic of the latest trends and findings in experimental psychology, Mach could hardly miss, for instance, a paper by W. Wundt, a leader of such studies, who analysed at length the merits of the idea of a finite universe.[141] Wundt's considerations constituted a detailed reply to an ingenious section of an otherwise rambling book by J. C. F. Zöllner,[142] the first to hold a chair of astrophysics at the University of Leipzig.

In 1872 Zöllner proposed nothing less revolutionary than abandoning the idea of an infinite universe consisting of an infinite amount of matter spread out evenly in an infinite space. As Zöllner correctly noted, the optical and gravitational paradoxes allowed but a finite total mass in the universe. Abandoning the idea of an infinite universe implied, however, parting with Euclidean space, as it was highly repugnant to conceive a finite universe with abrupt boundaries beyond which there could only lie the absolute nothing. The solution of this difficulty lay, according to Zöllner, in the acceptance of non-Euclidean geometry for cosmology as suggested in 1854 in a now historic paper by B. Riemann. But Zöllner went far beyond Riemann. In an astonishing paragraph he explained that in such a finite, non-Euclidean world the particles of matter would produce an oscillation on a cosmic scale between the relatively highest and lowest concentration of matter. The particles of matter must "after finite intervals of time, the magnitude of which depends on the velocity of particles and on the curvature of space, approach again one another and in such a manner that the pendulum-like, periodic, living force would change into attraction during the contraction [Annäherung] and into a repulsion during the expansion [Entfernung]."[143]

Half a century later Zöllner's ideas were still startlingly novel when Einstein outlined his General Theory of Relativity and when soon afterwards the modern idea of an oscillating universe was first derived from Einstein's original proposals. Until then scientists, who felt it their sacred duty to vindicate the infinity of the universe in time, fell back unsuspectingly on the inexhaustible energy reservoirs of an infinite universe. The efforts in cosmology of S. Arrhenius,[144] L. Zehnder,[145] and others, who tried to come up with something more plausible than Rankine's ether-walls, provided only fresh instances of the basic contradictoriness of such an enterprise. In their schemes not only physics failed to receive its due, but also the impartiality of scientific reasoning, which above all should be a respecter of facts, no matter what their wider implication may be.

The most palpable of those wider implications is that the various moments of time have an ineradicable uniqueness of their own, that time is an irreversible flow, that the historical process both on the human and on the cosmic level is arranged, as far as experience shows, as a straight arrow. True, no specific experience, including common-sense experience, is necessarily the ultimate form of truth. The human mind is able to imagine constructs that by running counter to hallowed human experiences can put them in deeper perspectives. The interpretation of physical effects on the basis of statistical mechanics is a good example. It has proved itself a major advance in physical science, but only so far as physicists looked for what it predicted as a most probable event. Yet, the success of the theory rests, paradoxically enough, on assumptions that make, theoretically at least, realizable even the most improbable. As Boltzmann, one of the founders of statistical mechanics, put it, a model of the universe consisting of atoms forever in random motion rested on equations in which "the positive and negative directions of time are equivalent."[146] The possible consequences of this postulate are, as can easily be guessed, rather drastic for man's interpretation of the physical world. Boltzmann himself phrased the problem in questions that hid nothing of the real nature of the new situation: "Is the apparent irreversibility of all known natural processes consistent with the idea that all natural events are possible without restriction? Is the apparent unidirectionality of time consistent with the infinite extent or cyclic nature of time?"[147]

The answer given by statistical mechanics is well known and can be summed up in a few words. Stars and galaxies should be considered as minute parts of the universe, and geological epochs should be viewed as mere fragments of a second on the cosmic scale. Such a universe, because of its immensely long past, is already in its most probable state, that is, in thermal equilibrium, or in a condition of complete inactivity. But from the mathematical viewpoint the possibility, however small, remains for a most improbable concentration of energy in small pockets of the universe. Thus, all of a sudden a large number of atoms, that represent merely a speck in the universe, may reverse their tendency to disperse, and as a result a new galaxy may begin to evolve out of cosmic ashes at a most unexpected moment and in a most unlikely corner of the universe. Or as Boltzmann painted the new cosmic picture: "In the entire universe, the aggregate of all individual worlds, there will however in fact occur processes going in the opposite direction. But the beings who observe such processes will simply reckon time as proceeding from the less probable to the more probable states, and it will never be discovered whether they reckon time differently from us, since they are separated from us by eons of time and spatial distances $10^{10^{10}}$ times the distance of Sirius – and moreover their language has no relation to ours."[148]

Boltzmann's claim that such a model of the universe was "at least a possible one" and that it was "free of inner contradictions," was true from the viewpoint of mathematics, but mathematical truth is not always enough even in physics, let alone in a broader understanding of the cosmos. The saving grace of the Boltzmannian cosmology was that its most special features were relegated to the realm of the unobservable, to the realm of the infinitely distant. Nor was anything genuinely scientific in the possibility of a world composed of parts in which the structure (and language which presumably grew out from the physical structure) of one part had no relation to the other, and could even be diametrically opposed to one another. But fascination with mathematical glitter and mistrust in traditional patterns of reasoning can trap a mind even as cautious as that of Boltzmann. The real thrust and preferences of his thinking came to the fore as he advised his reader about the difference between time as a straight, infinite line, or as a closed circle: "In any case, we would rather consider the unique directionality of time given to us by experience as a mere illusion arising from our specially restricted viewpoint."[149] This choice he seemed to take seriously.

In view of this one wonders whether Boltzmann really meant what he wrote about his world model: "Obviously no one would consider such speculations as important discoveries or even – as did the ancient philosophers – as the highest purpose of science. However, it is doubtful that one should despise them as completely idle. Who knows whether they may not broaden the horizon of our circle of ideas, and by stimulating thought, advance the understanding of the facts of experience."[150] Whatever the success of statistical theory in the laboratory, or even in astronomy, the universe failed to provide any evidence for those most improbable cosmic reverse processes. More importantly, Boltzmann's chiding of "ancient philosophers" shows the proverbial insensitivity of the natural philosopher (as physicists were still called in his time) for the philosophy of nature on which ultimately physics, too, rests. In a crucial measure physical science itself owes its only viable emergence to the concern of some "ancient philosophers" whether the flow of events, human and cosmic, was linear or cyclic. In Boltzmann's century, and Boltzmann could have very well seen this, the problem still produced agitated and startling reflections. It would have been within his easy reach to see in the writings of some of his well-known contemporaries that a consistent espousal of the cyclic patterning of time yields a mawkish picture of the cosmos and of human history as well. Boltzmann gave no sustained attention to this possibly because of the mental reserve that usually accompanies thorough scientific training. Others in his century who lacked that training had become, as the next chapter will show, hapless captives to the inner logic of eternal reversals as well as returns. The murky backwaters they produced is a striking vindication of the clear vision generated by the notion of a universe conceived as the Creator's handiwork.

NOTES

1. Subsequent references will be to the translation by Stillman Drake (Berkeley: University of California Press, 1962).
2. *Ibid.*, p. 53.
3. *Ibid.*, p. 57. Hardly an easy task, because only God can bring it about, adds Salviati with devout emphasis.
4. *Ibid.*, p. 59.
5. *Ibid.*
6. *Ibid.*, p. 60.
7. *Ibid.*, p. 61.

8. *Ibid.*, p. 62.
9. *Ibid.*, p. 185.
10. *Ibid.*, p. 265.
11. *Ibid.*, p. 234.
12. *Ibid.*, p. 421.
13. *Ibid.*, p. 104.
14. *Ibid.*
15. *Ibid.*, pp. 328, 334, 339.
16. *Ibid.*, p. 268.
17. Quoted in C. C. Gillispie, *The Edge of Objectivity: An Essay in the History of Scientific Ideas* (Princeton, N.J.: Princeton University Press, 1960), p. 53.
18. *Dialogue*, p. (v).
19. Subsequent references will be to the edition in *Œuvres de Descartes*, edited by Charles Adam and Paul Tannery (nouvelle présentation; Paris: J. Vrin, 1967), vol. XI.
20. Translated in *The Philosophical Works of Descartes*, by E. S. Haldane and G. R. T. Ross (rev. ed.; Cambridge: Cambridge University Press, 1931), vol. I, pp. 81–130.
21. *Ibid.*, p. 81.
22. *Ibid.*, p. 105.
23. *Ibid.*
24. *Ibid.*, p. 107. Here Descartes merely repeated a claim frequently voiced on the pages of his *Le Monde*.
25. *Le Monde*, chap. vii, p. 47.
26. The original Latin text published in 1644 was followed by the French version in 1647.
27. *Le Monde*, chap. vii, pp. 38, 41, 44.
28. *Ibid.*, p. 47.
29. *Ibid.*, p. 39.
30. *Ibid.*, p. 43.
31. *Discourse on the Method*, p. 109.
32. Letter to Mersenne, July 27, 1638, in *Œuvres*, vol. II, p. 268.
33. Letter to Mersenne, May 17, 1638, in *Œuvres, vol. II*, p. 142.
34. *Discourse on the Method*, p. 99.
35. *Ibid.*, p. 121.
36. *Ibid.*, p. 122.
37. A rather natural attitude on the part of one about whom Huygens later could say that he "put forward his conjectures as verities, almost as if they could be proved by his affirming them on oath. . ."; see "Remarques de Huygens sur *La vie de Descartes* par Baillet," in M. V. Cousin, *Fragments philosophiques* (5th ed.; Paris: Didier, 1865–66), vol. III, p. 118.
38. Letter to Mersenne, Dec. 23, 1630, in *Œuvres*, vol. I, pp. 195–96.
39. As E. Gilson pointed out (*René Descartes' Discours de la méthode: Texte et commentaire* [Paris: J. Vrin, 1925, pp. 456–57]) Bacon saw in the "instantia crucis" a means to ascertain the cause of a perplexingly complex set of appearances, whereas for Descartes the "experimentum crucis" merely helped to decide which among various explanations, equally probable *in se*, agreed with the course of nature.
40. In *The Works of Francis Bacon*, edited by J. Spedding, R. L. Ellis and D. D. Heath (Boston: Taggard and Thompson, 1863), vol. VI. All subsequent references to Bacon's writings will be to this edition. At this point mention should be made of the often quoted essays of M. B. Forster, "The Christian Doctrine of Creation and the Rise of Modern Natural Science," *Mind*, 43 (1934): 446–68; and "Christian Theology and Modern Science of Nature," *Mind*, 44 (1935): 439–66 and 45 (1936): 1–27. He made detailed analysis of the thought of Descartes, Galileo and Bacon but on the basis of a total neglect of their medieval forerunners and also in the conviction that the Christian notion of God (Creator) was at least as much Greek as it was Jewish. See on this p. 465, note 1 in his first article. Such a position was rather dated even in the 1930's.
41. *The Works of Francis Bacon*, vol. VI, p. 212.
42. *Ibid.*, p. 223.

43. *Ibid.*, p. 224.

44. See, for instance, F. H. Anderson, *The Philosophy of Francis Bacon* (Chicago: University of Chicago Press, 1948), chaps. iv and v.

45. *Works*, vol. VI, p. 224.

46. Originally written in Latin; for the English version see *Works*, vol. X, pp. 341–99.

47. *Ibid.*, p. 386.

48. *Ibid.*

49. *Ibid.*

50. Subsequent quotations will be from the English translation of the Latin version in *Works*, vol. VIII.

51. *Ibid.*, p. 510.

52. *Ibid.*

53. *Ibid.*, p. 511.

54. *Ibid.*, p. 510.

55. *Ibid.*, p. 511.

56. In his Essay, "Of Vicissitude of Things," in *Works*, vol. XII, p. 275. In speaking in the "Descriptio globi intellectualis" about the various theories concerning the composition of stars, Bacon raised the question harking perhaps back to the notion of Great Year, "whether stars are in long revolutions of ages created and dissipated"; *Works*, vol. X, p. 456.

57. On Mersenne the indispensable modern source of information is the massive monograph by R. Lenoble, *Mersenne ou la naissance du mécanisme* (Paris: J. Vrin, 1943).

58. Paris: sumptibus Sebastiani Cramoisy, 1623. See Art. VI of Qaestio IV of cap. xliv (col. 715), where critical evaluation is made of the various meanings of *bereshit* ("in the beginning").

59. *Ibid.*, cols. 588–90; the Great Year is again mentioned in col. 1749, where metempsychosis is rejected.

60. Originally published in 1624. In the edition of 1630, published under the title, *Questions rares et curieuses* etc. (Paris: chez Pierre Billaine, MDCXXX), the precession of equinoxes with its 28,000 years is referred to in connection with other specific data of the solar system that, according to Mersenne, can only be explained by the sovereign choice of the Creator. See chap. vi, pp. 96–120.

61. See Pensée 374 in *Œuvres complètes*, edited by J. Chevalier (Paris: Bibliothèque de la Pléiade, Éditions Gallimard, 1954), p. 1187.

62. See *The Excellency of Theology: or the Pre-eminence of the Study of Divinity above that of Natural Philosophy*, in *The Works of the Honourable Robert Boyle* (new ed.; London 1772), vol. IV, p. 11.

63. *Ibid.*, pp. 10–11.

64. *Ibid.*, p. 11.

65. In Book III, chap. xvii, pp. 246–51 in *A Collection of Several Philosophical Writings of Dr. Henry More* (4th ed.; London: Joseph Downing, 1712). Originally published in 1659.

66. As a source of his information on the views of various Stoics and Epicureans about the Great Year, More gives Lipsius' *Physiologia Stoicorum*, a work discussed in the previous chapter. See p. 266.

67. Subsequent references are to the 3rd edition, London: W. Innys, 1713.

68. In Part III, chap. iv, entitled "The Opinions of the Ancient Heathen Philosophers and Other Writers concerning the Dissolution," pp. 326–37.

69. *Ibid.*, p. 337.

70. *The Annals of the World* . . . (London: E. Tyler, 1658), p. 1. The original Latin was published in 1650. Ussher's modern detractors have not, of course, consulted the original text. It would have made it clear to them that Ussher did not propose his universal timetable as a tenet of the Church. Nor did Ussher fix the date of Creation for 9 o'clock in the morning as some modern scientists claimed. It is not known on whose authority Ussher's dates were added as marginal notes in the reference editions of the Authorized Version Half a century later Bossuet's *Discours sur l'histoire universelle* carried similar marginal note

starting from 4004 B.C. See p. 15 in the critical edition containing the variants of the several revised editions of that work (Paris: Emler Frères, Libraires, 1829).

71. On the dominance of that idea on Newton's whole life and thought, see F. E. Manuel, *A Portrait of Isaac Newton* (Cambridge, Mass.: The Belknap Press of Harvard University Press, 1968).

72. The claim that Newton believed in a cyclic cosmos was made by D. Kubrin in his "Newton and the Cyclical Cosmos: Providence and the Mechanical Philosophy," *Journal of the History of Ideas*, 28 (1967): 325–46. The principal weakness of Kubrin's claim rests on his distinctly unbalanced manner of weighing the evidences pro and con, and on his failure to specify sufficiently the meaning of "cyclic cosmos."

73. For a convenient reference, see the Dover edition of the *Opticks* (New York, 1952), pp. 400, 402.

74. *Ibid.*, p. 402.

75. See the translation by A. Motte, revised by F. Cajori, *Principia or Mathematical Principles of Natural Philosophy* (Berkeley: University of California Press, 1960), p. 541.

76. See the text in D. Kubrin, *art. cit.*, p. 343.

77. *Newton's Philosophy of Nature: Selections from his Writings* edited by H. S. Thayer (New York: Hafner, 1960), p. 54.

78. As Kubrin agrees with M. Mandelbaum, *art. cit.*, p. 331 note.

79. *Principia*, p. 544.

80. In *The Works of Richard Bentley*, edited by A. Dyce (London: Francis Macpherson, 1838), vol. III; see especially Sermon iii, pp. 51–72.

81. *Ibid.*, pp. 65–66.

82. *Ibid.*, pp. 66–68.

83. This was discussed in detail in my *The Paradox of Olbers' Paradox* (New York: Herder & Herder, 1969), pp. 60–66.

84. As extensively documented in the same work.

85. Written under the title, "De rerum originatione radicali," in 1697, but not published until 1840. Modern English translation is available in *Monadology and Other Philosophical Essays*, translated by P. Schrecker and A. M. Schrecker (Indianapolis: Bobbs-Merrill Co., 1965), pp. 84–94.

86. *Ibid.*, p. 93.

87. *Ibid.*, p. 91.

88. *Ibid.*, p. 94.

89. "Critical Remarks concerning the General Part of Descartes' *Principles*," *ibid.*, pp. 22–80.

90. First published in 1755. Subsequent references are to the English translation of the first two Parts in W. Hastie, *Kant's Cosmogony* (Glasgow: James Maclehose and Sons, 1900). For further details, see my *The Milky Way: An Elusive Road for Science* (New York: Science History Publications, 1972), pp. 197–98.

91. Although Kant realized the lack of meaning of a centre in an infinite Euclidean space, he argued that a physical universe must have a centre which he identified with the point of largest density where the world-formation started following the creation of matter. See *Kant's Cosmogony*, pp. 140–42.

92. The Third Part of Kant's work is devoted to the physical and moral characteristics of the inhabitants of each planet.

93. *Ibid.*, p. 154.

94. *Ibid.*, p. 152.

95. *Ibid.*, pp. 152–53.

96. *Cosmologische Briefe über die Einrichtung des Weltbaues* (Augsburg: Eberhart Kletts Wittib., 1761).

97. As stated by Lambert in a letter to Kant. See Hastie, *Kant's Cosmogony*, p. lxx.

98. *Cosmologische Briefe*, p. 220.

99. Written in 1761, but published only in 1779, three years after Hume's death. References are to the edition by H. D. Aiken (New York: Hafner Publishing Co., 1948). For quotation, see p. 52. Hume's discussion of cosmology is mainly in Part VIII of the work.

100. *Ibid.*, p. 53.
101. *Ibid.*, pp. 52–53.
102. *Ibid.*, p. 54.
103. *Ibid.*
104. *Ibid.*, p. 56.
105. Note VII in the fifth edition published in 1824 (Paris: Bachelier), pp. 409–18. The actual and considerably more complicated story of Laplace's nebular hypothesis is discussed in my forthcoming book, *The Puzzling World of Planets: A History of Theories on the Evolution of Planetary Systems.*
106. Except in the form of a discussion of Newton's Scholium to the *Principia*; see *ibid.*, pp. 393–94.
107. *Ibid.*, p. 393.
108. *The Monthly Review or Literary Journal* (London), 18 (1795): 13–16.
109. *Ibid.*, p. 15.
110. *Ibid.*, p. 16. The Scottish philosopher, Dugald Stewart, mentioned these speculations together with the ancient idea of Great Year, as examples of the sometimes overpowering and misleading influence of mathematical patterns on human thought, in his *Elements of the Human Mind* (3 vols., 1792–1827); see *The Collected Works of Dugald Stewart*, edited by Sir William Hamilton (Edinburgh: Thomas Constable, 1854), vol. III, pp. 168–69. In another section of that work, Stewart asked with tongue in cheek whether those who admired the idea of Great Year should not also consider recurring decimals as physical events. See *Works*, vol. IV, p. 205.
111. See, for instance, the latter work, translated by June Barraclough (New York: The Noonday Press, 1955), pp. 4–5.
112. *Ibid.*, p. 72. ("The Fifth Stage").
113. Quoted in C. A. Lubbock, *The Herschel Chronicle* (Cambridge: University Press, 1933), p. 197.
114. In *The Edinburgh Review* 87 (1848): 170–229. The review covered only the English translation of the first volume published in 1846, a year after the publication of the first volume of the German original.
115. *Ibid.*, p. 178.
116. *Ibid.*
117. *Ibid.*, pp. 178–79.
118. "An Account of Carnot's Theory of the Motive Power of Heat" (1849), in *Mathematical and Physical Papers* (Cambridge: Cambridge University Press, 1882), vol. I, p. 119.
119. "On a Universal Tendency in Nature to the Dissipation of Mechanical Energy," in *Mathematical and Physical Papers*, vol. I, pp. 511–14.
120. "On the Interaction of Natural Forces" (1854), in *Popular Lectures on Scientific Subjects*, translated by E. Atkinson (New York: D. Appleton, 1873), p. 193.
121. "Ueber verschiedene für die Anwendung bequeme Formen der Hauptgleichungen der mechanischen Wärmetheorie," *Annalen der Physik und Chemie* 125 (1865): 400.
122. *The Unseen Universe* (9th rev. ed.; London: Macmillan 1880), pp. 127–28.
123. "On the Age of the Sun's Heat" (1862), in *Popular Lectures and Addresses* (London: Macmillan, 1891), vol. I, pp. 349–50.
124. *The Scientific Papers of James Clerk Maxwell*, edited by W. D. Niven (Cambridge: Cambridge University Press, 1890), vol. II, p. 226.
125. "On the Reconcentration of the Mechanical Energy of the Universe," in *Miscellaneous Scientific Papers*, edited by W. J. Millar (London: Charles Griffin & Co., 1891), p. 202.
126. Subsequent references will be to the American printing of the fourth edition (New York: D. Appleton, 1881).
127. In the *British Association Report, 1874* (London: Murray, 1875), pp. lxvi–xcvii.
128. *Heat a Mode of Motion*, p. 467.
129. *First Principles* (3rd ed.; London: Williams and Norgate, 1870), p. 253, note. The historian, John William Draper, echoed Spencer as he closed his discussion of the main conceptual acquisitions of the "European Age of Reason" with a statement of the cyclic characteristic

of every process, historic and cosmic alike. In his words "the multiplicity of worlds in infinite space leads to the conception of a succession of worlds in infinite time." He spoke, of course, of the creation (of which he mentioned only its kind of very recent date) and approaching end of all things, as "unworthy hypotheses." See his *History of the Intellectual Development of Europe* (New York: Harper and Brothers, 1876), vol. II, p. 336.

130. *First Principles*, p. 271.

131. *Ibid.*

132. *Ibid.*, p. 537.

133. *Ibid.*, p. 535.

134. *Ibid.*, pp. 172–73. Immediately afterwards. Spencer singled out contemporary theology as still being dominated by the same confusion in its advocacy of a beginning and end of the world.

135. *British Association Report, 1874*, p. xcvii.

136. *Ibid.*, pp. xcvi–xcvii.

137. "Über statistische Mechanik," (1904), in *Populäre Schriften* (Leipzig: J. A. Barth, 1905), p. 362.

138. *Art. cit.*, p. 202.

139. 3rd ed.; London: Longmans, Green & Co., 1872, p. 308.

140. *History and Root of the Principle of the Conservation of Energy*, translated by P. E. B. Jourdain (Chicago: Open Court, 1911), p. 62.

141. "Über das kosmologische Problem," in *Vierteljahrschrift für wissenschaftliche Philosophie* (Leipzig), 1 (1877): 80–136.

142. *Über die Natur der Cometen: Beiträge zur Geschichte und Theorie der Erkenntniss* (Leipzig: Wilhelm Engelmann, 1872), pp. 299–312.

143. *Ibid.*, pp. 308–09.

144. In various, once most popular, but today deservedly forgotten works, and especially in an article aimed at evading the conclusiveness of Olbers' Paradox about the finiteness of the universe. See the English translation of that article, "Infinity of the Universe," *The Monist*, 21 (1911): 161–73.

145. The most complete form of Zehnder's ideas on this point is in his *Der ewige Kreislauf des Weltalls* (Braunschweig: F. Vieweg, 1914).

146. *Lectures on Gas Theory*, translated by S. G. Brush (Berkeley: University of California Press, 1964), p. 446. The German original was published in two parts in 1896 and 1898.

147. *Ibid.*

148. *Ibid.*, pp. 447–48.

149. *Ibid.*, p. 446.

150. *Ibid.*, p. 447.

CHAPTER THIRTEEN

On Murky Backwaters

The history of classical physics from Galileo to Kelvin is a process with many a lesson. It illustrates, for instance, the tenacious ability of science to purge itself, in the long run at least, of some false assumptions and illusory expectations. The scientific movement also displays an effectiveness which prevents its mighty stream from being overly polluted by the backwaters which keep forming along its main course. Murky and repulsive as these backwaters may be, a more than cursory look at them may well repay the effort. Much about health is learned from the analysis of disease, and the outlines of the rules of rationality can be more sharply discerned against the dark background of obscurantist aspirations.

Not too surprisingly, some of the willful trends of thought running counter to science succeeded in putting on the veneer of scientific repute. Their chief attractiveness usually consisted in some deceivingly simple approach toward unlocking fundamental puzzles of nature. The simplicity of the approach meant that one could dispense with the hard-to-learn scientific method, provided one possessed by nature an intuitive, almost magic power to look at things in the "right" way. Another attractive feature of such pseudo-scientific trends usually consisted in their partially justified insistence that "official" or quantitative science was not doing justice to the whole man, that its method could not account for purpose and beauty, these essential ingredients of proper understanding.

Most modern forms of pseudoscience have their roots in the rise of Romanticism. Various features of Romanticism are all too evident, for instance, in the basic assumptions as well as in the vagaries of Naturphilosophie. Naturphilosophie looked for its theoretical justification in the works of Fichte, Schelling, and Hegel, chief representatives of German Romanticism and Idealism. Common to their thinking was a theological orientation with distinctly pantheistic colouring, and this also meant a heavy emphasis on the priority of subjective consciousness over the external world. For Fichte, Schelling, and Hegel the external world was ultimately the emanation of the subjective Spirit, or Ego, which unfolded itself through the circular process of thesis, antithesis, and synthesis.

Fichte, Schelling, and Hegel were, of course, eager to find a reasonable escape from the clutches of an eternal cosmic treadmill, which was a logical consequence of some of their basic premises. Thus, one of Fichte's most widely read works, *The Vocation of Man*,[1] ends on the hopeful reflection that "the universe is to me no longer what it was before – the ever-recurring circle, the eternally-repeated play, the monster swallowing itself up only to bring itself forth again."[2] He took great care to impress the idea on his reader that "The sun rises and sets, the stars sink and reappear, the spheres hold their circle-dance, – but they never return again as they disappeared, and even in the bright fountain of life itself there is life and progress. Every hour which they lead on, every morning and every evening, sinks with new increase upon the world; new life and new love descend from the spheres like dew-drops from the clouds, and encircle nature as the cool night the earth."[3]

As proofs in support of his contention he offered a scientific and a philosophical assertion. The former suggested that in an infinite nature the same configuration of factors and circumstances can never occur twice. "The same circumstances can never return unless the whole course of Nature should repeat itself, and two Natures arise instead of one; hence the same individuals, who have once existed, can never again come into actual being. Further, the *man-forming* power of Nature manifests itself, during the same time in which I exist, under all conditions and circumstances possible in that time. But no combination of such circumstances can perfectly resemble those through which I came into existence, unless the universe could divide itself into two perfectly similar but independent worlds. It is impossible that two perfectly similar individuals can come into actual existence at the same time. It is thus determined what I, this definite person, must be; and the general law by which I am what I am is discovered."[4] The philosophical proof consisted in Fichte's spirited presentation of human will as the magic lever which lifts man to the plane of eternal destiny in union with the World Spirit: "And now the Eternal World rises before me more brightly, and the fundamental law of its order stands clearly and distinctly apparent to my mental vision. In this world, *will* alone, as it lies concealed from mortal eye in the secret obscurities of the soul, is the first link in a chain of consequences that stretches through the whole invisible realms of spirit; as, in the physical world, *action* – a certain movement of matter – is the first link in a material chain that runs through the whole system of nature. The will is the efficient, living principle of the world of reason, as motion is the efficient, living principle of the world of sense."[5]

If will was the ultimate wellspring for man to achieve his full self-realization in the universe, it was natural that Fichte accounted in terms of *will* for the creation of an intelligible world. By creation and Creator he meant the unfolding of the Eternal Will. The following passage should clearly convey the pantheistic and voluntaristic features of Fichte's thought on this point: "That Eternal Will is thus assuredly the Creator of the World, in the only way in which He can be so, and in the only way in which it needs creation: – in the finite reason. Those who regard Him as building up a world from an everlasting inert matter, which must still remain inert and lifeless, – like a vessel made by human hands, not an eternal procession of His self-development, – or who ascribe to Him the production of a material universe out of nothing, know neither the world nor Him."[6]

This flat rejection of the Christian idea of a Creator is even clearer in Fichte's crudely immanentist and subjectivist account of the reduction of the Creator to the awakening of the human mind to the level of complete self-consciousness: "Only in our minds has he created a world; at least that *from which* we unfold it, and that *by which* we unfold it; – the voice of duty, and harmonious feelings, intuitions, and laws of thought . . . In our minds He still creates this world, and acts upon it by acting upon our minds through the call of duty as soon as another free being changes aught therein. In our minds he upholds this world, and thereby the finite existence of which alone we are capable, by continually evolving from each state of our existence other states in succession."[7]

In view of this, one may almost expect that strange narrowing of the concept of reason evident in Fichte's description of a "merely rational" creation. For him it was equivalent to a smoothly operating machine in which all results were automatically obtained, including all human actions and enterprises. Freedom in that "rational" universe would have been, according to Fichte, the continual source of "unskillful direction," "wasteful extravagance and circuitous byways."[8] Consequently, there is not much convincing power in Fichte's spirited account of the gradual conquest by man of the method and results of science: "Science, first called into existence by the pressure of necessity, shall afterwards calmly and deliberately investigate the unchangeable laws of Nature, review its powers at large and learn to calculate their possible manifestations; and while closely following the footsteps of Nature in the living and actual world, form for itself in thought a new ideal one. Every discovery which Reason has extorted from Nature shall be maintained throughout the ages, and become the ground of new knowledge for the common possession of our race. Thus shall Nature ever become more and more intelligible and transparent, even in her most secret depths; human power, enlightened and armed by human invention, shall rule over her without difficulty, and the conquest, once made, shall be peacefully maintained. This dominion of man over Nature shall gradually be extended, until, at length, no farther expenditure of mechanical labour shall be necessary than what the human body requires for its development, cultivation, and health; and this labour shall cease to be a burden; – for a reasonable being is not destined to be a bearer of burdens."[9]

The Utopian ring on which the foregoing passage ends is as indicative of the inconsistencies and willfulness of Fichte's thought as is his remark on Schelling's *Naturphilosophie*. For it was hardly in line with Fichte's own thinking about nature to warn that "we must not be blinded or led astray by a philosophy assuming the name of *natural* which pretends to excel all former philosophy by striving to elevate Nature into Absolute Being and into the place of God."[10] Still, the remark was to the point. Schelling's philosophy of nature was imbued with pantheism and with its natural ally, idealistic subjectivism.

Here only a few samples can be given[11] of the shocking vagaries of Schelling's scientific account of the universe. He presented gravitation as corresponding to the male sex, and light as reflecting the female principle. Gravitation was also connected by him to animal instinct, and he tied plant life to magnetism. Human irritability manifested, according to him, outbursts of electricity. In the same vein Schelling set up a one-to-one correspondence between the five senses and the allegedly five fundamental forces of nature. It was with the same arbitrariness that he offered an a priori proof based on animistic considerations about the two foci which make the planetary orbits elliptical. The same animistic bipolarity was, in his opinion, at the root of the complementary processes of condensation and evaporation. After all this Schelling's description of man as a "perfect cube" may only come as a mild shock. Nor should it be surprising that a thinker who took such liberties with nature also believed that no secret of nature would ultimately 'remain hidden' to man, whom he described as the par excellence embodiment of the 'world spirit': "All motions of the great or little nature are concentrated in him, all forms of actuality, all qualities of earth and heavens. He is in a word the system of the universe, the fullness of infinite substance on a small scale – that is the integrated being, man become God."[12] Such a man knew not the difference between science and omniscience.

Schelling's manhandling of science will be better understood upon recalling that his thinking was wholly dominated by attention to circular patterns. He saw them everywhere in nature, in the life of society as well as in the history of thought. The ominous spectre of eternal recurrences as an inevitable consequence of his basic presuppositions could hardly escape him. In fact, it was with obvious desperation that he tried to dissipate the prospect of an eternal treadmill. Actually, an entire book of Schelling, *The Ages of the World*,[13] was devoted to that task. Still, to the question, "How or by what was life delivered from the cycle ["Umtrieb"] and led into freedom?",[14] Schelling gave no convincing reply. For him, "life did not begin at any time but began from all eternity never truly to end, and ended from all eternity to begin again and again."[15] He saw the very essence of life or existence in that endless circular pattern: "By that continual return to the beginning and the eternal recommencing, that life makes itself substance in the real sense of the word (*id quod substat*), into the always abiding; it is the constant inner mainspring and clockwork, it is time which is eternally beginning, eternally becoming, always devouring itself and always giving birth to itself again."[16]

The idea of circular repetitions is clearly implied in Schelling's description of existence as the perennial alternation of systole and diastole: "It is a completely involuntary movement which, once begun, automatically repeats itself. The beginning again, rising again, is a systole, is tension, which reaches its acme in the third potency; the returning to the first potency is diastole, relaxation, upon which, however, new contraction immediately follows. Consequently this is the first pulsation, the beginning of that alternating movement which goes through all visible nature, of the eternal contraction and eternal expansion of the universal ebb and flood."[17] According to him it was the emergence of an "eternal longing" within that pulsating life plasma that marked the possibility of rising above the perennial circularity: "This entrance of longing in eternal nature indicates a new moment, which we must therefore hold fast in our consideration. This is that moment which the divining primitive world designated by the breaking apart of the world-egg, by which it intimated precisely that closed wheel, that impenetrable motion which could not be stopped. This is the moment when the earthly and heavenly separate for the first time."[18]

Yet, in Schelling's reasoning, the realization of the "longing to escape from the eternal cycle"[19] merely transported the enslavement to cyclic processes from the unconscious level to the fully conscious participation in them. This is why Schelling kept insisting on the difference between the unconscious and conscious states of the universal being. "If there were nothing except that blind necessity, then life would remain in this dark, chaotic condition of an eternally and therefore never-beginning, eternally and therefore never-ending, movement. But the sight of eternal freedom raises that highest [power] of nature to freedom, too, and all other powers together with the highest come to stability and reality, since each power attains the place proper to it. And thus each shares the higher influence of which it is in immediate need, while indirectly all share the divine influence."[20] The divinity as conceived by Schelling was, however, a far cry from the Judeo-Christian God creating with an absolute beginning. Schelling claimed that succession in God is real,[21] that He has a life which rotates in an uninterrupted circle.[22] Commenting on the biblical name of God, Yahweh, Schelling phrased it as "I am who I was, I was who I shall be, I shall be who I am," and added: "Consciousness of such an eternity is impossible without a distinction of periods."[23] In the same vein Schelling insisted that the words of Genesis, "In the beginning . . .," designated the first of endless periods.[24] This truly baffling exercise in biblical interpretation did not only form an appalling instance of inconsistency in thinking. More importantly, it showed the extent to which Schelling's mind was dominated by the idea of a periodicity encompassing every facet and level of existence.

The connection between a pantheistic, anti-creationist metaphysics or theology and an interpretation of the world in terms of perennial cycles resulted in Schelling's hands in a revolting type of pseudoscience. This is also in evidence when one's attention is turned to Hegel. He expressed his belief in the fundamental nature of circular patterns whether writing about logic, philosophy of law, history of philosophy, or theology.[25] He defined the essential point to be observed by science as the circular return of the whole into itself where the first is the last and the last is the first.[26] This he explicitly contrasted with the concept of an absolutely given beginning. The Hegelian version of exact science illustrates only too well the morass of arbitrariness which seems to be generated when the pattern of perennial circularity is accepted as the supreme framework of thought and existence.[27]

The same inner logic reveals itself in the scientific writings of those who took as their guide one or the other of the leaders of German Idealism. Schelling's influence was, for instance, very strong on Lorenz Oken, the most important figure of Naturphilosophie. The theoretical and cosmological parts of his most extensive work, the *Elements of Physiophilosophy*,[28] are a strange mixture of pantheistic and Christian metaphysics with a distinct emphasis on the former. A case in point is Oken's treatment of the idea of creation out of nothing. While he seemed to assert the Judeo-Christian formulation of it,[29] he also stated that "the creation of the elements is none other than a representation of the three divine ideas in a finite sphere. Creation is a process of formation *of the* nothing."[30] (Italics added.)

This last phrase, which may sound senseless at its worst or naively poetical at its best, is, however, very logical in Oken's interpretation of nature where from the "Creator" to the lowest form of "created" being every process is characterized by a circular movement based on the tension of polarities. Thus, Oken stated that "for God to become real, he must appear under the form of the sphere. There is no other form for God. God manifesting [himself] is an infinite sphere." And so is the universe: "The universe is a globe, and everything, which is a Total in the universe, is a globe."[31] To be sure, not a motionless one. And here again Oken drew the closest parallel between God and the universe: "God is a rotating globe. The world is God rotating."[32] Clearly, then, the merits of a straightforward linear pattern with an absolute beginning became highly questionable, and this is precisely what Oken suggested: "All motion is circular, and there is everywhere no straight motion any more than there is a single line or straight surface. Everything is comprehended in ceaseless rotation. Without rotation there is no being and no life; for without it, there is no sphere, no space and no time."[33] Oken therefore was very consistent when he declared about time that "it is not a continuous stream, but a repetition of one and the same act, namely, the primary act, like as it were a rolling ball, which constantly returns upon itself."[34]

In order to do justice, however slight, to real processes which very often display no circularity, Oken made some concession about the exclusiveness of spherical and circular patterns. But he did not go so far as to admit without further ado the reality of straight lines: "There is no mathematically straight line in the world: all real lines are polar: they are all rooted in God by one extremity, by the other in finitude."[35] The key word in the passage was "polar." It gave Oken the chance to find everywhere foci of an ellipse, the pattern which seemed to satisfy both actual periodicity and the principle of thesis-antithesis-synthesis. He wrote: "No finite is absolutely spherical. As the real universe can only exist in a bicentral condition, so is there in this respect also no universal central body. It is there, but under the form of bicentrality, as sun and planet. God only is monocentral. The world is the bicentral God, God the monocentral world, which is the same with monas and dyas. The primary polarity, the dyas, the radiality, the light establishes itself in nature as bicentrality, which is the cosmogenic expression for self-manifestation or self-consciousness. Self-consciousness is a living ellipse."[36]

The last phrase should give a good insight into the nomenclature with the help of which Oken unfolded his views about gravitation, planetary motion, the comets, the ether, electricity, magnetism, and other topics in physics and astronomy. What he actually did represented a hopeless play on words, as for instance when he declared about the formation of planets from the ether: "The ether is not . . . absolutely imponderable, but only so in relation to the heavenly bodies. Light and heat are therefore ponderose substances, though they are not ponderable."[37] Such abandonment of the basic rules of consistency was the price exacted from those who cast their intellectual lot with the notion of an ultimate being which "is fluctuating in his eternity," to recall a startling phrase of Oken.[38] For them the world, too, became a fluctuating, amorphous entity, an embodiment of vagueness which does not yield to quantitative or experimental verification.

If one is to place Naturphilosophie within the trends generated by German Idealism it undoubtedly should be grouped with the ones forming the so-called Hegelian right. It was no coincidence that Naturphilosophie saw a strong revival a century or so later among the exponents of Nazi ideology. Nor was the swastika chosen without full recognition of the fact that throughout classical antiquity it served as the chief symbol of eternal recurrence. What occurred on the Hegelian right also evidenced itself in virtue of an inner logic on the Hegelian left. There, too, scientific sanity became a principal victim of willfully posited principles.

A classic and early illustration of this is the interpretation of the laws of physical science by F. Engels, a chief architect of Marxist dialectic. His admiration for Hegel was unbounded. Engels' Dialectics of Nature[39] contains well over fifty references to Hegel but only a few of them reveal a touch of criticism. By and large Engels readily subscribed to Hegel's astonishing dicta whether they concerned the laws and findings of physical science or the facts of its history. Thus, he saw merits in Hegel's stultifying pronouncements about what the relative distances of the planets ought to be.[40] He even found something basically good in Hegel's definition of electricity as "the anger, the effervescense, proper to the body," or "the angry self" of matter.[41] Hegel's supremacy in physics was again voiced by Engels when he wrote: "If nature itself proceeds exactly like old Hegel, it is surely time to examine the matter more closely."[42] Part of that matter concerned Hegel's claim about the presence of a repulsive force everywhere in nature as an indispensable counterpart of universal attraction. "Hegel is quite right in saying," Engels wrote, "that the essence of matter is attraction and *repulsion*."[43] The principal proof of this consisted of the basic laws of Hegelian dialectic on the basis of which it could be established *a priori*, so Engels insisted, that "the true theory of matter must assign as important a place to repulsion as to attraction, and that a theory of matter based on mere attraction is false, inadequate, and onesided."[44] Engels' arbitrary handling of scientific theory was fully matched by his evaluation of experimental proofs. According to him, the universal repulsion in nature evidenced itself in the pressure of light which bends, for instance, the tail of comets away from the sun.[45]

It was with similar highhandedness that Engels spoke of great men of physics. He called Newton "an inductive ass"[46] and dismissed one of Helmholtz's opinions as "pure childishness."[47] His comment on Thomson's and Tait's *Treatise on Natural Philosophy* was that "not only the ability to think, but also to calculate, has come to a standstill in the two foremost mechanicians of Scotland."[48] He felt no misgivings about brushing aside either a Maxwell or a Carnot. Faraday was spared of strictures only because his views were classified by Engels as closely duplicating those of Hegel.[49] But there could be no mercy for Clausius of entropy fame. In Engels' eyes Clausius was the bogeyman-scientist whom he tried to discredit, ridicule, or dismiss whenever opportunity arose. The reason for Engels' animosity toward Clausius is not difficult to identify. Clausius, entropy, and the heat-death of the universe meant one and the same thing for Engels. They represented the most palpable threat to the materialistic pantheism of the Hegelian left for which the *material* universe was and still is the ultimate, ever active reality. Engels made no secret about the fact that the idea of a universe returning cyclically to the same configuration was a pivotal proposition within the conceptual framework of Marxist dialectic. He saw the whole course of science reaching in Darwin's theory of evolution the final vindication of the perennial recurrence of all, as first advocated by the founders of Greek philosophy. Or as he registered the final phase of the conquest of science: "The new conception of nature was complete in its main features; all rigidity was dissolved, all fixity dissipated, all particularity that had been regarded as eternal became transient, the whole of nature shown as moving in eternal flux and cyclic course."[50]

Such a contention depended, of course, on the ability of dissipated energy to reconcentrate itself. This question, an insoluble enigma to the best minds in physics, represented no problem for Engels. While he admitted that radiating heat disappeared, so to speak, into infinite space, he felt sure that the cold bodies of defunct stars must, sooner or later, collide with one another. The enormous heat resulting from such collisions could then provide the energy necessary for the restarting of the whole process of physical and biological evolution in a particular section of the universe. Absolutely certain as such a conclusion appeared to Engels, he assigned as a future task for science to demonstrate that "the heat radiated into space must be able to become transformed into another form of motion, in which it can once more be stored up and rendered active."[51] About the positive outcome of such research he felt no doubt: "The eternally repeated succession of worlds in infinite time is only the logical complement to the co-existence of innumerable worlds in infinite space."[52] His concluding paragraph on the question should be quoted in full because of its explicitness on the eternal recurrence of all as the fundamental pattern of nature:

It is an eternal cycle in which matter moves, a cycle that certainly only completes its orbit in periods of time for which our terrestrial year is no adequate measure, a cycle in which the time of highest development, the time of organic life and still more that of the life of beings conscious of nature and of themselves, is just as narrowly restricted as the space in which life and self-consciousness come into operation; a cycle in which every finite mode of existence of matter, whether it be sun or nebular vapour, single animal or genus of animals, chemical combination or dissociation, is equally transient, and wherein nothing is eternal but eternally changing, eternally moving matter and the laws according to which it moves and changes. But however often, and however relentlessly, this cycle is completed in time and space, however many millions of suns and earths may arise and pass away, however long it may last before the conditions for organic life develop, however innumerable the organic beings that have to arise and to pass away before animals with a brain capable of thought are developed from their midst, and for a short span of time find conditions suitable for life, only to be exterminated later without mercy, we have the certainty that matter remains eternally the same in all its transformations, that none of its attributes can ever be lost, and therefore, also, that with the same iron necessity that it will exterminate on the earth its highest creation, the thinking mind, it must somewhere else and at another time again produce it.[53]

Engels' deep attachment to the idea of eternal succession of worlds is also evidenced in the manner in which he vindicated it as an integral part of the socialist, that is Marxist, Weltanschauung. His principal target in this respect was Eugene Dühring, privatdozent at the University of Berlin, who in 1875 declared himself a convert to socialism. Unfortunately for Dühring, he did not rest content with a general statement about his new allegiance, but published in the same year a lengthy justification of it, covering the fields of philosophy, science, sociology, ethics, economics, and history.[54] With his customary air of infallibility Engels immediately spotted Dühring's views as unorthodox, which could not be permitted to pollute the clear atmosphere of genuine socialism.[55] Tellingly enough what Engels found deficient in Dühring's "socialist" interpretation of natural science concerned among other things the perennial succession of worlds. On this score Dühring showed some hesitation as he found it unjustifiable, in view of the Second Law of thermodynamics, to equate matter, motion, and eternity. Thus, he left the door open, so Engels accused him, to the possibility of a cosmic beginning and end, beyond which the First Cause, or God, lurked. This represented in Engels' eyes the most fundamental and pernicious error from the viewpoint of socialism.

Whatever faults Engels did find with another socialist, Louis Auguste Blanqui,[56] the latter was certainly not remiss in elaborating the prospects of eternal recurrences. As a matter of fact Blanqui's *L'éternité par les astres*[57] is the lengthiest assertion of the topic written by a modern author. In view of the highly activist, revolutionary career of Blanqui, which covered a good part of nineteenth-century French history, his startling excursion into cosmology and philosophy of science should come as a surprise to anyone thinking of Marxism as a purely economico-social theory. Blanqui's "flights of astronomical fancy" bespeak an all-encompassing Weltanschauung and spell out in full, as was aptly noted, "the assumptions upon which rests his philosophy of being and change."[58] That philosophy was built around the need and justification of a merciless revolutionary programme aimed at an immediate and total transformation of society. The world picture of that programme matched fully its concept of society as composed of purely material units to be rearranged, transformed, or eliminated so that the momentarily best equilibrium state of society could be achieved in the shortest possible time.

The chief contention of the *L'éternité par les astres* is that any phrase and even any single moment of human civilization and history which has already occurred is actually occurring and will occur again on an infinite number of earth-like planets everywhere in the universe, and with the exact replicas of any number of human individuals, both famous and unknown. "Our earth," Blanqui wrote, "as well as the other celestial bodies, is the *repetition* of a *primordial* combination, which reproduces itself always the same and which exists simultaneously in billions of identical copies. Each copy is born, lives and dies in its turn. It is being born and is dying by the billions in each passing moment. On each of these 'earths' there succeed to one another all the material bodies, all the organic beings, in the same order, in the same place, at the same moment as they follow one another on the other earths, twin replicas of this earth. Consequently, all the facts already accomplished on our globe, or to be accomplished before its death, are accomplished in exactly identical form on the billions of its likes. And since the same holds for all the stellar systems, the whole universe is the permanent reproduction, without end, of a material and of a personnel always renewed and always the same."[59]

"As a result," Blanqui reasoned, "each man possesses in cosmic spaces an infinitely large number of duplicates that live his life in exactly the same manner as he lives it himself. He is infinite and eternal in the person of his alter egos, not only regarding his actual age, but all his ages. He has, by the billions, in each actual second, twin brothers who are being born, others who are just dying, others whose age is increasing from second to second, from his birth to his death."[60] Or putting it more concretely in Blanqui's words: "If someone interrogates the celestial regions, billions of his twins raise their eyes at the same time, with the same question in their thoughts, and all these glances cross one another invisibly. And it is not only once that these silent questions traverse the space, but always. Each second of eternity has seen and will see the situation of today, that is, the billions of twins of our earth inhabited by our twin egos. Thus, each of us has lived, is living, and will live without end in the form of billions of alter egos."[61]

Fantastic and obscurantist as such thoughts may appear, Blanqui could not be more emphatic in saying that in all this "there is no part of any revelation, or of any prophet, but the whole is a simple deduction from spectral analysis and from Laplace's cosmology."[62] In other words it was on science and science alone that Blanqui rested his case. Obviously, he leaned on a number of other presuppositions as well, to some of which reputable scientists in his time readily subscribed. Such was, for instance, the concept of an infinite space filled with an infinite number of stars. That Blanqui, a dilettante in matters scientific, did not perceive the grave contradictions in such a world picture should not be surprising. Before 1872, when the L'éternité par les astres was published, most scientists failed to take notice of the fact that in such a universe the gravitational potential should be infinite at any point and that the skies should be blazing at each point day and night with an infinite brilliance. The relatively few scientists who showed awareness of the gravitational and optical paradox of the infinite universe took the matter in stride mainly because of the conviction that intelligibility and infinite meant one and the same thing.[63] At any rate, of the two scientists, Laplace and Arago, who were referred to by name in Blanqui's work,[64] the first never touched on the problem, while the latter considered, and in a rather perfunctory manner, only the optical part of the problem, known as Olbers' paradox.[65]

Whatever Blanqui learned from Laplace and Arago, it was not the modest amount of cautious reserve that can be found in the writings of these two. The nebular hypothesis which Laplace offered as a tentative explanation of the origin of planetary systems was taken by Blanqui as final word in the matter. There was also a similar wilfulness in Blanqui's acceptance of each element as an ultimate form of matter.[66] As his theory needed such elementary bodies restricted in number, he readily overlooked the fact that the opinions of scientists were anything but unanimous about the ultimate nature of the elements. Blanqui's insistence on density as *the* clue to the physical processes unfolding in each celestial body[67] also reveals the arbitrary technique of dilettante cosmologists. The only point where Blanqui correctly read the science of his day was the principle of the homogeneous composition of the cosmos, a tenet which had just obtained its principal support from the spectral analysis of the sun's light. He displayed his amateurish arrogance at its rawest when he came face to face with the grave cosmological implications of the dissipation of energy. To account for the necessary reconcentration of energy in his perpetually recurring universe Blanqui postulated the continual collision of defunct stars. Obviously, once the energy of motion had been transformed into heat and dissipated, it could not be available again, but Blanqui saw gravitational attraction as an inherent property of matter constituting an inexhaustible source of energy.[68]

The most telling evidence about the dilettantish mind of Blanqui, the cosmologist, comes from his lengthy discussion of the nature of comets. As the topic has no logical place in his efforts to prove the eternal recurrences of all, it can only be taken as evidence of his obsession with some little detail, a pattern so characteristic of the thinking of scientific dilettantes. For only an obsessed mind could reject, as Blanqui did, any and all scientific opinions and data about the nature of comets, and keep insisting that in the comets one had at hand the strict evidence of a unique state of matter which sets them wholly apart from the rest of the universe. Such a procedure was all the more curious on Blanqui's part as his theory rested on the homogeneity of the material composition of the universe to which he made the comets constitute a flagrant exception. All bodies, stars, planets, and living beings were formed alike, according to him, from "simple bodies" or elements, but he described hydrogen as a granite when compared with the almost infinite tenuousness of the material composing the comets.[69]

One is indeed in the presence of an almost morbid mental phenomenon on finding Blanqui offer variations of his claim about the "otherworldly" nature of comets. "The comets are truly fantastic beings"; "they are scientific nightmares," "desperate enigmas," "an obstacle almost insurmountable to knowing the universe," and "an indecipherable and indifferent myth."[70] Such had to be truly the case if there was even a modicum of truth in Blanqui's words: "The comets are neither ether, nor gas, nor light, nor solid, nothing similar to that which constitutes the celestial bodies; they are rather an indefinable substance, that does not seem to have any property of ordinary matter, and has no existence outside the domain of the sun's rays which pulls them out of nothing at one moment and lets them fall back there in the next. Between this astral enigma [the comets] and the stellar systems which form the universe [there is] a radical separation. These are two modes of isolated existence, two categories of matter wholly distinct, and without any tie except that of a disordinate if not foolish gravitation. In the description of the world, the comets have no right to be taken into account. They are nothing, they do nothing, their only role is to be an enigma."[71] The real enigma consisted in the fact that a considerable part of a short book on eternal recurrences was taken up by a baffling discourse aimed at restoring some scientific importance to the comets, which was lost, so Blanqui argued, after their superstitious character had been exploded. In that biased treatment of the comets, commented Blanqui acidly, man showed his true mien. Actually, it was Blanqui who gave a revealing insight into the inner recesses of his mind.

That mind was that of a fanatical social prophet and activist, and not that of a sober scientist. For there is a distinct fanaticism in the manner in which Blanqui outlined the details of his "scientific" argument of the infinite and endless repetition of everything in the universe. The very core of his proof of eternal recurrences can be stated in a relatively few words. According to Blanqui there is only a restricted number of the main types of physical entities in an infinite universe. These are the large gas clouds, the stars, and the planets. Furthermore, there is only a limited number of elements, or fundamental building blocks, of which everything else is composed. Consequently, in an infinitely large universe in which an infinite number of stars and planetary systems are always in the process of formation and dissolution, it is inevitable on the basis of probability that there should be identical developments and formations. Or to paraphrase Blanqui's argument in the usual parlance of probability games: given a die with a large number of sides, each of which represents an actual planet, any sequence of throws, however variegated, will be obtained time and again if the throws continue throughout eternity.[72]

This is indeed elementary and requires no special familiarity with science, let alone some rare "feeling" for the scientific. What requires genuine scientific spirit concerns the application of the foregoing simple consideration about probability to the *actual* universe. It is there that Blanqui's reasoning went astray. The universe even a hundred years ago had to appear far more complicated than to fit into the simplest "scientific" cubbyholes construed by Blanqui. True, classical physics was based on the idea of absolute determinism in physical processes, that is, on the decisive importance of the initial conditions for the rest of the process in question. But on all the other contentions of Blanqui about the main features of the physical universe and its laws, the discrepancy between the physics and astronomy of the 1860's and the science of Blanqui could not have been wider.

That planets, stars, and nebulae formed the main classes of physical entities was at best a convenient classification. Actually, Blanqui spoke of nebulae as gas clouds which may give rise to individual planetary systems and he simply overlooked the true meaning of nebulae as immense systems of stars. He failed to consider the double stars whose large number was well known at that time, or the scant likelihood for the emergence of planetary systems around them. His assumption, that density was the clue to the actual manner in which a particular star or planet developed, represented a gross over-simplification. Blanqui's claim that gravitational attraction was an inexhaustible source of energy simply flew in the face of basic laws of thermodynamics. His most ironical mishandling of the scientific data occurred as he faced the objection based on the total lack of evidence of those cosmic catastrophies, the collision of defunct stars. He could have seized with some plausibility on the novae of 1572 and 1604, but he treated them as of secondary importance.[73] In fact, he went so far as to dismiss the whole idea of a need for observational evidence on behalf of the pivotal point in his cosmology: "The rejuvenating conflagrations never have witnesses," he wrote. As a result, no conviction could be generated by his subsequent assertion that the small nebular patches, of which thousands were revealed by a good telescope, represented those gigantic collisions.[74]

If Blanqui's account of cosmology evidenced an occasional mixture of flippancy and cynicism, these two traits became dominating as he faced in detail the problem of individual recurrences of the countless incidents of human history. Clearly, if everything was predetermined by ironclad physical necessity, what merit could there be in trying to prove any point? What could be the meaning of writing a book if Blanqui himself had to recognize with gripping candor the inevitable: "What I am writing this moment in a dungeon of the fortress of Taureau, I have already written and shall write again, for eternity, on a table, with a pen, in quite similar manner and circumstances."[75] If this held true of everybody, as Blanqui hastened to add, and if events repeated themselves in an inexorable sequence, could there be any sense in attempting to influence their course? To a radical revolutionary like Blanqui, such questions must have presented themselves with particular urgency, but he blandly skirted the issue. About the question of the freedom of choice in man, he simply stated that men apparently kept choosing between various possibilities. Placed at a bifurcation man could go either way, one of which "would lead to misery, disgrace and slavery. The other would lead to glory and freedom."[76]

Blanqui had, of course, an insurmountable task on hand as he tried to specify the meaning which could be assigned in a consistent manner to that last word, freedom, in a universe with room only for inflexible and immutable laws.[77] He could not help showing his true colour as he defined human will, the source of choices, as mere caprice,[78] the impulse of fantasies, the effervescence of temperament.[79] He did not attempt to conceal that such a situation was hardly encouraging from the viewpoint of cultural efforts. Actually, he took sardonic comfort in the fact that human freedom (caprice) could make hardly a dent in the global processes of nature: "The most gigantic efforts of men do not create as much as a molehill, though this does not stop them from posing as conquerors and from falling in ecstasy on account of their genius and power."[80] Recalling the fate of great ancient cultures now buried under layers of sand, Blanqui noted with satisfaction that man can disturb only to a very slight extent the realm of matter. But, he added in a cynical vein, men can play havoc with themselves: "Under the influence of passions and conflicting interests, the human species is agitated with a greater violence than the stormy ocean."[81]

Blanqui was unable to come up with a satisfactory reason why "one should foresee this subversive influence which changes the course of individual destinies, destroys or modifies the animal life, tears apart nations and ruins empires."[82] The urgency of cultural, social, and scientific endeavour remained hopelessly alien in the picture of cosmic and human history drawn by Blanqui. What encouragement could be derived from Blanqui's morbid outlines of both identical and slightly different returns of major historical events? Was it not an exercise in cynicism to describe the English losing the battle of Waterloo on other earths where Grouchy, one of Napoleon's generals, failed to make his slight but fatal mistake? What was the profit in conjuring up a Napoleon defeated at Marengo because of the unpredictable caprice of human volition, if the workings of that caprice were, after all, absolutely predetermined? Could there be anything constructive in spinning fantasies with Blanqui about the chance reversals of historic battles on innumerable replicas of this very earth of ours?[83]

One is indeed at a complete loss when trying to fathom the reasons of Blanqui for writing this book. A sworn enemy of the Church, Christianity, and of anything savouring of the supernatural and spiritual, Blanqui possibly intended his book to be another blow at religion by portraying human existence along "strictly scientific lines" which deny to man any sense of lasting purpose.[84] But such a tactic also works in the opposite direction. It effectively prevents "scientific socialism" from claiming science as the road toward a secure betterment of the human condition by revealing man a hapless prisoner of the forces of nature. The motivation of the L'éternité par les astres may forever remain a riddle, but it is certainly a most revealing book. It shows in its real nature that frame of thinking which tries to come to grips with the universe while resolutely rejecting its created character. Blanqui was very explicit on this point. One of the first affirmations of his book is that "matter does not issue out of nothing, nor can it return there. Matter is eternal, imperishable. Although on a perpetual course of transformations, it cannot diminish nor increase by as much as an atom."[85]

Blanqui's work palpably shows that the logic of such a statement, allegedly based on facts alone, leads to the disappearance of all rationality in one's world picture. With expressive consistency Blanqui noted that the world could not be considered a clockwork,[86] the traditional epitome of rationality, as the world was in reality an infinite chaos with infinite variations, of which many were identical repetitions. In the cosmic eternity as pictured by Blanqui there was no direction but only resignation if not despair. It is by showing this, perhaps unintentionally, that Blanqui performed a cultural and scientific service. The rambling and agitated pages of the *L'éternité par les astres* certainly convey the hardly intended message that the notion of a creation out of nothing is the only logical foundation of a confidently rational attitude toward the cosmos and man's place in it. This lesson is also embodied in a short passage of the concluding part of the book, where Blanqui painted in broad strokes a portrait of the universe: "The universe is at the same time life and death, destruction and creation, change and stability, tumult and rest. It comes about and dissolves without end, it is always the same, with all beings forever rejuvenated. In spite of its perennial becoming it is cast in bronze, and prints incessantly the very same page. Both in its details and entirety, it is transformation and immanence for eternity."[87] Hegel, no doubt, would have been pleased with this withdrawal from the domain of scientific clarity into the shadowy realms of pantheistic poetizing.

Hegel's influence on Blanqui is largely a matter of educated guessing based on the affinity of minds engrossed with the supremacy of dialectic or cyclic patterns. In Nietzsche's case the evidence of Hegel's impact is obvious.[88] As one would expect in view of Nietzsche's capricious mind, his remarks about Hegel are not without barbs, but these mainly concern incidental points. With the pivotal idea of Hegel's system of thought Nietzsche expressed full agreement. He wrote in 1886 in the Preface to the second edition of *The Dawn of Day* that "we Germans . . . catch the scent of truth, a *possibility* of truth, at the back of the famous fundamental principle of dialectics with which Hegel secured the victory of the German spirit over Europe."[89]

What is so significant in this reference to Hegel is that it comes from the second and final phase of Nietzsche's intellectual journey. It represents, therefore, Nietzsche's own judgment of the true trend of a book finished in 1881, the very same year when he had undergone the most decisive experience of his life. This consisted in his being captivated with a dramatic suddenness by the idea of eternal recurrences. In his autobiographical *Ecce Homo* written in 1888 Nietzsche gave the following account of the event: "I now wish to relate the history of *Zarathustra*. The fundamental idea of the work, the *Eternal Recurrence*, the highest formula of a Yea-saying to life that can ever be attained, was first conceived in the month of August, 1881. I made a note of the idea on a sheet of paper, with the proscript: 'Six thousand feet beyond man and time.' That day I happened to be wandering through the woods alongside the Lake of Silvaplana, and I halted not far from Surlei, beside a huge rock that towered aloft like a pyramid. It was then that the thought struck me."[90]

The work to which Nietzsche referred in the foregoing passage is his famous *Thus Spake Zarathustra*.[91] Its four books written in four very short periods of time between 1883 and 1885 bespeak not only the extraordinary vehemence of inspiration by which the idea of eternal recurrence kept Nietzsche's mind and moods in a permanent grip. They also give some clue about Nietzsche's familiarity with the idea of eternal recurrences prior to August 1881. Zarathustra (Zoroaster), the ancient religious leader of the Iranian highland, is mentioned next to Heraclitus in Nietzsche's essay on "Philosophy during the Tragic Age of the Greeks," written in 1873.[92] A year earlier Nietzsche published his *Birth of Tragedy* in which he interpreted Greek drama in a Hegelian manner as the expression of a dialectic experience of perennially contending forces. As he remarked in 1888, the work "smells shockingly Hegelian."[93] It is from the same year that dates the only admission of Nietzsche about his possible encounter with the doctrine of eternal recurrences well before his "moment of truth" at Silvaplana. Once more he referred to Heraclitus: "The doctrine of the 'eternal recurrence,' that is, of the unconditioned and infinitely repeated cycle of all things—this doctrine of Zarathustra's *might* after all have been already taught by Heraclitus. At any rate the portico [the Stoa], which inherited well-nigh all its fundamental conceptions from Heraclitus, shows traces thereof."[94]

Clearly, a professor of classics, as Nietzsche had been for eight or so years, could not have remained ignorant of the ancient Greek formulations of the doctrine. As a lover of literature, he might have been reminded of it while reading Hölderlin[95] and Heine.[96] His early advocacy of materialism almost certainly turned his attention to F. A. Lange's *History of Materialism* of which the second and vastly revised edition from 1873 discussed the question of eternal returns with references to Epicurus, Lucretius, and Blanqui.[97] The publicity given to the address of C. W. von Nägeli, the leading biologist of Germany, before the annual gathering of German scientists in Munich in 1877, also might have been instrumental in further nurturing an idea already planted in the deeper recesses of his mind.[98] It is more difficult to trace the steps of the process that led to the emergence of Zarathustra as a central symbolic figure in Nietzsche's thinking. At any rate, almost immediately after his experience at Silvaplana, Nietzsche penned the first plans of *Thus Spake Zarathustra* on a sheet of paper which contains under the headings, "Midday and Eternity," "Guide-Posts to a New Way of Living," the following two short reflections: "Zarathustra born on Lake Urmi; left his home in this thirtieth year; went into the province of Aria, and, during ten years of solitude in the mountains, composed the Zend-Avesta. – The sun of knowledge stands once more at midday; and the serpent of eternity lies coiled in its light – : It is *your* time, ye midday brethren."[99]

The question of the origins of Nietzsche's acquaintance with the concept of eternal returns and with the figure of Zarathustra is only of secondary importance when compared with the sweep, originality, and persistence by which Nietzsche articulated the doctrine of eternal recurrence.[100] His portrait of Zarathustra has little to do with the mythical prophet of Iranian religion. What Nietzsche says about the eternal recurrence, in a large part through Zarathustra's mouth, provides, however, a unique illustration in modern setting of the full range of intellectual, moral, and emotional consequences which arise when the doctrine is espoused with utmost seriousness and consistency as the very foundation of one's world view and as the supreme norm for one's course of action. It certainly was not an exercise in trite clichés of rhetoric when Nietzsche jotted aphoristic phrases like: "The doctrine of the Eternal Recurrence is the turning point of history," and "The moment in which I begot recurrence is immortal, for the sake of that moment alone I will endure recurrence."[101] The latter phrase is a clear indication of the dramatic intensity with which he looked into the face of a doctrine capable of evoking the darkest perspectives.

He called unabashedly the idea of eternal circuits "the most oppressive thought,"[102] "the most abysmal thought,"[103] and "the heaviest burden."[104] It was, in his eyes, the very denial of the notion of an ultimate peace, of an ultimate reason, of an ultimate trust, of an abiding love: "There is no longer any reason in that which happens, or any love in that which will happen to thee – there is no longer any resting place for the weary heart, where it has only to find and no longer to seek."[105] Such an outlook may easily take on a demonic character, against which human nature would react violently: "Wouldst thou not throw thyself down," Nietzsche asked himself, "and gnash thy teeth, and curse the demon," if he "crept after thee into thy loneliest loneliness some day or night and said to thee: 'This life, as thou livest it at present, and has lived it, thou must live it once more, and also innumerable times; and there will be nothing new in it, but every pain and every joy and every thought and every sigh, and all the unspeakably small and great in thy life must come to thee again, and all in the same series and sequence – and similarly this spider and this moonlight among the trees, and similarly this moment, and I myself. The eternal sand-glass of existence will ever be turned once more, and thou with it, thou speck of dust'. "[106]

This demonic character of the idea of eternal returns comes to surface on more than one occasion in the weird monologues of Zarathustra. There is, for instance, the howling of a terrified dog symbolizing the spontaneous horror of Zarathustra himself when he looks into the consequences of the dwarf's dark utterance: "Everything straight lieth . . . All truth is crooked; time itself is a circle."[107] Or one may recall that "fatally weary, fatally intoxicated sadness which spake with yawning mouth" about the return of even the smallest, of even the most insignificant man, – a prospect that conjured up for Zarathustra the vision of an earth crushed under human dust and bones.[108] Zarathustra had indeed to recognize that assuming the role of the teacher of eternal returns would be both his "great fate" and also his "greatest infirmity." It is in an ominous silence that he listens to the words outlining his mission:

> Thou teachest that there is a great year of Becoming, a prodigy of a great year; it must, like a sand-glass, ever turn up anew, that it may anew run down and run out: –
> – So that all those years are like one another in the greatest and also in the smallest, so that we ourselves, in every great year, are like ourselves in the greatest and also in the smallest.[109]

Zarathustra's objection that he, body and soul, will totally dissolve in death is brushed aside:

> 'Now do I die and disappear,' wouldst thou say, and in a moment I am nothing. Souls are as mortal as bodies.
> But the plexus of cases returneth in which I am intertwined, – it will again create me! I myself pertain to the causes of the eternal return.
> I come again with this sun, with this earth, with this eagle, with this serpent – *not* to a new life, or a better life, or a similar life:
> – I come again eternally to this identical and selfsame life, in its greatest and its smallest, to teach again the eternal return of all things, –

Y

– To speak again the word of the great noontide of earth and man, to announce again to man the Superman.[110]

The last phrase leads us to the psychologically most critical question of the doctrine of eternal returns. Is it indeed possible, as Nietzsche claimed, to turn the mesmerizing picture of eternal returns into a vision of noontide, into the laughter of a transfigured being surrounded by light?[111] Is there any plausibility in the claim that one could be truly ardent "for the marriage-ring of rings – the ring of the return"?[112] Could it somehow be possible for any human to greet that demonic whisper as the voice of God and "*to long for nothing more ardently* than for this last eternal sanctioning and sealing"[113] in the bleak prospect of endless returns? Nietzsche's affirmative reply is shockingly blunt: the well-nigh impossible can be achieved though only at an exorbitant price, the complete abdication of reason. This is why Nietzsche took pains to emphasize that the cyclic pattern should not be thought of as something expressive of rationality in contrast to the universal chaos: "The chaos of the universe, inasmuch as it excludes any aspiration to a goal, does not oppose the thought of the circular process: the latter is simply an irrational necessity, absolutely free from any formal ethical or aesthetical significance."[114] In the same vein he denounced any speculative inquiry into a possible time-lag between the death and rebirth of the same individual: "Ye fancy that ye will have a long rest ere your second birth takes place, – but do not deceive yourselves! 'Twixt your last moment of consciousness and the first ray of the dawn of your new life no time will elapse, – as a flash of lightning will the space go by, even though living creatures think it is billions of years, and are not even able to reckon it. Timelessness and immediate re-birth are compatible, once intellect is eliminated!"[115] With critical reason out of the way, Nietzsche could advocate an interpretation of existence independent of the categories of both rationality and irrationality: "We must guard against ascribing any aspiration or any goal to this circular process: Likewise we must not, from the point of view of our own needs, regard it as either monotonous or foolish, etc. We may grant that the greatest possible irrationality, as also its reverse, may be an essential feature of it, but we must not value it according to this hypothesis. Rationality or irrationality cannot stand as attributes of the universe."[116]

Clearly, Nietzsche used the term "scientific" in a most peculiar sense when he made the staggering claim that the idea of eternal recurrence was "the most scientific of all possible hypotheses."[117] The meaning intended by him is contained in the context where "the most scientific hypothesis" is described as a "European form of Buddhism." The characterization was fitting well beyond Nietzsche's intentions. Neither in its original setting, nor in its Nietzschean paraphrase, did Buddhism prove itself an inspiration for scientific inquiry. Not that Nietzsche had not been seized shortly after his electrifying experience of August 1881 by the urge to justify scientifically the idea of eternal returns on a cosmic scale. In the fall of the same year he planned to embark on a rigorous ten-year long study of physical science so that he could turn his claim about the eternal return of all into a scientifically unassailable demonstration.[118]

These plans never materialized. From early youth Nietzsche disliked the systematic and abstract intellectual exercise demanded by scientific studies. Although his personal library contained well over a hundred books on science, most of them were popularizations. About half of these show some evidence of having been perused by him.[119] But whatever information he had gathered there, it fell on the unfriendly soil of a highly impulsive, restlessly emotional mind. Obsessed as he was by the soundness of the idea of eternal returns he could not refrain from constructing a "scientific" proof of it, which he kept submitting for the rest of his life without ever catching a glimpse of the futility of the enterprise.

The proof rested on a most arbitrary handling of some of the fundamental tenets of late nineteenth-century physics. These were the principle of the conservation of energy, the law of entropy, and the infinity of the material universe. Of these the last rested on less solid ground than it appeared to be. At any rate, Nietzsche had no inkling of either of the two great puzzles, the gravitational paradox and the optical or Olbers' paradox implied in the notion of an infinite universe.[120] His rejection of the infinity of space, matter, and energy derived from his instinctive realization that were such the case, the number of possible configurations would not be finite and consequently the same configurations would not return with inevitable regularity. He formulated this point with great explicitness in one of his earliest "scientific" reflections on the idea of eternal recurrence: "The extent of universal energy is limited; it is not 'infinite': We should beware of such excesses in our concepts! Consequently the number of states, changes, combinations, and evolutions of this energy, although it may be enormous and practically incalculable, is at any rate definite and not unlimited. The time, however, in which this universal energy works its changes is infinite – that is to say, energy remains eternally the same and is eternally active: – at this moment an infinity has already elapsed, that is to say, every possible evolution must already have taken place. Consequently the present process of evolution must be a repetition, as was also the one before it, as will also be the one which will follow. And so on forwards and backwards! Inasmuch as the entire state of all forces continually returns, everything has existed an infinite number of times."[121] His refusal to recognize the full weight of the law of entropy, or the steady degradation of energy, reflected the same determination to evaluate the laws of physics in terms of the "fundamental truth" of eternal recurrences. It should not, therefore, be surprising to find Nietzsche repeating the wholly arbitrary contention that the principle of the conservation of energy meant that the finite universe would forever remain active by going through the same cycles and self-manifestations throughout eternity.

The following passages should forcefully intimate the measure of willfulness in Nietzsche's efforts to create the impression that physics supported his basic claim according to which the universe ultimately was a "Will to Power." "Formerly," he offered his interpretation of the principle of conservation of energy, "it was thought that unlimited energy was a necessary corollary to unlimited activity in time, and that this energy could be exhausted by no form of consumption. Now it is thought that energy remains constant and does not require to be infinite. It is eternally active but it is no longer able eternally to create new forms, it must repeat itself: that is my conclusion."[122] The only truth in this passage is contained in the last words. The conclusion was Nietzsche's alone and not of any of the scientists responsible for the formulation of the principle. His assertions about the finiteness of the universe, energy, and space were equally lacking in scientific clarity and cogency, as amply revealed, for instance, in the following passage: "That a state of equilibrium has never been reached, proves that it is impossible. But in infinite space it must have been reached. Likewise in spherical space, the *form* of space must be the cause of the eternal movement, and ultimately of all 'imperfection'."[123]

The passage contains what for Nietzsche loomed large as the incontrovertible proof on behalf of his world conception: if the universe could reach a final stage it would already have reached it in all certainty during its infinite past. For him this consideration constituted the only certainty in cosmology: "If the universe were in any way able to congeal, to dry up, to perish; or if it were capable of attaining to a state of equilibrium; or if it had any kind of goal at all which a long lapse of time, immutability, and finality reserved for it (in short, to speak metaphysically, if becoming could resolve itself into being or into nonentity), this state ought already to have been reached. But it has not been reached: it therefore follows . . . This is the only certainty we can grasp, which can serve as a corrective to a host of cosmic hypotheses possible in themselves."[124] The circularity of such an argument comes through clearly in Nietzsche's following statement: "If energy had ever reached a stage of equilibrium that stage would have persisted; it has therefore never reached such a stage. The present condition of things contradicts this assumption. If we assume that there has ever been a state absolutely like the present one, this assumption is in no wise refuted by the present state. For, among all the endless possibilities, this case must already have occurred, as an infinity is already behind us. If equilibrium were possible it would already have been reached."[125] Evidently, Nietzsche kept asserting possible consequences of what he should have established in the first place, the eternity and finiteness of the universe.

It would, however, be somewhat mistaken to press Nietzsche on points of scientific merit. Ultimately he wanted a notion of the world with which the spirit of scientific inquiry had little if anything to do. He was willing to show what the universe looked like when viewed in the mirror of his mind and, characteristically enough, the lengthy description started with a reference to the universe as a "monster of energy" and ended in a Dyonisian reveling over the utter aimlessness of its twofold "voluptuousness," the source of perennial returns. Such a universe was, as he put it, "*The Will to Power – and nothing else!*" Humans, too, were put by him into the same category: "And even ye yourselves are this will to power – and nothing besides."[126]

It throws a sharp light on Nietzsche's wholly unscientific way of thinking that he never came to grips with the obvious conflict between the rigorous determinism on which rested his claim of identical returns and the complete willfulness in which he saw the ultimate feature of the universe. At any rate, he was unable to do justice to science. What he stated about scientific laws, about the aim and achievements of physics, shows that he considered exact science an arbitrary projection of one's mind and will. His stricture of those who claimed universally valid laws of ethics ended with the exhortation: "Let us leave this nonsense and this bad taste to those who have nothing else to do, save to drag the past a little distance further through time, and who are never themselves the present, – consequently to the many, to the majority! We, however, *would seek to become what we are*, – the new, the unique, the incomparable, making laws for ourselves and creating ourselves! And for this purpose we must become the best students and discoverers of all the laws and necessities of the world. We must be *physicists* in order to be *creators* in that sense, – whereas hitherto all appreciations and ideals have been based on *ignorance* of physics, or in *contradiction* to it. And therefore, three cheers for physics! And still louder cheers for that which *impels* us to it – our honesty."[127] It is very unlikely that physics had ever been given a more insulting compliment.

Nietzsche's discussion of the merits of the atomic theory provides a good illustration of his evaluation of physics as a set of basically arbitrary laws. He derided physicists for believing in a "true world." According to Nietzsche the "true world" of physicists consisting in a "fixed systematizing of atoms to perform necessary movements" was merely a projection of their minds and of their peculiar way of thinking. In the closing decades of the nineteenth century Nietzsche was not, of course, the only critic of the atomic theory. Positivist scientists and philosophers of science, like Mach, Ostwald, and Stallo, turned their opposition to atoms into the touchstone of the truth of their contentions. Their radical views had often been seized upon by thinkers like Nietzsche, who saw in the doubts besetting the conceptual foundations of late classical physics a confirmation of their own attitude extolling the subjective and the irrational. Nietzsche was hardly a disciplined thinker sensitive to the fine points of the anti-atomism of positivism, and he was certainly poles apart from the thinking of the great majority of physicists about physical reality and its lawfulness. By and large physicists always did their work in the belief that it was possible to ferret out, to some extent at least, the real laws governing the physical universe, and that those laws made sense only if they could be considered universally valid. Such convictions were wholly alien to Nietzsche's thinking, as shown by his criticism of the scientific belief in atoms: "The atom, which they [the physicists] postulate is arrived at by the logic of that perspective of consciousness; it is in itself therefore a subjective fiction. This picture of the world which they project is in no way essentially different from the subjective picture: the only difference is, that it is composed simply with more extended senses, but certainly with our *senses* . . . And in the end, without knowing it, they left something out of the constellation: precisely the necessary *perspective factor*, by means of which every centre of power – and not man alone – constructs the rest of the world *from its point of view*, – that is to say, measures it, feels it, and moulds it according to its degree of strength . . ."[128]

The passage also gives several glimpses into Nietzsche's inability to escape the confines of a crude voluntarism. It was not abstract reasoning which gave rise to another statement of his about atoms: "Even in the inorganic world all that concerns an atom of energy is its immediate neighbourhood: distant forces balance each other. Here is the root of *perspectivity*, and it explains why a living organism is 'egotistic' to the core."[129] His criticism of the use in mechanics of such notions as "attraction" and "repulsion" shows equally well that science and its philosophy were totally misunderstood by him: " 'To attract' and 'to repel,' in a purely mechanical sense, is pure fiction; a word. We cannot imagine an attraction without a *purpose*. – Either the will to possess one's self of a thing, or the will to defend one'

self from a thing or to repel it – *that* we 'understand': that would be an interpretation which we could use."[130] The same confusion about scientific reasoning and
manner of speech is revealed in Nietzsche's appraisal of the quantitative method.
Because it yielded only quantitative relations, a point clear to most physicists,
though not to some philosophers, he asked indignantly: "The possibility of expressing all phenomena by means of formulae – is that really 'understanding'? What
would be understood of a piece of music, if all that were calculable in it and capable
of being expressed in formulae, were reckoned up?"[131] Undoubtedly music was
far more than quantities of sound or levels of frequencies. But could there be
any music if man cared not for rhythm, a strict sequence of quantities, which is
the very soul of music?

That such and similar questions had never been pondered by Nietzsche makes a
very poor reflection on the man, who, if one adopts the most benevolent interpretation of his often violent and garbled statements, wanted to become the herald
of a higher, truly human culture. His picture of man and of culture was hopelessly
one-sided. In the ideal culture, as outlined by Nietzsche, science, technology, and
the domination of nature by man, had no serious roles to play. His ideas about
science gave him a badly distorted image of the real portrait of scientific quest,
of its motivations and possibilities. Zarathustra's description of science as a pro-
longed, ancient fear, which at long last became subtle, spiritual, and intellectual,[132]
is perhaps too enigmatic to set great store by it, but Nietzsche could also be very
argumentative about science. He claimed that science denied freedom of will in
man, because science assumed material processes to be independent of volition:

> Science does *not* inquire what impels us to will: on the contrary, it *denies* that
> *willing* takes place at all, and supposes that something quite different has happened
> – in short, that the belief in "will" and "end" is an illusion. It does not inquire
> into the motives of an action, as if these had been present in consciousness
> previous to the action: but it first divides the action up into a group of pheno-
> mena, and then seeks the previous history of this mechanical movement – but
> *not* in terms feeling, perception, and thought; from this quarter it can never
> accept the explanation: perception is precisely the matter of science, which has
> to *be explained.* – The problem of science is precisely to explain the world,
> *without* taking perceptions as the cause: for that would mean regarding *perceptions*
> themselves as the *cause* of perceptions. The task of science is by no means
> accomplished.
>
> Thus: either there is *no* such thing as will, – the hypothesis of science, – or the
> will is *free*. The latter assumption represents the prevailing feeling, of which
> we cannot rid ourselves, even if the hypothesis of science were *proved*.[133]

Nietzsche's strange defence of the freedom of will in the teeth of mechanistic
physics was all the more so as for him everything in the universe was a manifestation
of will. This is why he could pay lip service at best to the merits of the application of
the "soulless" method of science to the material universe. He could not say something unreservedly praiseworthy about science, about the atmosphere of scientific
enterprise, about its practitioners, about its relation to culture. In the same breath
he also voiced disdain or took back much of what he grudgingly conceded.

Illustrations of this are everywhere in the works which he wrote following his total commitment to the idea of eternal recurrences. He both praised and deplored the severe and manly atmosphere of science.[134] While acknowledging the role of science in culture he hastened to add that this held true of a future type of science wholly transcending the basic misconceptions of the present one.[135] He admitted the beauty of science but only with the proviso that it was a savage beauty.[136] It is rather easy to sense that it was not sympathy that inspired his description of science as "the exactest humanising."[137] He spoke of science as an enterprise without faith in itself and in need of further justification.[138] His characterization of physics as "a boon for the mind" was not only qualified by the claim that "science (as the road to knowledge) acquires a new charm after morality has been laid aside,"[139] but it was largely offset by his contemptuous declaration that "science is laying the road to *sovereign ignorance*, to a feeling that 'knowledge' does not exist at all, that it was merely a form of haughtiness to dream of such a thing."[140]

Evidently, his deep-seated disdain of science was voiced when he spoke of science "as a half-way house, at which the mediocre, more multifarious, and more complicated beings find their most natural gratification and means of expression: all those who do well to avoid action."[141] To grasp the full significance intended by Nietzsche in making such utterances one should recall his ideal of man, the Superman. Among his features one would look in vain for traits characteristic of scientists. As if compelled by a subconscious aversion toward men of science, it was always the mediocrity of the average scientist that Nietzsche liked to belabour. The unique qualities of creative scientists were largely left unmentioned by him. He liked to poke fun at the average scientist, whom he carefully distinguished from the genius. The scientific man was for him an old maid, a person of commonplace virtues, of unimposing attitudes, but always in covert search for recognition, for "the sunshine of good name," for the "perpetual ratification of his values and usefulness." According to Nietzsche, a chief motivation of the man of science was "the inward *distrust* which lies at the bottom of the heart of all dependent men and gregarious animals." The learned, or rather, the scientific man was, in his words, "full of petty envy," and had "a lynx-eye for the weak points" in those whom he could not emulate. The man of science was a person who grew cold and reserved because of his inferiority complex with respect to the Superman, or "the man of the great current." By contrast, the appearance of a scientist evoked for Nietzsche the image of a "smooth and irresponsive lake, which is no longer moved by rapture or sympathy."[142]

Nietzsche's myopia about scientists was matched by his vision of the scientific movement. He viewed it with heavy blinders indeed. He was unable to distinguish convincingly between classical or mechanistic physics, and the mechanistic dogmas which enjoyed undue popularity in his time. Like Goethe, whose revulsion for the mechanistic nightmare of the Enlightenment he approvingly recalled,[143] he failed to see the profound difference between physical science and the mechanistic philosophies foisted upon it by some of its non-scientist interpreters. This myopia of his only increased as time went on. From his early writings to the posthumous *Will to Power* there is a steady increase of venom in Nietzsche's comments on classical physics as a mechanistic science.

What irritated Nietzsche about science was not primarily its allegedly "lifeless," mechanistic character. Underlying the often superhuman efforts of the great men of physics there was the unshakable conviction in the progressive value of scientific research. Copernicus' vision of the heliocentric arrangement of planets was born of his belief in a universe structured on an immutable, once-and-for-all established pattern. Kepler's heroic struggles to unfold the laws of planetary motion were sustained by his faith in the eternal character of the fundamental patterns of the universe. Galileo, in turn, had already seen the chief glory of scientific work in its cumulative character. The great masters of physics in Nietzsche's century were animated by the very same convictions. Nietzsche possibly sensed the enormous contrast between the spirit of science and the atmosphere of eternal recurrences: not even the most shocking pages of *Thus Spake Zarathustra* contain a hint about the same scientific discoveries being made all over again an infinite number of times. Could a Galileo derive any comfort from the prospect of going through his trial not once but endlessly? Would he or any other man of science have been willing to care for a world in which no ignorance and error could be definitively displaced by the results of science? Clearly, anyone with Nietzsche's absolute commitment to the idea of eternal recurrence could only look askance at the fact that more than one scientific achievement and conclusion evidenced the unmistakable characteristic of permanency. Nietzsche's preferences leaned towards a pre-scientific age when fables and fairy tales served as clues to nature and existence: "The wonderful did so much good to those men, who might well get tired sometimes of the regular and the eternal. To leave the ground for once! To soar! To stray! To be mad! That belonged to the paradise and the revelry of earlier times; while our felicity is like that of the shipwrecked man who has gone ashore, and places himself with both feet on the old firm ground – in astonishment that it does not rock."[144]

The image of a shipwrecked man completely at a loss over the stability of dry land conveys poignantly the extreme to which Nietzsche disowned confidence in the ability of man to recognize certainties and hold onto them. He scanned the various fields of man's inquiry and attacked the evidences of man's trust in lasting values and in lasting conclusions. He searched with passionate resolve for the logically most fundamental formulation of that trust and, not surprisingly, he identified it with the concept of a world issued from the hands of a Creator! It was with a crusading spirit that he kept depicting the belief in creation as the most fatal aberration of the human mind and held up to ridicule the Biblical story of creation.[145] As one would expect, he saw an absolute opposition between a created world and an eternally recurring one. "He who does not believe in the circular process of the universe," he wrote, "must pin his faith to an arbitrary God – thus my doctrine becomes necessary as opposed to all that has been said hitherto in matters of Theism."[146] In the same vein he urged: "Let us not indulge our fancy any longer with unthinkable things in order to fall once more before the concept of a Creator (multiplication out of nothing, reduction out of nothing, absolute arbitrariness and freedom in growth and qualities)."[147] He certainly showed no eagerness to form for himself a correct idea about Creator and creation out of nothing as he brushed both aside as intrinsically valueless concepts: "We need not concern ourselves for one instant with the hypothesis of a *created* world. The concept 'create' is today utterly indefinable and unrealisable; it is but a word which hails from superstitious ages; nothing can be explained with a word."[148] He characterized the notion of absolute beginning as absurd.[149] The moment of scientific truth coincided for him with the rejection of the doctrine of creation.[150] He showed a remarkable measure of consistency as he spelled out a logical implication of his reading of intellectual history: pantheism, he wrote, "or the doctrine of 'everything perfect, divine, eternal,' also leads to the belief in Eternal Recurrence."[151]

Such an intellectual bravado was not without its moments of agonizing doubt. There were moments of sober awakening in the frantic effort to exorcise that "half-Christian, half-German narrowness and stupidity" or the remnants of other-worldliness, which, as Nietzsche claimed, was still influencing modern life. It was one thing to hold high defiantly the opposite ideal, that "of the most world-approving, exuberant, and vivacious man," who not only accepted reality, but also wanted "to have it again as it was and is, for all eternity, insatiably calling out da capo, not only to himself, but to the whole piece and play." It was another to endure the ever-recurring chill of fear that all this defiance was but a tactic in self-defeat to replace an inescapable God with an inexorable one, the deity of eternal recurrences, and face the question: "What? And this would not be – circulus vitiosus deus?"[152] The true spirit of Nietzschean optimism was the demon of vicious circles.

In the psychological abysses to which Nietzsche carries his reader there is no irony but only tragedy. The futility of Nietzsche's efforts also derived from his failure to come to terms with the historical record. History was the crown witness that the fate and fortune of the scientific spirit and enterprise depended on theological beliefs, but in a sense diametrically opposite to the one claimed by Nietzsche. The only point where he succeeded was hardly intended by him. Unwittingly he painted a genuine picture of the intellectual morass that envelops man once he falls victim to the mirage of eternal returns. Nietzsche had certainly shown that beliefs, or ultimate assumptions, have an inner logic and an intrinsic capability of producing either light or darkness. In his agitated comments on the respective merits of creation versus eternal recurrence one finds a tragically expressive paraphrase of the decisive issue in the great drama of human culture which witnessed several still-births of science before its final and viable emergence in minds firmly committed to the Christian message about creation.

The sadly mistaken discourses of Nietzsche on science have an eloquence of their own. The sickly lingering of science within the ambience of a Naturphilosophie inspired by the pantheism and dialectic of German Idealism is an open book only to the professional student of the history of science. The setbacks suffered by science in institutionalized Marxism constitute a less esoteric matter. But in both cases the harmful consequences of a more or less overt acceptance of eternal, cosmic returns can only be seen if one is willing to take an exacting look beyond the horizons of stereotyped accounts of the principal factors at work in the history of science. Readers of Nietzsche should be in a far easier position. From the pages of the tragic prophet of eternal returns there blares forth in shocking directness the principal lesson of a fateful choice made against the clear evidence provided by the long and difficult progress of science. The miasmatic atmosphere of some murky backwaters of nineteenth-century intellectual history should be a powerful reminder of the fact that choices in cosmology have an impact far beyond a particular science. They touch not only science at its very roots but have a decisive influence on man's existential energies. The truth of this cannot be pondered carefully enough in an age in which questions about human and cosmic destiny loom menacingly large because of the highly accelerated advance of science, including the science of cosmology.

NOTES

1. In *The Popular Works of Johann Gottlieb Fichte*, translated by William Smith (4th ed.; London: Trübner & Co., 1889), vol. I, pp. 319–478.
2. *Ibid.*, p. 476.
3. *Ibid.*, pp. 476–77.
4. *Ibid.*, p. 337.
5. *Ibid.*, p. 441.
6. *Ibid.*, p. 462.
7. *Ibid.*
8. *Ibid.*, p. 440.
9. *Ibid.*, pp. 426–27.
10. *On the Nature of the Scholar and its Manifestations: Lectures delivered at Erlangen 1805*, in *The Popular Works*, vol. I, p. 224.
11. For further details and documentation, see my work, *The Relevance of Physics* (Chicago: University of Chicago Press, 1966), pp. 45–46.
12. *System der gesammten Philosophie und der Naturphilosophie insbesondere*, in *Sämmtliche Werke* (Stuttgart: J. G. Cotta'scher Verlag, 1860), vol. VI, p. 492.
13. Translated with introduction and notes by Frederick de Wolfe Bolman Jr., (New York: Columbia University Press, 1942). Schelling began writing this work in 1811, but it saw publication only shortly after Schelling's death in 1854. On the importance which Schelling attached to this work, see Horst Fuhrman, *Schellings Philosophie der Weltalter: Schellings Philosophie in den Jahren 1806–1821. Zum Problem des Schellingschen Theismus* (Düsseldorf: Verlag L. Schwann, 1954), pp. 191–218. On Fuhrman's discussion of Schelling's views about a cyclic patterning in God's life, see pp. 325–30.
14. *The Ages of the World*, p. 119.
15. *Ibid.*, p. 116.
16. *Ibid.*, p. 117.
17. *Ibid.*, p. 118.
18. *Ibid.*, p. 129.
19. *Ibid.*, p. 119.
20. *Ibid.*, p. 140.
21. *Ibid.*, p. 149.
22. *Ibid.*
23. *Ibid.*, p. 152.
24. *Ibid.*, p. 221.
25. This is well brought out in the *Hegel-Lexicon* compiled by H. Glockner (Stuttgart: Fr. Frommanns Verlag, 1938), vol. III, pp. 1297–99.
26. See *ibid.*, p. 1298.
27. For details, see my *Relevance of Physics*, pp. 47–48.
28. Translated by A. Tulk (London: Printed for the Ray Society, 1848). Oken's own definition

of "physiophilosophy" is worth noting: "Physiophilosophy is the science of the genesis of the world, or Cosmogony" (p. 17), and "It has to portray the first periods of the world's development from nothing" (p. 1).

29. *Ibid.*, p. 39.
30. *Ibid.*, p. 66.
31. *Ibid.*, p. 29.
32. *Ibid.*, p. 33.
33. *Ibid.*
34. *Ibid.*, p. 20.
35. *Ibid.*, p. 30.
36. *Ibid.*, p. 54.
37. *Ibid.*, p. 50.
38. *Ibid.*, p. 30.
39. The following references are to the translation by Clemens Dutt, accompanied with the preface and notes of J. B. S. Haldane (New York: International Publishers, 1940; fourth printing, 1960).
40. *Ibid.*, p. 246.
41. *Ibid.*, p. 86.
42. *Ibid.*, p. 127.
43. *Ibid.*, p. 259.
44. *Ibid.*, pp. 257–58.
45. *Ibid.*, p. 258.
46. *Ibid.*, p. 155.
47. *Ibid.*, p. 195.
48. *Ibid.*, p. 75.
49. *Ibid.*, p. 86.
50. *Ibid.*, p. 13.
51. *Ibid.*, p. 23. In notes added to the main text of the *Dialectics of Nature*, Engels admitted that the question of how the dissipated heat would once more turn into a useful source of energy had not yet been solved: "The cycle has not been traced and will not be until the re-utilisation of the radiated heat shall have been discovered" (p. 202). But the prospect of a positive outcome was a certainty for him: "It will be solved, just as surely as it is certain that there are no miracles in nature and that the original heat of the nebular ball is not communicated to it miraculously from the outside of the universe" (*ibid.*).
52. *Ibid.*, p. 24.
53. *Ibid.*, pp. 24–25.
54. *Cursus der Philosophie als streng wissenschaftlicher Weltanschauung und Lebensgestaltung* (Leipzig: Erich Koschny, 1875); see chap. ii, in Abschnitt II, "Grundgesetze des Universums," especially pp. 84–85.
55. Engels' critique of Dühring appeared three years later. Its English translation by Austin Lewis was published under the title, *Landmarks of Scientific Socialism: "Anti-Dühring"* (Chicago: Charles Kerr & Co., 1907), see especially pp. 74, 78, 84, and 88.
56. The most informative book on Blanqui in English is *The Theories of Louis Auguste Blanqui* (New York: Columbia University Press, 1957) by Alan B. Spitzer. On Engels' criticism of Blanqui's socialism, see p. 21. Nothing shows better Blanqui's prominence in the socialist camp than Marx's description of him and of his immediate group, "as the real leaders of the proletarian party, the revolutionary communists" (*ibid.*, p. 21). It is possible that Marx moved the headquarters of the First International to New York because he feared a possible domination of it by Blanqui's followers in London (*ibid.*, p. 15).
57. Written during Blanqui's imprisonment in the island-fortress of Taureau (off the coast of Brittany) during 1871 and published early next year with the subtitle *Hypothèse astronomique* (Paris: Librairie Germer Baillière, 1872). About two-thirds of the book, or the last three chapters, were also printed on the large quarto pages of the *Revue scientifique* in its issue of Feb. 17, 1872, pp. 797–803, with the note: "This article is an excerpt from the book of Mr. Blanqui, which will be published tomorrow. It seemed to us interesting to

show to our readers how the famous socialist agitator treated a scientific question." This might have very well been a back-handed compliment. The text published in the *Revue scientifique* contains entire paragraphs that are not in the book, a point apparently unnoticed so far.

58. As was aptly noted by A. B. Spitzer, *op. cit.*, p. 35.
59. *L'éternité par les astres*, p. 61.
60. *Ibid.*, p. 70.
61. *Ibid.*, pp. 70–71.
62. *Ibid.*, p. 75.
63. As shown in my *The Paradox of Olbers' Paradox* (New York: Herder & Herder, 1969), chap. viii.
64. *L'éternité par les astres*, p. 17. The older Herschel is also mentioned but only as the source of Laplace's nebular hypothesis, a rather erroneous claim.
65. See *The Paradox of Olbers' Paradox*, pp. 97–100 and 156–57.
66. *L'éternité par les astres*, p. 12.
67. *Ibid.*, p. 11.
68. *Ibid.*, p. 35.
69. *Ibid.*, p. 25.
70. *Ibid.*, pp. 23, 30, 44.
71. *Ibid.*, pp. 26–27.
72. *Ibid.*, pp. 51–52 and 64–65. Spitzer claims (*op. cit.*, pp. 36–37) that "many contemporaries were impressed by Blanqui's grasp and exposition of modern science," but his only proof consists in a reference to two French newspapers which took note of the book. There is a noncommittal reference to the science of the *L'éternité par les astres* in F. A. Lange's *The History of Materialism* (2d ed., 1873), translated by E. C. Thomas (New York: The Humanities Press, 1950), vol. I, p. 151 note.
73. *L'éternité par les astres*, p. 43.
74. *Ibid.*
75. *Ibid.*, p. 73.
76. *Ibid.*, p. 56.
77. *Ibid.*, p. 63.
78. *Ibid.*
79. *Ibid.*, p. 64.
80. *Ibid.*, p. 63.
81. *Ibid.*, p. 64.
82. *Ibid.*
83. *Ibid.*, p. 59.
84. Such is the opinion of A. B. Spitzer, *op. cit.*, p. 37.
85. *L'éternité par les astres*, p. 6.
86. *Ibid.*, p. 33.
87. Blanqui's influence may be in evidence in M. Guyau's elaboration on the bearing of spectroscopy on the interpretation of cosmic existence in his philosophical poem, *Vers d'un philosophe* (Paris: Librairie Germer Baillière, 1881), where the next to last section in the concluding part of the work is entitled, "L'analyse spectrale," (pp. 195–200). Against the debilitating vision of worlds following one another to no end, Guyau sought a moral philosophy, independent of science and also free of the categories of obligations and sanctions, in his *Esquisse d'une morale sans obligation, ni sanction* (4th ed.; Paris: Felix Alcan, 1896). – Another French intellectual, who was led by Blanqui to the idea of eternal recurrence, was the physician-turned-sociologist, Gustave Le Bon, who in his *L'homme et les sociétés* (Paris: J. Rotschild, 1881) took the circular cosmic process as the foundation of his cyclic theory of the dynamics of society and history. Forty some years later, the almost octogenarian Le Bon was still preaching the message of periodic recurrences of religious, moral, and intellectual "certainties" in his *La vie des vérités* (Paris: Flammarion, 1925). During the middle period of his intellectual career Le Bon's interest largely centered on the philosophy of physics. In once very popular, but now largely forgotten books, *L'évolution*

de la matière (Paris, Flammarion, 1908) and *L'évolution des forces* (Paris: Flammarion, 1912), he showed not only a praiseworthy awareness of the importance of the newly discovered radioactivity, but also put forward many obscurantist and wholly mistaken ideas about physics. His case seems to be one more illustration of that strange connection between a commitment to the idea of eternal returns and an abusive interpretation of science.

88. See on this W. A. Kaufmann, *Nietzsche: Philosopher, Psychologist, Antichrist* (Princeton, N.J.: Princeton University Press, 1950), pp. 289–91.

89. *The Dawn of Day*, translated by J. M. Kennedy, in *The Complete Works of Friedrich Nietzsche*, edited by Dr. Oscar Levy (Edinburgh: T. N. Foulis, 1909–11), vol. IX, p. 6. Subsequent references to Nietzsche's writings are to that edition.

90. *Ecce Homo*, translated by A. M. Ludovici, *Works*, vol. XVII, pp. 96–97.

91. Translated by T. Common, *Works*, vol. XI.

92. Translated by M. A. Mügge, *Works*, vol. II, see p. 77.

93. Translated by W. A. Haussmann, *Works*, vol. I, p. 190.

94. *Ibid.*, pp. 193–94.

95. In a composition, which he wrote as a schoolboy, Nietzsche called Hölderlin his favourite poet and during 1870–71 he toyed with the idea of writing his own Empedocles-drama. See on this *Der Tod des Empedokles*, edited by F. Beissner, in *Hölderlin: Sämtliche Werke* (Stuttgart: W. Kohlhammer, 1961), vol. IV, p. 348.

96. "It was Heinrich Heine," wrote Nietzsche in his *Ecce Homo* (p. 39), "who gave me the most perfect idea of what a lyrical poet could be." Nietzsche owned a copy of the *Letzte Gedichte und Gedenken von H. Heine* (Hamburg, 1869) which contained a draft of chapter xx of Heine's *Reise von München nach Genoa*, originally published in 1828. In that draft the poet stated, à propos of a touching encounter, that because of the laws governing chance combinations all configurations, things and persons alike, will reappear again in identical forms: "And so it will happen that a man, just like me, will be born again, and a woman, exactly like Maria, will be born . . . and the woman will finally give her hand to the man and say in a soft voice: let us be good friends." See *Heinrich Heines Sämtliche Werke*, edited by E. Elster (Leipzig: Bibliographisches Institut, n.d.), vol. III, p. 542. The stylistic elements of the whole section are closely paralleled in a passage of Nietzsche's *Joyful Wisdom*, quoted below (note 105).

97. On Lange's work see note 72 above. Nietzsche's library contained a copy of the 1887 edition of Lange's work. It is also known that Nietzsche had already perused in 1869 the edition of the same work from 1865. See on this A. Mittasch, *Friedrich Nietzsche als Naturphilosoph* (Stuttgart: Alfred Kröner Verlag, 1952), p. 366.

98. Nietzsche, an ardent opponent of Darwinism, had in his library several of Nägeli's works, among them the one carrying the text of his address of 1877. See on this Mittasch, *op. cit.*, pp. 367 and 371.

99. Quoted in the Introduction by Nietzsche's sister, Mrs. Förster, to *Thus Spake Zarathustra*, p. xvi.

100. More or less detailed discussions of Nietzsche's doctrine of eternal returns can be found in practically every book on Nietzsche. The most exhaustive treatment of the question is by K. Löwith, *Nietzsches Philosophie der ewigen Wiederkunft des Gleichen* (Berlin: Verlag Die Runde, 1935), but he too failed to put the problem in the perspective of science in general and of the history of science in particular. See also Appendix II, "Nietzsche's Revival of the Doctrine of Eternal Recurrence," in Löwith's *Meaning of History* (Chicago: University of Chicago Press, 1949), pp. 214–22.

101. Aphorisms 41 and 62 in "Explanatory Notes to *Thus Spake Zarathustra*," in *The Twilight of the Idols, or How to Philosophise with the Hammer*, translated by A. M. Ludovici, *Works*, vol. XVI, pp. 267 and 274.

102. *The Will to Power*, translated by A. M. Ludovici, *Works*, vol. XV, p. 424; aph. 1059.

103. *Thus Spake Zarathustra*, in *Works*, vol. XI, p. 264.

104. *The Joyful Wisdom*, translated by T. Common, *Works*, vol. X, p. 270; aph. 341.

105. *Ibid.*, p. 220; aph. 285.

106. *Ibid.*, pp. 270–71; aph. 341.

107. *Thus Spake Zarathustra*, in *Works*, vol. XI, p. 190.
108. *Ibid.*, p. 268.
109. *Ibid.*, p. 270.
110. *Ibid.*, pp. 270–71.
111. *Ibid.*, p. 193.
112. *Ibid.*, p. 282.
113. *The Joyful Wisdom*, in *Works*, vol. X, p. 271.
114. "Eternal Recurrence," in *The Twilight of the Idols*, in *Works*, vol. X, pp. 270–71; aph. 341.
115. *Ibid.*, p. 253; aph. 33.
116. *Ibid.*, p. 247; aph. 20.
117. *The Will to Power*, in *Works*, vol. XIV, pp. 48–49; aph. 55. It is equally staggering that a noted Nietzsche-scholar could remark: "It may seem odd that this most scientific hypothesis has found no acceptance among scientists." (W. A. Kaufmann, *Nietzsche*, p. 287).
118. As reported by his confidant, Lou Andreas-Salome, in her *Friedrich Nietzsche in seinen Werken* (Dresden: Carl Reiszner, n.d.; reprint of the first edition of 1894), pp. 128–29. The failure of most Nietzsche-interpreters to call attention to the glaring lack of scientific merit in Nietzsche's advocacy of eternal returns is well exemplified in *What Nietzsche Means* by G. A. Morgan Jr., (Cambridge, Mass.: Harvard University Press, 1943), see especially p. 286.
119. See on this the work of A. Mittasch quoted above in note 97. It is a storehouse of information on every small detail that could possibly be gathered about Nietzsche and science, but its general trend is wholly misleading. Only inveterate admirers of Nietzsche would hold that his utterances on scientific theories and science in general have scientific value. Thus, the chapter, "The Cosmologist," in F. A. Lea, *The Tragic Philosopher: A Study of Friedrich Nietzsche* (London: Methuen, 1957), displays no grasp whatever of scientific cosmology.
120. The two problems were referred to with sufficient frequency during the latter part of the 19th century. J. C. F. Zöllner, professor of astrophysics at Leipzig, to whom Nietzsche made numerous references, and whose book, *Über die Natur der Cometen* (Leipzig: W. Engelmann, 1872), Nietzsche had in his library, pointed out in detail in that book, long before the advent of relativistic cosmologies, that a four-dimensional, closed universe with a finite mass, provides a logical solution to the problem. See on this my work, *The Paradox of Olbers' Paradox*, pp. 158–60.
121. "Eternal Recurrence," *Works*, vol. XVI, p. 237, aph. 1.
122. *Ibid.*, p. 238; aph. 2. It should not be surprising that G. Batault, who defended Nietzsche in his "L'hypothèse du retour éternel devant la science moderne," (*Revue philosophique*, 57 [1904]:158–67), also ignored the dissipation of energy.
123. *The Will to Power*, in *Works*, vol. XV, p. 427; aph. 1064.
124. *Ibid.*, pp. 429–30; aph. 1066.
125. "Eternal Recurrence," in *Works*, vol. XVI, pp. 241–42; aph. 14.
126. *The Will to Power*, in *Works*, vol. XV, p. 432; aph. 1067.
127. *The Joyful Wisdom*, in *Works*, vol. X, p. 263; aph. 335.
128. *The Will to Power*, in *Works*, vol. XV, p. 120; aph. 636.
129. *Ibid.*, p. 121; aph. 637.
130. *Ibid.*, p. 114; aph. 627.
131. *Ibid.*, p. 112; aph. 624.
132. *Thus Spake Zarathustra*, in *Works*, vol. XI, p. 371.
133. *The Will to Power*, in *Works*, vol. XV, p. 140; aph. 667.
134. *The Joyful Wisdom*, in *Works*, vol. X, pp. 227–28; aph. 293.
135. *Ibid.*, p. 159; aph. 113.
136. *The Dawn of Day*, in *Works*, vol. IX, p. 312; aph. 427.
137. *The Joyful Wisdom*, in *Works*, vol. X, p. 158; aph. 112.
138. *The Genealogy of Morals: A Polemic*, in *Works*, vol. XIII, pp. 192 and 197.
139. *The Will to Power*, in *Works*, vol. XV, p. 99; aph. 594.
140. *Ibid.*, p. 104; aph. 608.

141. *The Will to Power*, in *Works*, vol. XV, p. 100; aph. 597.
142. *Beyond Good and Evil: Prelude to a Philosophy of the Future*, translated by H. Zimmern, *Works*, vol. XII, pp. 138–39; aph. 206.
143. *Thoughts out of Season*, translated by A. Ludovici, *Works*, vol. IV, p. 58.
144. *The Joyful Wisdom*, in *Works*, vol. X, pp. 82–83; aph. 46.
145. "Has anybody ever really understood the celebrated story which stands at the beginning of the Bible, – concerning God's deadly panic over *science*?" – asked Nietzsche with a jeer in *The Antichrist: An Attempted Criticism of Christianity*, in *Works*, vol. XVI, pp. 197–98; aph. 48.
146. "The Eternal Recurrence," in *Works*, vol. XVI, p. 244; aph. 16.
147. *Ibid.*, pp. 243–44; aph. 15.
148. *The Will to Power*, in *Works*, vol. XV, pp. 428–29; aph. 1066.
149. "The Eternal Recurrence," *Works*, vol. XVI, p. 239; aph. 6.
150. See *ibid.*, p. 239; aph. 5, and *The Will to Power*, in *Works*, vol. XV, p. 427; aph. 1062.
151. *The Will to Power*, in *Works*, vol. XIV, p. 49; aph. 55.
152. *Beyond Good and Evil*, in *Works*, vol. XII, p. 74; aph. 56.

CHAPTER FOURTEEN

Oscillating Worlds and Wavering Minds

"It is hardly necessary to say that if we add to these theories of the future death of the universe the next stage, namely, that after its dematerialized form it again rematerializes somehow, we find resurrected the old Babylonian, Hindu, Greek, and Roman theories of the *annus magnus*, with all their traits. The above theories of the death of the universe are but a step toward that end. And it is not improbable that in the near future some of the astrophysicists will make this new discovery."[1] So wrote in the mid-thirties the famed Russian-American sociologist, Pitirim A. Sorokin, in a work devoted to the patterns of social and cultural processes. Among those patterns, the cyclic ones commanded special interest on his part. He took great pains to present his topic within the broadest possible framework, or the findings and speculations about geophysical and cosmic cycles. As far as the idea of a universal heat-death was concerned, it was already in the air in the late twenties largely because of the attention drawn to it in the classic popularizing works of such leading astronomers as Eddington and Jeans. But Sorokin failed to learn by 1935 or so, that what he had envisioned as a future development in cosmology had already been a well-established aspect of Einstein's General Theory of Relativity.

True, comparatively few people were privy at that time to the concept and intricacies of an oscillating type of universe which was shown in 1922 by A. Friedman[2] as a possible solution of the cosmological equations of Einstein implying a universal heat-death was concerned, it was already in the air in the late twenties, remained largely esoteric even among cosmologists during the twenties,[3] a decade or so later there was something baffling in Sorokin's unfamiliarity with the idea of an oscillating universe which would have provided a most satisfying capstone to his astonishingly well documented discussion of cyclic patterns in the various aspects of social and cultural life.[4]

While scientific cosmology has become in this century an increasingly forbidding field of investigation because of its highly complex mathematics, word about the oscillating universe made headlines in various newspapers on more than one occasion during the early thirties.[5] Enthusiastic newspaper coverage of a scientific novelty is not, however, a sure index of the theory's popularity in scientific circles. In the case of Einstein's cosmological ideas, admiration on the part of scientists was mixed with some deep-seated uneasiness. The idea of a finite universe was a bitter pill to swallow after three hundred years of classical physics during which intelligibility was equated with actual Euclidean infinity. Parting with that hallowed tenet was not made much easier by the beautifully consistent account by Einstein of a finite though unbounded universe.[6]

The concept of finiteness along the parameter of matter (and space) was the target of spirited objections on the part of prominent scientists. Thus, William D. Mac-Millan devoted his lengthy address to the American Mathematical Society on April 10, 1925, to the defence of Euclidean infinity as the true structure of actual physical reality, and the only logical framework of cosmological speculations.[7] The same year witnessed the publication of Bertrand Russell's *The ABC of Relativity*.[8] He found the concept of a finite universe highly unpalatable and, in order to take the sting out of it, he resorted to an argumentation which flew in the face of a tenet held by him inviolable up to that time. The tenet was the eternal truth of at least the major propositions of science. In *The ABC of Relativity* Russell kept, however, insisting on the basic revisability of any scientific proposition and conclusion, and specifically singled out the newly established truth of the finiteness of the total amount of matter constituting the universe.[9]

An even more disturbing factor for many scientists was the growing realization that the universe may very well be finite in time. While Condorcet still could base his rosy description of the future on Laplace's conclusions about the stability of the solar system, its basically transitory character became clearly recognized as the nineteenth century went on. The amount of energy of any closed system, small or large, could only be considered as finite, but the unquestioned infinity of the universe provided the magic panacea for the eternalists. With the sudden realization of the finiteness of the universe a new significance was added to the estimates of the age of the earth and stars, already in progress since Rutherford's pioneering work on radioactivity in the opening years of this century. In a finite universe which had to be conceived as being constituted of stars with basically similar characteristics, the idea of a finite past inevitably offered itself. At any rate, the energy-reservoir of a finite universe could not be imagined as inexhaustible, unless one were willing to part with the fundamental laws of physics and in particular with the proven law of entropy. Eddington and Jeans in their classic popularizing works did indeed their literary best to impress upon their readers the idea of an inescapable heat-death for the universe. The "rundown" of the universe could not help prompting speculations about the extent of time that had elapsed since the process of the degradation of energy started, but reliable estimates could not be obtained. The total mass and the total energy available in a finite cosmos represented no easy target for any calculation, however circumspect.

The situation had taken on an entirely new aspect when, in 1927, G. Lemaître published his now historic paper[10] written with no awareness of Friedman's articles. Lemaître proposed an "intermediate solution" between Einstein's original, static cosmological model and W. de Sitter's interpretation of it. Actually, the interpretative model constituted a separate class of its own as the "abstract universe" of de Sitter contained no matter.[11] Yet, an indirect feature of de Sitter's model consisted in the fact that there the properties of gravitational field implied a certain rate of recession for nebulae which had been actually observed in a number of cases. Lemaître's treatment of the problem could hardly be more impressive with respect to specific results. These were a logical explanation of the energy used in the expansion of the universe; a simple relationship between the total mass of the universe and the so-called cosmological constant; a formula and a table of values for the red-shift of receding galaxies in fine agreement with the actually observed

z

data; and the conclusion that "the radius of the universe increases without limit from an asymptotic value R_o for $t = -\infty$".[12]

Recounting all this is in a sense anticipating the actual story. Lemaître's paper, published in a scientific journal of rather restricted circulation, did not come to the attention of cosmologists and of the world of science in general until its "rediscovery" by Eddington in 1930. If no one appreciated Lemaître's paper more than Eddington, no one was more surprised either. When he stumbled on Lemaître's paper, he had already been working for several months with G. C. McVittie on the problem of whether Einstein's spherical universe was really stable. Lemaître's paper, as Eddington put it, gave "a remarkably complete solution of the various questions connected with the Einstein and de Sitter cosmogonies," and made all too apparent the instability of the world model of Einstein. Forestalled "by Lemaître's brilliant solution" to contribute "some definitely new result," Eddington took upon himself to bring out in detail the bearing of that instability to the understanding of "the behavior of spiral nebulae."[13]

The behaviour of spiral nebulae to which Eddington referred was their recession. The scientific world had learned shortly after the publication of Lemaître's paper that this recession matched a simple linear function: the distance-velocity relationship. In the papers of E. Hubble and M. L. Humason[14] which disclosed this remarkable feature of the universe, there was no suggestion whatever about the possibility of estimating on the basis of that relationship the time already elapsed since the beginning of the recession of nebulae from a common point, or from an original, extremely high condensation of matter. Cosmological inferences were conspicuously absent in their papers. They were in fact so much startled by the implications of their finding that they refused to see in the red-shift of galaxies an evidence of receding motion. Eddington showed, therefore, both insight and boldness as he set out to unfold the vast range of consequences of the reality of an expanding universe which was stated in all explicitness in the opening phrase of the last paragraph of Lemaître's paper.[15]

Eddington proceeded about the task with his customary incisiveness and philosophical sensitivity. He took pains to emphasize that to "every expanding solution there corresponds a contracting solution,"[16] that "an infinite variety of solutions can be found representing spherical worlds which are not in equilibrium," and that "whilst remaining spherical they expand or contract."[17] Eddington did not profess to know which of these two processes was the very first on the purely theoretical level. But theory when coupled with the data of observation quickly brought him face to face with the idea of an absolute cosmic beginning. He hastened to note not so much exactly as whimsically that "the initial small disturbance can happen without supernatural interference."[18] Another point that seemed to disturb Eddington was the apparently "young age" of the universe. Since he believed that the expansion of the "embryonic" universe might have been much slower than the expansion of a universe already differentiated into nebulae, he felt himself justified in calculating the "age of the universe only from the time when the universe had reached, say, 1.5 times its initial radius to the present day." On such a basis it was not possible to assign more than 10 billion (10^{10}) years to the universe. This, however, implied for him the "oddity" that the sun which has existed for 5 billion years "should have waited so long and then formed its system of planets just at the time the universe toppled into a state of dispersion."[19]

About one point, he offered no serious disclaimer. The process of expansion appeared to him irreversible for all practical purposes. He conjured up a future where astronomers would look in vain for nebulae: they all will have recessed beyond the ken of observation. He also stated in a phrase, highly significant in retrospect at least, that the last conclusion rested on the observational evidence of the recession of nebulae and "that none are coming in to replace them." And he added: "Unless a theory is invented which provides some force opposing this recession, there is no evading the rapid departure of nebulae from our neighbourhood."[20] When a few months later, at the meeting of the Mathematical Association, on January 5, 1931, Eddington delivered his presidential address,[21] he articulated anew much the same points and concerns. As to the notion of a beginning of "the present order of Nature," by which he undoubtedly meant an absolute beginning, he candidly remarked that it was "repugnant" to him.[22] On the other hand he also admitted the impossibility of finding a reliable loophole in the considerations leading inescapably to the reality of an absolute beginning. In addition to the one-way process of an expanding universe there was also the one-way increase in the amount of entropy, or an increase in the disorganization of matter constituting the universe. In other words, atoms tended to get into more and more probable configurations. Thus, the undeniable organization present in the universe could only be assigned to what Eddington non-committally called "anti-chance," which loomed larger and larger on the horizon as one traced backward the process of cosmic and geological evolution.

Eddington was fully aware that most men of science were anything but eager to take a serious look at the increasingly crucial role played by anti-chance as one retraced the steps of the cosmic timetable. Their reluctance was, in Eddington's eyes, a fundamental inconsistency which he laid bare in a graphic style: "We are unwilling to admit in physics that anti-chance plays any part in the reactions between the systems of billions of atoms and quanta that we study; and indeed all our experimental evidence goes to show that these are governed by the laws of chance. Accordingly, we sweep anti-chance out of the laws of physics – out of the differential equations. Naturally, therefore, it reappears in the boundary conditions, for it must be got into the scheme somewhere. By sweeping it far enough away from our current physical problems, we fancy we have got rid of it. It is only when some of us are so misguided as to try to get back billions of years into the past that we find the sweepings all piled up like a high wall and forming a boundary – a beginning of time – which we cannot climb over."[23] It was with his gentle wit and searing logic that Eddington made shambles of the major counter-argument based on the idea of statistical fluctuations. He termed it a blind alley to assume that since there is an infinite time ahead, very rare but sufficiently large reversals in the increase of entropy should take place with the result, he noted with tongue in cheek, that the present meeting of the Mathematical Society should occur by chance an infinite number of times while time flows on endlessly.

This is not the place to enter into the details of Eddington's reasoning which emphasized a little appreciated aspect of the concept of physical law, namely, its most intimate connection with man's consciousness of time. He kept a studied silence about the idea of an oscillating universe as a conceivable means for salvaging infinity along the parameter of time. In all likelihood, he saw no reason why the oscillating universe should be considered a possibility of any real merit. But he did not conceal another misgiving about the vision of a universe going through endless cycles. Already in his famous Gifford Lectures in 1927, he resolutely upheld the once-and-for-all unfolding of the cosmos as something far more meaningful than the prospect of endless repetitions: "At present we can see no way in which an attack on the second law of thermodynamics could possibly succeed, and I confess that personally I have no great desire that it should succeed in averting the final running-down of the universe. I am no Phoenix worshipper. This is a topic on which science is silent, and all that one can say is prejudice. But since prejudice in favour of a never-ending cycle of rebirth of matter and worlds is often vocal, I may perhaps give voice to the opposite prejudice. I would feel more content that the universe should accomplish some great scheme of evolution and, having achieved whatever may be achieved, lapse back into chaotic changelessness, than that its purpose should be banalised by continual repetition. I am an Evolutionist, not a Multiplicationist. It seems rather stupid to keep doing the same thing over and over again."[24] Such was a sharply drawn contrast between two basic options.

Five years later, in the first book written on the expanding universe, Eddington drew a similar contrast. Appalling as could appear the "state of dead sameness," or maximum entropy and cosmic finale, he described it as "a rather happy avoidance of a nightmare of eternal repetition."[25] The following year, in 1933, lecturing at Cornell University, he was even more explicit on the deeper ramifications of a universe repeating itself forever: "From a moral standpoint the conception of a cyclic universe, continually running down and continually rejuvenating itself, seems to me wholly retrograde. Must Sisyphus for ever roll his stone up the hill only for it to roll down again every time it approaches the top? That was a description of Hell. If we have any conception of progress as a whole reaching deeper than the physical symbols of the external world, the way must, it would seem, lie in escape from the Wheel of things. It is curious that the doctrine of the running-down of the physical universe is so often looked upon as pessimistic and contrary to the aspirations of religion. Since when has the teaching that 'heaven and earth shall pass away' become ecclesiastically unorthodox?"[26]

The foregoing quotations are rich in allusions, which any reader of this book may now be tempted to consider as almost trite, the wording by Eddington notwithstanding. The fact is, however, that Eddington displayed in his time an unusual insight by the remarks he made. While Whitehead's emphasis on the role of the Christian faith in the Creator in the rise of modern science[27] made some inroads into public awareness, the sinister role of the ancient addiction to the idea of an eternal and cyclical cosmos was hardly mentioned, let alone elaborated, in the standard presentations of the prospects of scientific enterprise in the various phases of human culture. A rare exception was a work by Victor Monod, entitled *Dieu dans l'univers*.[28] Its author, characteristically enough, was a perceptive theologian. He did not turn the tide, but at least he exemplified for a smugly complacent historiography of science that theology can help the man of science to spot facts and factors of utmost importance for the understanding of the rise and evolution of science, to say nothing of its future prospects.

The vocal prejudice among men of science in favour of a "never-ending cycle of rebirth of matter and worlds" evidenced, indeed, a baffling oversight of the historical background of the question. No wonder that only a cursory allusion was made to it at the Centenary Meeting of the British Association for the Advancement of Science in September 1931, where a distinguished panel of scientists discussed the evolution of the universe in the light of the freshly proposed idea of an expanding universe.[29] The author of the remark was Ernest W. Barnes, Lord Bishop of Birmingham, and the controversial figure of his Church during much of the first half of the century. With scholarly attainments in both science and theology, he sounded more often than not patently unorthodox from the theological viewpoint. He must have also appeared very unorthodox to the typical scientist as he decried the concept of infinite space as "simply a scandal to human thought."[30] It was in much the same way that he spoke of infinite time. Not that he was particularly happy with the idea of the expanding, or as he put it, "inflating" universe. For him "a steady and continued inflation, either of a currency or of a universe," was disquieting.[31] Still, it represented in his eyes the lesser of two evils. Reluctance to postulate an absolute beginning, he noted, left one with the alternative of "an infinite regress, a never-ending sequence of alternate periods of world-building and world-destruction, the rise and fall of universes without end."[32]

The panel also included the three leading authorities and advocates of the expanding universe, Lemaître, Eddington, and de Sitter. As for all three the idea of the expanding universe represented a fairly well established conclusion, they felt free to go one step farther. Thus, Lemaître disclosed his speculations about the original shape of the universe as one huge "primeval atom,"[33] while Eddington offered one of the first of his daring speculations about numerical correlations between certain atomic and cosmic quantities.[34] De Sitter's most interesting remark concerned the bearing of the known density of matter in the universe on the choice between expanding and oscillating models. The latter possibility, he noted, would have required a density considerably higher than actually observed.[35]

Another prominent cosmologist, E. A. Milne, also on the panel, restricted himself to a special detail in the evaluation of stars which had nothing to do, immediately at least, with the great endpoints of cosmic evolution. He wryly called attention to the fact that with respect to the universe "its future evolution is a matter of speculation and its past evolution a matter of inference."[36] But Milne hardly meant to give thereby the impression that fundamental issues in cosmology were of little interest to him. By 1931 he had already been working on one of the major achievements of modern cosmology, his *Relativity, Gravitation and World-Structure*.[37] There he warned about a formidable if not insuperable difficulty to be faced by oscillating models of cosmology: the gravitational collapse at a given degree of the cosmic compression of matter. In addition to grave scientific difficulties, the idea of an oscillating universe also implied for him something akin to a conceptual monstrosity: "It has sometimes been held that these oscillating universes of general relativity provide for cycles of evolution, and for this reason they have been favourably considered. Actually they have not been previously analysed in . . .

detail as regards their appearance to observation, so that it has not been previously realized how remarkably and irretrievably they stand in conflict with experience ... They must be totally and absolutely rejected as irrational. They are the fantastic weavings of the mathematical loom, orgies of mathematical licence, divorced from experience. They are possible only in the sense that in a dream everything is possible."[38] Once more, dreams appeared more attractive to many than facts.

The panel did not lack in voices trying to shore up the cause of a universe infinite in time. Such was certainly the gist of the comments of Jeans, who emphasized the never-ending process of expansion. "The universe can never come to rest; it is destined to go on changing for ever – continually swelling in size, and continually dissolving into nothing but size, yet never attaining either complete dissolution or truly infinite size."[39] His argument that "matter turns into energy and energy into mere bigness of space" was, of course, conclusive only if the rate of transformation of matter (the total amount of which had to be finite) corresponded to an asymptotic function. Even more questionable were Jeans' allegations of a failure of the conservation of matter and energy on a cosmic scale. Nor did R. A. Millikan make himself sufficiently clear as to what he really intended to prove by attributing an atom-building ability to high energy cosmic rays in outer space.[40] However, those familiar with other pronouncements of Millikan along the same lines, could have no misgiving about his real aim. It consisted in furthering the credibility of the idea of a forever active universe.

The scientific foundation of Millikan's contention rested on the hypothesis of W. D. MacMillan[41] that energy readily transformed itself into atoms, and on Millikan's belief that high energy cosmic rays kept turning into atoms. Or as Millikan stated following his acceptance of the gold medal of the British Society of Chemical Industry in 1928: "With the aid of this assumption one would be able to regard the universe as in a steady state now, and also to banish forever the nihilistic doctrine of its ultimate 'heat-death'."[42] About the notion of a cosmic heat-death Millikan made further statements which throw strong light on his ulterior convictions in the matter. Those convictions made him misrepresent the historical record, on the one hand, and laid bare his real and hardly scientific concern, on the other. Concerning the first point, Millikan falsely presented the body scientific as having always objected to "the extravagant and illegitimate extrapolation" of the Second Law of thermodynamics to the universe as a whole. Some scientists did so, but certainly not their overwhelming majority. As to the second point, Millikan named as his allies the "modern philosophers and theologians" who made the same objection on the ground that the idea of a universal heat-death "overthrows the doctrine of Immanence and requires a return to the middle-age assumption of a *Deus ex machina*."[43] In the medievals Millikan might have found better allies.

Millikan seemed to succeed in creating the impression that cosmic-ray studies provided the long-sought escape hatch from the cosmic relevance of the Second Law of thermodynamics. But Eddington, who noted the credence given to Millikan's claim, saw it as a balloon easily deflatable by a pointed remark. The ironical tone was all too obvious in Eddington's comment: "Millikan has sometimes called the atom-building process a 'winding-up' of the universe; but 'up' and 'down' are relative terms, and a transformation of axis may be needed in comparing his description with mine."[44] Other defenders of the eternity of the universe, expanding though it may be, made it only clear by their often rash remarks how desperate had become the predicament of eternalists. Probably the classic incident in this respect was the heated reaction of Walter Nernst, a Nobel-laureate, to a paper read by the young F. von Weizsäcker[45] at a physics colloquium at the University of Berlin in 1938. The paper dealt with the remarkable harmony between the age of the sun, which Weizsäcker computed on the basis of the carbon-cycle, and the age of the universe derived from the red-shift of the nebulae. In his comment on the paper Nernst angrily contended that "the view that there might be an age of the universe was not science." According to him "the infinite duration of time was a basic element of all scientific thought, and to deny this would mean to betray the very foundations of science." He was not even willing to consider the notion of an "age of the universe" as a hypothesis suggested by some hints of experimental evidence. Nothing, he insisted, had claim to the status of "scientific hypothesis which contradicted the very foundations of science."

Obviously, Nernst's convictions harked back to a bygone age in science the spirit of which was hard a-dying. In the distinctly programmatic work of Abel Rey[46] on the idea of eternal returns as the supreme and most valid embodiment of scientific reasoning, only some very brief references to quanta and relativity indicated that it was not actually written around the turn of the century, but in the mid-twenties. The scientific considerations marshalled by Rey in support of his contention offered nothing new, though he waxed prolific on many occasions. On the one hand he insisted that the concept of energy was merely a concept to which nothing corresponded in nature;[47] on the other hand he claimed that irreversibility would not be observed in nature if science were able to follow each individual atom in its course as stars can be taken individually in celestial mechanics.[48] It was on such premises that he elaborated various details of the kinetic theory of gases. He hardly guessed that the prospect of an inexorable realization of extremely small probabilities in a sufficiently (infinitely) long time was merely an abstract number game, but not physics which is the investigation of physical reality. He certainly failed to realize the full weight of the fact that, kinetic theory or not, no experimental evidence could be invoked against the law of entropy, whereas the range of evidence in its support was simply immense and embraced all branches of physical science.

For all that, Rey's work is very important and it undeservedly gathers dust in the relatively few libraries listing a copy of it. As a philosopher of science Rey showed sharp awareness of the ultimate origins of certain intellectual choices and attitudes, and he spelled them out at the very outset. Half of the three-page long "avant-propos" to his book consisted of quotations from Proclus, Nietzsche, Le Bon, and Blanqui glorifying the idea of the eternal return of all. "The Reality of the Eternal Return" was the title of the last section of the last chapter in his book and it was all Nietzsche. On the basis of second-hand quotations from Nietzsche, Rey emphatically declared Nietzsche's extraordinary greatness as a scientist.[49] A professor at the Sorbonne, who twenty years earlier incisively analyzed the perplexity of scientists over the collapse of the foundations of classical physics,[50] undermined by modern physics, Rey should have known better. But he was unable to part with the belief that the world was a perfect machine in the Laplacian sense. He kept insisting that the universe was not only a perfect machine,[51] but also "a blind machine constructed in such a manner that it might pass through the same states in an infinite number of times."[52] Perfect machine and unending cyclic motion meant one and the same thing for Abel Rey.

No wonder that he presented the notion of eternal returns as the supreme embodiment and safeguard of objectivity in thinking.[53] For him it was the foundation of identity in a world of apparent flux. "It presents," he wrote, "a maximum of stability and guarantees the maximum of certitude to the laws [of science]. It secures to the mind the maximum of peace."[54] Such an utterance would amply justify the foregoing remark about the programmatic character of Rey's book. He indeed treated his topic in a truly metaphysical if not evangelistic vein. He made no secret about the fact that what he tried to convey consisted in the most explicit espousal of the train of thought, the classic expression of which was the ancient doctrine of the Great Year.[55] In this respect one cannot help admiring his utter consistency. His book ended in proclaiming the doctrine of pantheistic eternity in the words of Spinoza: "We have an inner experience of being eternal."[56]

As it happened all too often in the past, pantheistic eternalism did not prove itself in Rey's case, either, a helpful beacon. It not only deprived him of misgivings about the completeness of his scientific information in the matter; it also gave him a badly distorted picture as to what had actually taken place in scientific history. According to him the whole conceptual history of physics corresponded to a steady convergence toward the idea of eternal returns, and that no appreciable countercurrent evidenced itself before Clausius unfolded the full significance of Carnot's speculations on the efficiency of steam engines.[57] Such was a rather indefensible statement on the part of a French philosopher of science who offered his book as an analysis of the history of physics,[58] but who merely produced a preconceived account of it. How could he, a French rationalist, be oblivious, for instance, of the jubilation of Condorcet who greeted with a loud sigh of relief Laplace's demonstration of the stability of the solar system? Was it not the first reassuring reply to the anxiety felt from Newton on about the gradual loss of the momentum of planets through the ether, minimal as its resistance could be? There was something patently willful in Rey's haughty dismissal of Duhem who in Rey's allegation "wanted to join hands with Aristotle and Scholasticism over the head of Descartes."[59] If this meant that Duhem's monumental volumes were not foreign to Rey then only his fervent pantheism and not documentary evidence could make him belittle the crucial role of a faith in the Creator in the scientific methodology of those who prepared the way for the creators of mechanistic physics, Descartes, Galileo, Newton, and others.

If the distinguished professor of the Sorbonne deserved any accolade it was only for his candid unfolding of some philosophical and humanistic implications of the acceptance of the doctrine of eternal returns. His list of pivotal notions rooted in the idea of eternal returns speaks for itself. They were unconsciousness, determinism, inertia, universal reductibility to quantities, multiplicity, spatiality, stability, the ready-made, the haphazard, the deducible, and functional causality. The contrary series of pivotal points drawn up by him consisted of the following:

consciousness, freedom, activity, qualities as irreducible to quantities, unity, duration, flow, self-making, finality, the intuitive, and efficient causality.[60] In another revealing statement of Rey "the idea of eternal returns characterized the object, or the objectivity given, whereas the subjectivity was characterized by the ideas of aging, of progressive activity, of specific and irreducible moments, of irreversible duration, of endless and contingent creation, of finality, and of orientation in all cases toward a terminus."[61]

If drawing up such a list of pairs of opposites represented utter frankness on Rey's part, it also contained the rebuttal of the very heart of all that he tried to prove throughout some three-hundred often belaboured pages. After all, was it not the appreciation of freedom, of consciousness, of goal-awareness, of creative intuition, and of an insatiable urge for self-development that was needed by the scientific enterprise as an indispensable part of its very soul? Or were these factors and attitudes necessary for science only to be proven in the end a cruel illusion? Was anything further needed in the way of contemporary articulation to show that within the wheels of eternal returns man was but flotsam and jetsam on an endless dark sea and so was his hard-won scientific achievement?

Undoubtedly, Abel Rey would have immensely relished the vision of an oscillating universe had it come to his attention. Others with similar convictions, who were more fortunate in this respect, failed to emulate his unusual consistency. A case in point is Jean Perrin, a Nobel-laureate for his experimental verification of the Brownian movement. In 1941, in one of his last communications to the French Academy of Sciences, of which he was president in 1938, Perrin declared the idea of an age of the universe of some hundred billions (10^{11}) of years as "almost derisory."[62] Although his principal aim was to prove the fundamental unacceptability of the so-called "short-age" of the universe of a few billion (10^9) years calculated from the red-shift of nebulae, he showed impatience even with a much extended variety of the "long-age" of the universe, or some thousand billion (10^{12}) years. His sympathies belonged to the idea of a universe where at the moment when the density of matter reached zero, a contraction came into play. "The photons, which streak through this expanded space standing on the threshold of its era of contraction, become more and more piercing [aigus], and the chances increase that the conditions which gave them birth should be reversed: once again matter will continually appear at the expense of light, until . . . when the universe, in which nebulae and then giant stars evolve anew, will again begin an era of expansion. And so on indefinitely."[63] He even saw likelihood in Arrhenius' "dream" that spores are being carried by light pressure through cosmic spaces. Thus, he concluded, "there would triumph the trend of Things toward a Thought continually more elevated, by means of an eternal balancing in the universe which has not seen birth, and shall see no death."[64]

To recall approvingly in 1941 Arrhenius' antiquated ideas was as inconsistent with the progress of science, as was Perrin's eager exploitation of the idea of an oscillating universe. As a prominent scientist, Perrin should have known that a truly scientific disclaimer of his reasoning had already been "in the books" for at least seven years, namely, since the publication of Richard C. Tolman's *Relativity, Thermodynamics and Cosmology*.[65] Its author was probably better qualified than anybody to place a firm *caveat* against presenting the idea of an oscillating universe as a "proof that the actual universe will always provide a stage for the future role of man,"[66] precisely because of its endless oscillations. Tolman did pioneering studies in relativistic thermodynamics and found that some highly idealized forms of material configurations may undergo processes with the decrease of entropy for the whole system. In particular, he showed that such possibility could obtain under very special conditions for an oscillating type of cosmological model based on the General Theory of Relativity. Tolman, however, warned about the enormous difference between theoretical models and physical reality. "We must be careful," he wrote, "not to substitute the comfortable certainties of some simple mathematical model in place of the great complexities of the actual universe."[67] To nip in the bud any wishful thinking on the part of latter-day admirers of Nietzsche's advocacy of eternal returns, it should also be recalled that Tolman found no justification for the possibility of "strictly-periodic" cosmic oscillations, that is, oscillations with the exact repetitions of the same states.[68] His findings justified nothing more than "some liberalizing action on our general thermodynamic thinking."[69] By this he hoped to discourage two extreme positions. One was the eagerness to conclude from the mere possibility of reversible cosmological models to the assertion that "the actual universe will never reach a state of maximum entropy, where further change would be impossible."[70] About the other he wrote: "At the very least it would seem wisest, if we no longer dogmatically assert that the principles of thermodynamics necessarily require a universe which was created at a finite time in the past and which is fated for stagnation and death in the future."[71] This also meant that science was not to become substitute dogmatics.

Such was a valuable and well balanced advice in spite of the fact that all available evidence suggested the reality of the model with one single start and expansion for the universe. Dedicated believers in the creation of the universe should have rested content with the realization that nothing in modern science and in cosmology in particular, evidenced the eternity of things and processes, let alone the impossibility of a creation in time. But eagerness is an age-old feature to which men of all persuasion can readily succumb. Identification of the beginning of the actual expansion with the creation of the world out of nothing and in time was going far beyond the confines of science. It was also a step for which theology could hold no real licence. Lemaître, both a scientist and a priest, carefully avoided presenting his hypothesis of the "primitive atom" as the state of the world in which it came out from the Creator's hands.[72]

It was otherwise with Sir Edmund T. Whittaker, the most articulate spokesman in the 1940's of the idea that the beginning of the present expansion of the universe corresponded to the start of all physical processes and by inference to their creation. In his Riddell Lectures,[73] delivered in 1942, Whittaker set great store by the radical difference between the Great Year of antiquity and the Christian belief in creation, and on the bearing of that difference on the past and future of science. He gave a short though incisive review of some of the outstanding landmarks of the clash between the two views: Origen, Augustine, Nietzsche, and "one distinguished physicist" in recent years, or Millikan. Of the oscillating universe as such he made no mention. He emphasized that beyond establishing the remarkable convergence of the various determinations of the "age" of the universe, science was completely at a loss to speculate about physical processes that might have preceded that primordial stage: "This is the ultimate point of physical science, the farthest glimpse that we can obtain of the material universe by our natural faculties. There is no ground for supposing that matter (or energy, which is the same as matter) existed before this in an inert condition, and was in some way galvanized into activity at a certain instant: For what could have determined this instant rather than all the other instants of past eternity? It is simpler to postulate a creation *ex nihilo*, an operation of the Divine Will to constitute Nature from nothingness."[74]

Whittaker's reasoning received wide publicity a decade later in an address by Pope Pius XII to the Pontifical Academy of Sciences on November 22, 1951. The allocution which interpreted radioactive decay, entropy, and expansion as clear indications of a creation in time, referred explicitly to Whittaker, and saw print in very large quantities and many translations.[75] The Pope's willingness to set the time of creation at the then estimated past length of cosmic processes, or at about 5 billion years, must have certainly puzzled some in his audience. They all could only be intrigued by a little aside in the allocution about hypotheses, "sometimes unduly gratuitous," forged for salvaging the eternity of physical processes.[76] The only such hypothesis identified by the Pope was "that of continued supplementary creation," by which he evidently meant the pivotal postulate of the Steady-State cosmology, first outlined in 1948 in a now famous paper by H. Bondi and T. Gold,[77] who later joined forces with F. Hoyle.

The choice of the Pope was justified for at least two reasons. First, the process of creation postulated by the Steady-State theorists was diametrically opposed to the Christian idea of Creation. According to the latter, Creation is the work of an absolute, transcendental Being. According to the former, it is an emergence out of nothing but without a Creator.[78] Second, part of the originality of this "creation" was the daring by which an assertion flying in the face of elementary logic was upheld by its proponents. Equally surprising was the ready acceptance of the newfangled idea of creation by many scientists usually very keen on logic in scientific matters at least. Thus it happened that from the early fifties on, it has become the mark of cosmological sophistication to talk about rival cosmologies.[79] The rivalry existed largely on the level of wishful thinking, and of wilfully created publicity, rather than on the level of an urgency created by experimental hints. As a matter of fact, the theory has failed so far to secure a single piece of experimental verification. In addition to this and to the preposterous concept of spontaneous creation, together with its overtly anti-theological, or rather anti-Christian motivations,[80] it should also be pointed out about the Steady-State Theory that it is but a subclass of cosmologies based on the idea of perpetual returns.

The so-called "perfect cosmological principle,"[81] on which the Steady-State Theory is based, would in itself allow both the expansion and the contraction of the universe, but the latter possibility is rejected for the reason that in such a case "there would be even more radiation compared with matter than in a static universe."[82] The expansion is accepted out of deference for the observational evidence and also because a basically static universe would have long reached an equilibrium state. It has, however, been generally overlooked that, according to the theory, within any "local" volume of the universe the phenomena would go through a cyclic process. As galaxies recede from the "local" area, in the empty space thus produced hydrogen atoms are supposed to come into existence ("created"), and turn eventually into that cosmic plasma out of which new nebulae, stars, and planetary systems are formed. These would again move out from the "local" area due to the expansion operating all the time in the universe. At a sufficiently great distance from the point of their original formation the galaxies and their constituents would then meet their dissolution and more importantly their final "annihilation" in order to keep the total mass of the universe at the same finite value. This last stipulation is necessary in order to forestall the gravitational and optical paradox.

The whole edifice of the Steady-State Theory received an acid but well deserved criticism in 1953 in H. Dingle's presidential address to the Royal Astronomical Society.[83] From that address, which is a masterpiece of scientific criticism, only one point can be recalled here. It concerns the domination which can be exercised by prejudices on reasoning about matters scientific. The prejudice in question was the reluctance to accept the uniqueness of an event, or of an entity, small or great, living or not. Dingle spoke only of the "scientific roots" of such prejudice and he named in particular the emphasis of general relativity on the equivalence of status for all co-ordinate systems. According to him it was undue admiration for that principle that prompted the Steady-State theorists to assume the equivalence of all events at all places and times, in the teeth of evidence which showed that, as Dingle put it: "Every process we know, on the small or the large scale, is a one-way process, showing a preference for one direction over the opposite. The system of nebulae expands and does not contract, gravitation is an attraction and not a repulsion, the entropy of a closed system increases and does not decrease, every chemical process tends towards a state of equilibrium from which the substances concerned do not of themselves depart, organic evolution proceeds in one direction and not the opposite; and so on. There is nothing whatever in nature that indicates that any course of events is reversible. Admittedly the evidence is small compared with the magnitude of the problem, but it is all we have."[84]

Not that Dingle had not shared the desire to find some escape hatch from the inescapable. Thirty years earlier he noted with a touch of sadness in connection with the problem of the re-concentration of radiating energy in a finite universe that "a never-ending cycle of changes seems impossible at present to establish."[85] The magic cycle was still missing in the mid-fifties and is still wanting. To give the impression of having turned the trick, nothing less was needed than to graft on one's mind the long-discredited tradition of alchemists and of constructors of perpetual-motion machines, feverishly dedicated either to the programme of perpetuating the present order of things, or to the Utopia of getting something out of nothing. Once more the inner logic of perpetual recurrence exacted its sad ransom: people captivated by it had to abdicate an essential part of the scientific spirit. In this case it was the sensitivity for the true weight of the facts of Nature that had to be sacrificed and in a most unscientific manner.

The future historian of science will undoubtedly be puzzled by the fact that Dingle's incisive and well justified critique failed to turn the tide running in support of the Steady-State Theory. With science-news writers, as well as with authors of serious cosmological essays, it became imperative, throughout the fifties and the sixties, to reserve a special paragraph or chapter for the Steady-State Theory as one of the *principal* advances in modern cosmology. The appeal of the theory certainly could not lie in a correspondence with observational evidences, as these were entirely lacking. Moreover, the suggested observational verifications of the theory stubbornly failed to materialize. But the theory obviously offered the vision of a *perpetual* order of things, and only a few in scientific circles and in the educated public remembered the lessons of history in this respect. Thus, many responding chords were struck by the theory in a decade which witnessed the beginnings of a radical revolt against the Judeo-Christian foundations of Western culture. In an age mesmerized by the vision of a "Second Genesis" or a wholesale redesigning of mankind by molecular biology, by behaviouristic psychology, by a physicalist sociology, and by "thinking machines," what else could be more in line with the intellectual preferences than a perpetual emergence out of nothing but without a cause, which is the Creator Himself?

This is indeed the main reason why it has become a standard claim that the Steady-State Theory has a strong esthetic appeal. Esthetic considerations have a rightful role in fashioning scientific theories and in determining scientific choices. Yet, a science dominated by "esthetics" is bound to end in raping one's intellect. The simplicity of the Copernican arrangement of planets, or the sweeping consistency of the General Theory of Relativity, exerted a powerful esthetic appeal, but their crucial value consisted in revealing a strikingly broader and more exact correspondence among the phenomena. The esthetic appeal of the Steady-State Theory was markedly barren in this respect. By a most ironical twist, its "esthetics" demanded parting not only with some basic aspects of general relativity, but also with the principle of conservation of matter, to mention only two genuinely esthetic instances of creative, scientific thought. The Steady-State Theory implied an esthetics more blinding than enlightening, exactly as was the case in those long centuries when the belief in the Great Year cast its deceiving glitter over men's minds depriving them of indispensable insights.

The lure of a doubtful esthetics is today more persistent and pervasive than it would appear. The Steady-State Theory may now be on its way out due to the crushing blows it has received through experimental investigations in the mid-sixties.[86] For all that, the craving for the "beauty" of an eternal universe does not seem to diminish in intensity. The place of the Steady-State Theory, as the "esthetically true" cosmology, now seems to be taken over by the oscillating model of relativistic cosmologies. There is much symptomatic in this respect in the following passage from a widely read article from 1967 on a phenomenon connected with the starting phase of the expansion of the universe.[87] The phenomenon is called the "primeval fireball" of radiation that must have survived from the time when the universe was in a highly contracted state and with an exceedingly high temperature. In discussing the ideas of R. H. Dicke, a chief investigator of this radiation, the authors of the article stated that Dicke arrived at the idea of a primeval fireball radiation as he was considering in the summer of 1964 the origin of the universe. The authors then added: "It is difficult to explain the apparently spontaneous creation of matter that is called for if one associates the beginning of the expansion of the universe with its actual origin. Dicke, therefore, preferred an oscillating model in which the present expansion of the universe is considered to have been preceded by a collapsing phase."[88]

The statement which reflects the thinking of the article's authors rather than that of Dicke, should well reveal that baffling type of thinking for which the "spontaneous" creation of matter "in the beginning" is unacceptable, but which finds much less alarming the "spontaneous and continual" creation of matter as proposed by the Steady-State Theory. A thinking for which the idea of "spontaneous creation" is acceptable in every single moment but not in a very first moment, in many small quantities but not in one large quantity, is strange enough even if no exception is taken to that curious "spontaneousness" of the emergence of matter out of nothing.[89] Clearly, what really underlies the now increasingly fashionable option for an oscillating universe, is not strict logic but an undying eagerness to purchase eternity by eliminating time from one's thinking. In the oscillating universe the true meaning of time is no more preserved than it is in the Steady-State Theory. It is indeed difficult to understand how G. J. Whitrow, the author of a most perceptive modern discussion on time, could claim that alternative phases of contraction and expansion would secure a "genuine infinity of past time" by eliminating at the same time the question of a "natural origin."[90] Regress to infinity has never been able to distract man's search for a truly ultimate origin, nor can a true notion of time be found in the Heraclitean flux which at least for Heraclitus, as Whitrow pointed out, was compatible with the idea of a Great Year of 10,800 years. Furthermore, the idea of an oscillating universe is to be considered a "sheer speculation" not only because science has no explanation for the "creation anew" of stars and galaxies from the material remains of the previous cycle. There is also the problem of entropy and radiation in the reverse, or the wholly mysterious physics of the contracting cycle.

The most palpable difficulty of verifying the idea of an oscillating universe lies in the great uncertainties connected with the only possible though very indirect observational evidence that has so far been specified in its favour. The evidence would consist in a highly specific distribution of galaxies with distance. Although it has been repeatedly stated during the last fifteen years or so that a reply to this problem would be soon forthcoming, the uncertainty in measuring the distance of galaxies increases too rapidly beyond a billion light-years to permit a definitive interpretation of the data available. There is still much improvement needed in the use of the best optical and radiotelescopes to explore in a satisfactory manner the distribution of galaxies within the range extending from two to eight billion light years. Yet, accomplishment of this task seems to be an absolute prerequisite for a reliable preference for an oscillating model of the universe. Even then there still would remain major problems. One of them is the question of extrapolation. While in 1922 Friedman estimated the period of oscillation at 2.4 billion years, today astronomers speak of a time-span thirty times longer. If the period is eighty some billion years,[91] then by observing galaxies at eight billion light years (the most optimistic limit of present-day observations) astronomy has covered only the very youth of the present expanding phase of the universe. It should appear somewhat preposterous to predict unhesitatingly on such a meagre basis not only the rest of that awesomely long expanding period but also that it would certainly turn into contraction at an immensely distant moment.

The contracting phase has its own additional mysteries and paradoxes, for which physical astronomy does not even have a tentative solution. Of these the chief is the gravitational collapse which should naturally come at the end of the contracting phase. But even if the universe should be spared this, although on the basis of general relativity it is difficult to see why it should, there still remains the crucial question which should be answered if the idea of an oscillating universe is to remain within the ken of respectable scientific speculations: can there be any evidence of the contracting phase, say in the form of some radiation, that could be detected in the following expanding phase or vice versa? The most likely answer is the negative. Or as E. J. Öpik, the astronomer-author of a book distinctly in favour of the idea of an oscillating universe, described the situation when the whole universe precipitated itself into a narrow space, almost into a point: "Everything will perish in a fiery chaos well before the point of greatest compression is reached. All bodies and all atoms of the world will dissolve into nuclear fluid of the primeval atom – which in this case is not truly primeval – and a new expanding world will surge from it, like Phoenix out of the ashes, rejuvenated and full of creative vigour. No traces of the previous cycle will remain in the new world, which, free of traditions, will follow its course in producing galaxies, stars, living and thinking beings, guided only by its own laws."[92]

If, however, such is the case, then the properties of one phase will remain an unknown entity for scientists living in the ensuing phase and can hardly form the object of consistent scientific discourse. That such elementary considerations could escape a distinguished astronomer merely illustrates the point once made by Einstein: "The man of science is a poor philosopher."[93] Öpik justified his preference for the oscillating model on the rather questionable reasoning that if Lemaître's primeval atom "had no predecessor, it must have been created on the spot. It would seem that here, in the absence of positive evidence, the acceptance of an Act of Creation could be left to the taste and esthetic judgment of those concerned with the problem."[94] A little reflection or theological research could have shown Öpik that the act of Creation is, by definition, not an object of observation nor can it have observational evidence as such. It was merely a borrowing from age-old theological truism when a cosmologist of fine repute aptly stated that "the act of creation of a particle is not a physical event and is just as mysterious as the legendary waving of the magician's wand."[95]

Equally indefensible is Öpik's claim that it was "on purely esthetic grounds" that in some quarters disapproval has been voiced of a universe repeating its general features. Opposition of serious scientists to the idea of an oscillating universe has been based on respect for the sound facts and conclusions of physics and of cosmological investigations, all of which run counter to the idea and implications of an oscillating universe behaving like a perpetual-motion machine. Öpik is right in claiming that esthetic considerations should be absolutely irrelevant when one is faced with the question of the true plan of the physical universe. There is, however, much that is puzzling in his inability to see "why the Great Repetition should claim a lesser esthetic value than, for example, the annual succession of seasons so praised by poet and layman."[96] The repetition of sunrise and the return of spring is one thing, the recurrence of a "cosmic year," which Öpik revealingly equated with the "days of Brahma" and its congeries, is another. Perhaps Öpik himself suspected the incongruity of this as he strongly disputed the alleged identity of the oscillating universe and "of the eternal recurrence of all things" advocated by Nietzsche.[97]

To be classed with Nietzsche was also an embarrassing prospect to the French astronomer, A. Dauvillier, another articulate advocate of a universe going through cycles without end.[98] But he must have been somewhat wary of the idea of an oscillating universe, as he failed, strangely enough, to mention it throughout his lengthy efforts to discredit the reality of the expanding universe, as a once-and-for-all process. As one would expect, for Dauvillier the law of entropy had no validity for the universe as a whole. Scientifically, he showed himself heir to the contentions of Boltzmann and Arrhenius, who pictured the universe as a system in which some parts were always in dissolution and others in reconcentration. Their philosophical presuppositions were shared by Dauvillier. Thus, he declared that "at all times men tried to construct systems of the world by taking their start in their own creeds, and the result was a rich but sterile flow of hypotheses."[99] With a curious inconsistency Dauvillier carefully exempted from this devastating indictment the notion of a universe conceived along cosmic cycles. The bias in his reasoning spoke for itself: "The reversibility, being the condition of eternity, should be based on the *cosmic cycles*. All the phenomena which we observe are cyclical, from the meteorological, biological, geochemical cycles to the geological cycles, and the same should hold true in everything."[100] All this received its crowning touch of irony in the fact that Dauvillier was led back to that long discredited cosmologist, Aristotle, whose dictum on the eternity of the world, as evidenced by the cycles of heavenly motions, he quoted with approval.[101]

Dauvillier had other revealing sources as well. He divided, and rightly so, the cosmological hypotheses into two main classes: one of them was that of a universe forever existing and evolving, or rather transforming. As the most ancient matrix of such a conception, Dauvillier named the Buddhist thought formulated in India and China during the sixth century B.C., and articulated further by Confucius and Lao-Tse.[102] The same view of the universe was taught, Dauvillier continued, by the Greek philosophers, among whom he singled out Democritus, Epicurus, and Heraclitus. He hastened to add that the latter had carefully disavowed the idea of eternal returns in the Nietzschean sense.

Dauvillier's account of the other main class of cosmological hypotheses was markedly different in tone. In reviewing it, he criticized from the very start instead of reporting first: "The Jewish and Christian thought postulated an unknowable world of supernatural powers, transcending ours beyond its space and time and having created it *ex nihilo*."[103] Nevertheless, the "architecture without an architect," Dauvillier insisted, can be found everywhere in nature. As an example Dauvillier pointed to the process of crystallization and a most palpable form of it, snowflakes, the marvels, as he put it, of the haphazard interplay of molecules. What could be truly hazardous for the unwary and uninformed was Dauvillier's outline of the more recent history of cosmological theories. Its arbitrariness was best illustrated by a phrase which claimed that in France "science liberated itself in

1530 from the shackles of scholasticism by the freedom of instruction in the Collège Royal de France."[104] The phrase is from a chapter which begins by Dauvillier's quoting none other than Pierre Duhem, the author of the *Système du monde.* Even the overwhelming documentation of its ten monumental volumes was not weighty enough to make as much as a ripple on convictions steeped in the ebb and flow of eternal cycles.

One may, of course, try to dodge the issue, and claim as H. Alfvén did that cosmology has to rest content with the analysis of the present expansion and of the contraction immediately preceding it.[105] Still, it is all too easy to profess reservations about a universe oscillating forever, and pour gentle scorn at the same time on the so-called Big-Bang theory, lampooning the idea of a creation out-of-nothing as well. It is anything but easy to provide proofs on behalf of a world-model that could only be saved if one were willing to heed Alfvén's request: "We beg leave to sidestep the question, 'What happened before then?'."[106] But this is precisely what cannot be done in cosmology, be it scientific or philosophical. Beyond the billions of years of slow contraction in the previous phase, and the billions of light years, the assumed radius of the "original" state of maximum expansion, there looms the inevitable choice between finiteness and infinity. As to an infinite past, Alfvén failed to consider its grave scientific problems. As to an infinity in space and matter, he sought refuge in that hierarchically structured universe proposed some sixty years ago by C. V. L. Charlier.[107] The price of that choice was not only the abandoning of the General Theory of Relativity, but more ironically perhaps, the option for a cosmic hierarchy without a Supreme Hierarchos. But is it not the very particularity of a structure that clamours most convincingly for the recognition of its architect?

Such and similar examples of inept evasiveness in the face of fundamental questions about the universe should command the utmost attention in an age in which the question about the interaction between science and culture can no longer be evaded. This concern, at least in its seriousness, is relatively new. For well over two centuries Western Man lived in the Belief that science was an infallible and unfailing road to Utopia based on the completely rational ordering of things and persons alike. The conviction had firmly entrenched itself that science, or rather *more* science, was the magic cure for all the ills and woes of mankind. Such an outlook on science received its classic expression in Herbert Spencer's once highly esteemed poetry in prose:

> Thus to the question with which we set out – What knowledge is of most worth? – the uniform reply is – Science. This is the verdict on all the counts. For direct self-preservation, or the maintenance of life and health, the all-important knowledge is – Science. For that indirect self-preservation which we call gaining a livelihood, the knowledge of greatest value is – Science. For the due discharge of parental functions, the proper guidance is to be found only in – Science. For that interpretation of national life, past and present, without which the citizen cannot rightly regulate his conduct, the indispensable key is – Science. Alike for the most perfect production and highest enjoyment of art in all its forms, the needful preparation is still – Science. And for purposes of discipline –

intellectual, moral, religious – the most efficient study is once more – Science. . . .
Necessary and eternal as are its truths, all Science concerns all mankind for all
time.[108]

Still, if Spencer's works began to gather dust shortly after his death in 1903, it
was only recently that the general public, and the world of science as well, began to
realize the utter hollowness of Spencer's panegyrics on science. Significantly, this
new awareness asserted itself at a time when the level and spread of science-oriented
education far surpassed anything witnessed before. Modern life has become
dependent even in its most trivial details on the application of some scientific
invention, but at the same time obscurantist and irrational trends too have asserted
themselves to a truly alarming degree. The political, social, and individual arena
is now crowded with symptoms that hardly evidence that hallmark of scientific
mentality which is respect for facts and rationality. Of these symptoms one has a
special relevance to a survey of the long history of man's sad flirtation with the
idea of eternal returns. The symptom is the audacity with which astrology is
re-establishing itself as a respectable enterprise. The zodiac which was aptly called
the "Rorschach-text of mankind in its infancy"[109] is now taking in man's thinking
the same prominent position that it held two thousand years ago and earlier. Then,
as now, the psychological motivations underlying its popularity show remarkable
similarity. As a declaration of the American Association of Social and Psychological Studies stated as early as 1940: "The principal reason which turns some men
to astrology and other superstitions is that they lack the necessary resources to solve
the serious problems with which they are faced. Frustrated, they give in to the
pleasant suggestion that there is a golden key within their reach, a simple solution,
an ever present help in times of trouble."[110] Since 1940 the groping of modern man,
frustrated by his science and tormented by his post-Christian emptiness, is becoming even more desperate as he tries to find tranquillity. Man once again frantically
wants to abdicate his responsibilities by trying to get immersed in the great cosmic
ebb and flow.

Science, which in some sense cast him into that aimless to and fro of endless
waves, has no right to encourage him to grab some rudder, as science can assign
no aim, no direction. Science has long ago excluded from its realm considerations
about goals and aims. Science can reintroduce these notions only through the back
door, but nowadays such inconsistencies should give themselves away rather
rapidly. It is clearly inconsistent to bring to a close a brilliant lecture series on time
and space by projecting on the screen a magnificent Chinese drawing of a wave
curling upon itself, as the epitome of the oscillating universe, and insist at the same
time on man's duty to navigate with courage on the all-engulfing cosmic waves.
As far as science can say, man is flotsam and jetsam on those unfathomable waves
and nothing else. At the same time science should have kept saying that it was not
within its competence to say that man was truly nothing else.

By and large, most prominent scientists did their duty in this last respect.[111]
However, their voices were drowned out by the crusade of those who saw in
science the invincible antidote for obscurantism, real or imaginary. The result was
a drastic erosion of Western man's confidence in value judgments, as a grave

suspicion has been cast on any type of reasoning which was not "scientific," that is, quantitative. It is hardly a coincidence that this wholesale wavering about basic values should approach a crisis point at a time when in the most comprehensive part of science, cosmology, one witnesses a pronounced option for the oscillating universe, this scientifically coated modern version of the Great Year. It seems indeed that once more the inner logic of things and processes vindicates itself. The scientific enterprise, this acme of man's rationality, which got its wings from belief in the Creator, finds them now transformed into dark omens. The possible global abuse of the tools of science looms frighteningly large on the horizon. Science, which only five hundred years ago got off the ground against all odds, now is turning into a runaway sophistication for its own and not for man's sake. Revealingly enough, the impotency of science to provide norms and goals in addition to tools is dawning on humanity at a time when modern scientific culture is defiantly discrediting the faith in the Creator as anachronism from a naive age. But while intellectual fashions can be effectively shaped, man's basic aspirations seem to resist even the most skilful manipulations. Those aspirations can, however, be robbed of some hard-won guidelines and supports. Deprived of the moral and intellectual benefits of faith in the Creator, these aspirations are now being driven toward pseudo-philosophies and pseudo-sciences which promise salvation in an identification with the celestial realms of the zodiac and its ever same revolutions.

In our century, Marxism and Nazism earned the dubious distinction of being the two most violent manifestations of such pseudo-philosophies and pseudo-sciences. Characteristically, both displayed most explicit ties to the dogma of cyclic recurrences. In this respect Nazi ideology was in a sad sense the more consistent of the two. Unlike the disciples of Engels who tried to imagine an end to the dialectical cyclic process in the establishment of the heaven of a classless society, Nazi propaganda organs restricted their mad ambitions to the "next thousand years." After that the rotation of the swastika was to put other races into the driver's seat. Such is, however, a detail of relatively minor significance. What should command far greater attention in both cases is the inaptitude of the scientific realm to set up effective barriers against the rise and spread of totalitarian trends steeped in the anti-science of dialectic and cosmic recurrences. Science, both in Nazi Germany and in Communist Russia, had been successfully harnessed by irrational ideologies. If there remained in both systems a largely clandestine re-affirmation of trust in man's rationality, purpose, and inalienable dignity, the credit for this should go to convictions reaching far deeper than purely scientific training and information.

Even in the Free World it has become increasingly clear, and in a large part because of the traumatic experiences of recent history, that science was rather the superstructure than the very foundation of rationality. In the face of a possible global self-destruction, it had to be noticed that science can provide only the tools of constructive endeavour, but not its goals, norms, and much less the all-important commitment on man's part to rationality and good will. Today, only some inveterate believers in science as a cure-all fail to recognize the fallacy of the illusion that information, especially scientific information, will automatically produce the good, upright, and dedicated man. The poignant remark of Einstein, that it was man's heart, not the uranium, that needed to be purified, has been echoed by other creators of modern physics, this most spectacular evidence of man's extraordinary abilities. Some of the most sensitive of these physicists have become so much appalled by the diabolical transformation of the products of science into tools of destruction as to hint that were it possible for them to start their career anew it would not have anything to do with science.

Out of such despair arose, however, a more comforting symptom as well. It is the steadily growing realization that the man of science, no less than his counterpart in religion, lives ultimately by faith. With the mirage of positivism now being unmasked, it is easier to recognize that the scientific enterprise rests on a conviction which presupposes far more on man's part than the mere juxtaposition and correlation of the data observation. The conviction in question is nothing short of a faith which, like religious faith, consists in the readiness of going beyond the immediately obvious.[112] The step is not simply a glib conjecture about a deeper layer. It is rather a recognition of the indispensable need of such a layer if the scientific enterprise is to make any lasting sense. It is in that deeper layer that notions like the intelligibility, simplicity, and lawfulness of nature are taking on a meaning which demands absolute, unconditional respect and acceptance. It is that deeper meaning which science must command if its laws should be considered not merely clever manipulations of terminology and data, but a concrete encounter with the real structure of nature.

That real structure is not an *a priori* construct. Efforts, ancient and recent, aimed at deriving the shape and structure of the cosmos from preconceived considerations have one thing in common: their miserable failure. The universe is an entity which is *given*, in the most ontological sense of this word. This feature of the universe, so commodiously ignored from Bruno on by a convenient recourse to the presumed infinity of the world, is again imposing itself on the inquiring mind with elemental force due to the rather recent but inevitable acceptance of the finiteness of world in matter and space. There will, of course, be many who keep trying to shore up their Spinozean pantheism and immanentism by desperately claiming infinity for the universe along the parameter of time. Neither science nor scientific history will be their ready allies. As the foregoing chapters tried to illustrate, science owes its only viable birth to a faith, according to which the world is a created entity, that is contingent in every respect of its existence on the creative act of God, and that its existence has an absolute origin in time, majestically called "in the beginning."

When a human enterprise is at a crisis point, the most helpful remedy consists in tracing the road to the present, as the diagnosis of the symptoms of a disease is basically dependent on a careful analysis of the past record. In the whole record of science the most fundamental and yet most ignored issue centres on the question why science failed to emerge until rather recently in mankind's history. If science does indeed represent in history the stir that was surpassed only by the birth in Bethlehem, then a studied neglect of that question is intellectual smugness at best and mental cowardice at worst. The question is not only momentous but also monumental. It is undoubtedly beyond the capabilities of one individual to answer it in a comprehensive fashion. A commanding view of so many different cultures is only one of the enormous tasks facing anyone who tries to come up with at least a moderately satisfactory answer. There is, in addition, the vastness of the history of science. The present work certainly does not claim to have entered into every aspect of such a vast problem. One point seems, however, to stand out very clearly. All great cultures that witnessed a stillbirth of science within their ambience have one major feature in common. They all were dominated by a pantheistic concept of the universe going through eternal cycles. By contrast, the only viable birth of science took place in a culture for which the world was a created, contingent entity. To make this point is not merely a reply to a past issue. The reply seems to have a direct relevance to the gravest perplexity of the modern, scientific world. The very roots of that perplexity form a mirror-image of the age-old need to make a choice between two ultimate alternatives: faith in the Creator and in a creation once-and-for-all, or surrender to the treadmill of eternal cycles. Such should indeed be the case, as the present is always the child of the past. The present and past of scientific history tell the very same lesson. It is the indispensability of a firm faith in the only lasting source of rationality and confidence, the Maker of heaven and earth, of all things visible and invisible.

NOTES

1. *Social and Cultural Dynamics*, vol. II, *Fluctuation of Systems of Truth, Ethics, and Law* (New York: American Book Company, 1937), p. 383, note.
2. "Über die Krümmung des Raumes," *Zeitschrift für Phyik*, 10 (1922): 377–86.
3. Einstein's reaction to Friedman's ideas was typical in this respect, as his two short communications on Friedman's paper dealt with such minor details as algebraic errors, etc. See *Zeitschrift für Physik*, 11 (1922): 326 and 16 (1923): 228. Einstein's second communication concluded with the somewhat apologizing remark that he considered Friedman's results "valid and enlightening." Still, it was not until 1931 that Einstein considered in any detail the contraction (see his "Zum kosmologischen Problem der allgemeinen Relativitätstheorie," *Sitzungsberichte der Preussischen Akademie der Wissenschaften, Phys.-Math. Klasse* 1931 [XII], pp. 235–37) which, as Friedman showed, was implied in Einstein's world model.
4. Sorokin's unawareness of the idea of an oscillating universe was all the more puzzling as he explicitly mentioned de Sitter and Lemaître, who referred repeatedly to the theoretical possibility of a contracting phase without endorsing it.
5. For some details on this, see P. Michelmore, *Einstein: Profile of the Man* (New York: Dodd, Mead & Co., 1962), p. 164.
6. See, for instance, the reaction of the French physicist, E. Borel, in his *Space and Time* (London: Blackie and Son, 1926), pp. 226–27, prompted largely by Einstein's insistence on the finiteness of the universe in his lectures given at the Sorbonne in March 1922.
7. "Some Mathematical Aspects of Cosmology," *Science*, 62 (1925): 63–72, 96–99, 121–27.
8. New York: Harper and Brothers, 1925.
9. *Ibid.*, pp. 163–68.
10. "Un univers homogène de masse constante et de rayon croissant, rendant compte de la vitesse radiale des nébuleuses extra-galactiques," *Annales de la Société Scientifique de Bruxelles*, 47A (1927): 49–59.
11. De Sitter first outlined his world-model in 1917 as an immediate elaboration on Einstein's ideas published a few months earlier. De Sitter's own, non-technical, account of his world-model is in his *Kosmos* (Cambridge, Mass.: Harvard University Press, 1932), pp. 120–22.

12. Quotation is from the English translation of Lemaître's paper, published mainly through Eddington's concern, under the title, "A Homogeneous Universe of Constant Mass and Increasing Radius accounting for the Radial Velocity of Extra-galactic Nebulae," *Monthly Notices of the Royal Astronomical Society*, 91 (1930–31): 483–90. For quotation see p. 489.

13. "On the Instability of Einstein's Spherical World," *Monthly Notices of the Royal Astronomical Society*, 90 (1930–31): 668–69.

14. See Selections 54, 55 and 56 in H. Shapley (ed.), *Source Book in Astronomy 1900–1950* (Cambridge, Mass.: Harvard University Press, 1960).

15. *Art. cit.*, p. 489. See also Lemaître's paper, "The Expanding Universe," dated March 7, 1931, and published together with the English version of his famed paper, *ibid.*, pp. 490–501.

16. "On the Instability of Einstein's Spherical World," p. 669 note.

17. *Ibid.*, p. 668.

18. *Ibid.*, p. 670.

19. *Ibid.*, p. 677.

20. *Ibid.*

21. "The End of the World: from the Standpoint of Mathematical Physics," *Nature*, 127 (1931): 447–53.

22. *Ibid.*, p. 450.

23. *Ibid.*

24. *The Nature of the Physical World* (Cambridge, Cambridge University Press, 1928), p. 86.

25. *The Expanding Universe* (Cambridge: Cambridge University Press, 1932), p. 125.

26. *New Pathways in Science* (Cambridge: Cambridge University Press, 1934), p. 59.

27. See on this chap. x, pp. 230–31.

28. Paris: Librairie Fischbacher, 1933.

29. "Discussion on the Evolution of the Universe," in *British Association for the Advancement of Science: Report of the Centenary Meeting: London, 1931, September 23–30.* (London: Office of the British Association, 1932), pp. 560–610.

30. *Ibid.*, p. 599. As a proof of the finiteness of the universe he mentioned the infinite gravitational potential arising in every point in a Newtonian universe, but he failed to recall Olbers' Paradox, or the optical counterpart of the gravitational paradox inherent in an infinite, homogeneous universe.

31. *Ibid.*, p. 600.

32. *Ibid.*, p. 599.

33. *Ibid.*, p. 607.

34. *Ibid.*, p. 587.

35. *Ibid.*, p. 584.

36. *Ibid.*, p. 579.

37. Oxford: Clarendon Press, 1935.

38. *Ibid.*, p. 337.

39. *British Association, Report of the Centenary Meeting*, p. 578.

40. *Ibid.*, pp. 594–605.

41. For instance, in his address quoted in note 7 above.

42. Reprinted under the title, "Available Energy," in his *Science and the New Civilization* (New York: Charles Scribner's Sons, 1930), pp. 108–09.

43. *Ibid.*, p. 110.

44. *The Expanding Universe*, p. 124.

45. The story was told by Weizsäcker himself in his *The Relevance of Science: Creation and Cosmogony* (New York: Harper & Row, 1964), pp. 151–52. In spite of his professed concern, expressed in the subtitle, for the notion of creation, Weizsäcker failed to do justice to the role played by that notion in the history of science.

46. *Le retour éternel et la philosophie de la physique* (Paris: Ernest Flammarion, 1927).

47. A point already emphasized in Rey's earlier work, *L'énergétique et le mécanisme* (Paris: Alcan, 1907).

48. *Le retour éternel*, pp. 44, 94, 298. Rey had no doubt that science would eventually succeed

in this respect. Still, his hopes can now appear doubly ironical, as they saw print in the same year (1927) when Heisenberg enunciated the uncertainty principle.

49. *Le retour éternel*, pp. 309–11.
50. *La théorie de la physique chez les physiciens contemporains* (Paris: Alcan, 1907).
51. *Le retour éternel*, p. 267.
52. *Ibid.*, p. 44.
53. *Ibid.*, p. 292.
54. *Ibid.*, p. 274.
55. *Ibid.*, pp. 10–11 and 273.
56. *Ibid.*, p. 316. (The passage is from Spinoza's *Ethics*, Part V, prop. xxiii, scholium).
57. *Ibid.*, p. 16.
58. *Ibid.*, p. 272.
59. *Ibid.*, p. 290.
60. *Ibid.*, p. 286.
61. *Ibid.*
62. "L'âge de l'univers," in *Comptes Rendus*, 213 (1941): 326.
63. *Ibid.*, p. 329.
64. *Ibid.*
65. Oxford: Clarendon Press, 1934. A year earlier H. P. Robertson published his classic review of previous investigations of relativistic cosmological models giving detailed attention to Tolman's studies on the question already in print. See *Reviews of Modern Physics* 5 (1933): 87–88.
66. *Relativity, Thermodynamics and Cosmology*, p. 488.
67. *Ibid.*, p. 487.
68. *Ibid.*, p. 431.
69. *Ibid.*, p. 444.
70. *Ibid.*
71. *Ibid.*
72. See his *L'hypothèse de l'atome primitif: essai de cosmogonie* (Neuchâtel: Du Griffon, 1946), a collection of six lectures given between 1929 and 1945.
73. Published under the title, *The Beginning and End of the World* (Oxford: Oxford University Press, 1942). Whittaker's Donnellan Lectures given at Trinity College, Dublin, in June 1946, had a similar thrust as can be seen from the title of the published text, *Space and Spirit: Theories of the Universe and the Arguments for the Existence of God* (Hinsdale, Ill.: Henry Regnery Co., 1948).
74. *The Beginning and End of the World*, p. 63.
75. One of the numerous publications of the English translation was by the National Catholic Welfare Conference: Washington, D.C., 1951, under the title, *The Proofs for the Existence of God in the Light of Modern Natural Science*.
76. *Ibid.*, p. 11.
77. "The Steady-State Theory of the Expanding Universe," *Monthly Notices of the Royal Astronomical Society*, 108 (1948): 252–70.
78. The idea of a creation out of nothing but without a Creator was most explicitly stated by H. Bondi in his *Cosmology* (Cambridge: Cambridge University Press, 1952), p. 144.
79. As can be seen from the title of the published text of a Symposium broadcast in 1959 by the BBC, *Rival Theories of Cosmology* (London: Oxford University Press, 1960). Rivalry could, of course, arise only if the validity of entropy was systematically underplayed, if the enormous difficulties of the physics of an oscillating universe were consistently minimized, and if the conceptual contradiction of a creation out of nothing but without a Creator was simply ignored.
80. This is fairly in evidence, for instance, in the various semi-popular expositions of cosmology by F. Hoyle.
81. Or the postulate that the large-scale features of the universe should appear the same at any point *and* at any moment.
82. "The Steady-State Theory of the Expanding Universe," p. 256.

83. See *Monthly Notices of the Royal Astronomical Society*, 113 (1953): 393–407.
84. *Ibid.*, p. 406.
85. *Modern Astrophysics* (New York: The Macmillan Co., 1927), p. 408.
86. The two most important of these are the recent counts on the distribution of very distant galaxies, and the failure to detect in outer space a specific radiation which should occur, according to the Steady-State Theory, if hydrogen atoms would indeed keep emerging out of nothing at a given rate.
87. See J. E. Peebles and D. T. Wilkinson, "The Primeval Fireball," *Scientific American*, 216 (1967): 28–37.
88. *Ibid.*, p. 36.
89. The inept footwork of some modern cosmologists around the notion of creation is best illustrated by the statement of G. Gamow, author of *The Creation of the Universe* (New York: The Viking Press, 1956), who felt it important to add a note to the second printing of the work, that he understood the term "creation" not in the sense of making something out of nothing, but rather as "making something shapely out of shaplessness," as meant, for example, in the phrase "the latest creation of Parisian fashion."
90. *The Natural Philosophy of Time* (London: Thomas Nelson, 1961), p. 24.
91. See, for instance, the report of The New York Times on May 23, 1965 (p. E6 col. 2): "Last week, Dr. Allan R. Sandage of the observatory on Mount Palomar said, in a telephone interview, that he has compared the brightness and velocities of the nine quasars on which such data are available. They are consistent, he said, with a universe that is closed and oscillatory at a rate of one explosion every 82 billion years."
92. *The Oscillating Universe* (New York: The New American Library, 1960), p. 123.
93. Actually, Einstein merely quoted with approval an old saying. See his "Physics and Reality" (1936) in *Out of my Later Years* (New York: Philosophical Library, 1950), p. 58.
94. *The Oscillating Universe*, pp. 120–21.
95. G. J. Whitrow in his *The Natural Philosophy of Time*, pp. 26–27. The rest of the quote, "for we cannot reason about it, any more than we can reason about the creation of the whole universe," is true, of course, from the viewpoint of the "natural philosopher."
96. *The Oscillating Universe*, p. 124.
97. *Ibid.*, p. 123.
98. *Les hypothèses cosmogoniques: théorie des cycles cosmiques et des planètes jumelles* (Paris: Masson & Cie, 1963), p. 83. See also his paper, "Les hypothèses cosmogoniques et la théorie des cycles cosmiques," *Scientia*, 98 (June–July 1963): 121–26.
99. *Les hypothèses cosmogoniques*, p. 78.
100. *Ibid.*, p. 120.
101. *Ibid.*
102. *Ibid.*, p. 83.
103. *Ibid.*, pp. 83–84.
104. *Ibid.*, p. 84.
105. *Worlds-Antiworlds: Antimatter in Cosmology* (San Francisco: W. H. Freeman, 1966), p. 70.
106. *Ibid.*
107. The basic inadequacies of Charlier's model were discussed in my *The Paradox of Olbers' Paradox* (New York: Herder & Herder, 1969), pp. 198–205. Curiously, Alfvén claims (*op. cit.*, p. 96), that in Charlier's universe the total mass is infinite. As a matter of fact, the whole purpose of the hierarchical distribution of mass in an infinite space was to make the total mass finite and avoid thereby the gravitational and optical paradox!
108. "What Knowledge Is of Most Worth?" (1850), in *Education: Intellectual, Moral, and Physical* (New York: D. Appleton, 1889), pp. 93–94.
109. G. Bachelard, *L'air et les songes: essai sur l'imagination du mouvement* (Paris: J. Corti, 1943), p. 25.
110. Quoted in P. Couderc, *L'astrologie* (Paris: Presses Universitaires Françaises, 1951), p. 52.
111. As documented with respect to physicists in my *The Relevance of Physics* (Chicago: University of Chicago Press, 1966), passim.
112. On this one may consult my paper, "The Role of Faith in Physics," *Zygon*, 2 (1967): 187–202.

Index of Names